High-Performance Liquid Chromatography in Forensic Chemistry

CHROMATOGRAPHIC SCIENCE

A Series of Monographs

Editor: JACK CAZES
Fairfield, Connecticut

Other Volumes in Preparation

High-Performance Liquid Chromatography in Forensic Chemistry

Edited by

IRA S. LURIE
Drug Enforcement Administration
Special Testing and Research Laboratory
McLean, Virginia

JOHN D. WITTWER, JR.
Drug Enforcement Administration
South Central Regional Laboratory
Dallas, Texas

Chem
Sep/ae

MARCEL DEKKER, INC.   New York and Basel

6960 1987
CHEM

LIBRARY OF CONGRESS CATALOGING IN PUBLICATION DATA
Main entry under title:

High-performance liquid chromatography in forensic
   chemistry.

      (Chromatographic science ; v. 24)
      Includes bibliographical references and indexes.
      1. Chemistry, Forensic--Addresses, essays, lectures.
2. Liquid chromatography--Addresses, essays, lectures.
I. Lurie, Ira S., [date]. II. Wittwer, John D.,
[date]. III. Series.
HV8073.H53 1983   363.2'56'015430894   83-7461
ISBN 0-8247-1756-2

MARCEL DEKKER, INC.

270 Madison Avenue, New York, New York 10016

Current printing (last digit):
10 9 8 7 6 5 4 3 2 1

PRINTED IN THE UNITED STATES OF AMERICA

To our wives, Linda and Juliet

Contributors

IRA S. KRULL, Institute of Chemical Analysis, Northeastern University, Boston, Massachusetts

IRA S. LURIE,* Northeast Regional Laboratory, Drug Enforcement Administration, New York, New York

ALBERT H. LYTER, III,† Scientific Services Division, Bureau of Alcohol, Tobacco, and Firearms, Rockville, Maryland

W. W. McGEE, Department of Chemistry, University of Central Florida, Orlando, Florida

MICHAEL A. PEAT, Center for Human Toxicology, University of Utah, Salt Lake City, Utah

JOHN D. WITTWER, JR., South Central Laboratory, Drug Enforcement Administration, Dallas, Texas

Present affiliations:
*Special Testing and Research Laboratory, Drug Enforcement Administration, McLean, Virginia
†Federal Forensic Associates, Los Angeles, California

Preface

In the field of analytical chemistry high-performance liquid chromatography (HPLC) is considered by many to be the most exciting and dynamic new technique of the past decade. Since its advent in 1969, tremendous improvements have been realized in pumping systems, sample introduction modes, column design, and detectors to make it a rapid, accurate, and precise technique for the analytical determination of compounds. The number of mobile phases in HPLC are infinite and thus the separation possibilities are limited only to the analyst's imagination. Nonvolatile, polar, and thermally degradable compounds that are difficult to analyze by gas chromatography are particularly suited for modern liquid chromatography. In addition, since the technique is nondestructive, compounds can be isolated for identification by spectrographic techniques. Thus for the forensic chemist HPLC is an invaluable tool.

Although improvements in this field will continue to be realized, the technology of HPLC has grown out of its infancy. Now is a good time to take a breath and examine the role of HPLC for the forensic chemist.

The emphasis in this book is on practice rather than basic theory. Until now no single text has treated HPLC with the forensic chemist in mind. Theoretical considerations, hardware, and qualitative, quantitative, and preparatory work are discussed. Practical applications are cited throughout the text. Detailed charts of retention indices are included as well. The authors are chemists from government and university laboratories whose expertise is in forensic analysis and who have a great deal of experience working with HPLC for both method development and everyday case loads. The purpose of this book is to provide a basic understanding of the technique of HPLC as well as to present applications relevant to forensic analysis for the analyst with or without experience in HPLC. Chemists in related disciplines such as pharmaceutical science and clinical chemistry will also benefit.

We are grateful to the Drug Enforcement Administration for providing us with the training, resources, and encouragement that was necessary to pursue the exciting technique of HPLC.

Ira S. Lurie
John D. Wittwer, Jr.

Contents

1 Principles of High-Performance Liquid Chromatography

Ira S. Lurie*

Northeast Regional Laboratory
Drug Enforcement Administration
New York, New York

I. INTRODUCTION

Chromatography is a technique that involves separation of solutes based on selective interactions between a stationary phase and a fluid (mobile phase). If the stationary phase is a solid support or a liquid-coated solid support and the mobile phase is a liquid, the technique is known as liquid chromatography. Depending on how the stationary phase is physically applied, there are various forms of liquid chromatography. If the stationary phase is applied as a layer, the technique is referred to as thin-layer and paper chromatography. When the stationary phase is placed in a column, the chromatographic system is liquid column chromatography. In gas chromatography the stationary phase is a solid support or a liquid-coated solid support and the mobile phase is a gas. An excellent historical treatment of chromatography is presented in a book edited by Ettre and Zlatkis [1].

---

*Present affiliation: Special Testing and Research Laboratory, Drug Enforcement Administration, McLean, Virginia.

II. CLASSIC VERSUS MODERN LIQUID CHROMATOGRAPHY

Let us look at the differences between liquid column chromatography
which was developed before 1969 (classic column chromatography) and
the recently developed technique high-performance liquid chromatography (modern column liquid chromatography). In classic column
chromatography we have the following features.

1. Columns with particle sizes 100 µm or greater are packed by the
    user, employed only once, and discarded.
2. Sample application is skill dependent and time consuming.
3. Mobile-phase flow is accomplished by gravity.
4. Sample detection is afforded by collecting and analyzing frac-
    tions manually.
5. Analysis time is typically several hours.

In contrast, modern column chromatography has the following
attributes:

1. Columns with particle sizes of 3-50 µm are generally commercially
    packed and used for hundreds and possibly over a thousand in-
    jections.
2. Precise sample injections are easily and rapidly achieved using
    syringes or injection valves.
3. Precise mobile-phase flow is obtained by utilizing pumps capable
    of operating under high pressure.
4. Sample detection is continuous utilizing sensitive detection
    systems.
5. Analysis time is typically 1 h or less, with resolution superior
    to that obtained over classic column chromatographic techniques

A comparison of classic column, thin-layer and paper, and
modern column chromatography is given in Fig. 1.

III. CHROMATOGRAPHIC MODES

For broad classification purposes high-performance liquid chromatography (HPLC) can be divided into five separation modes.

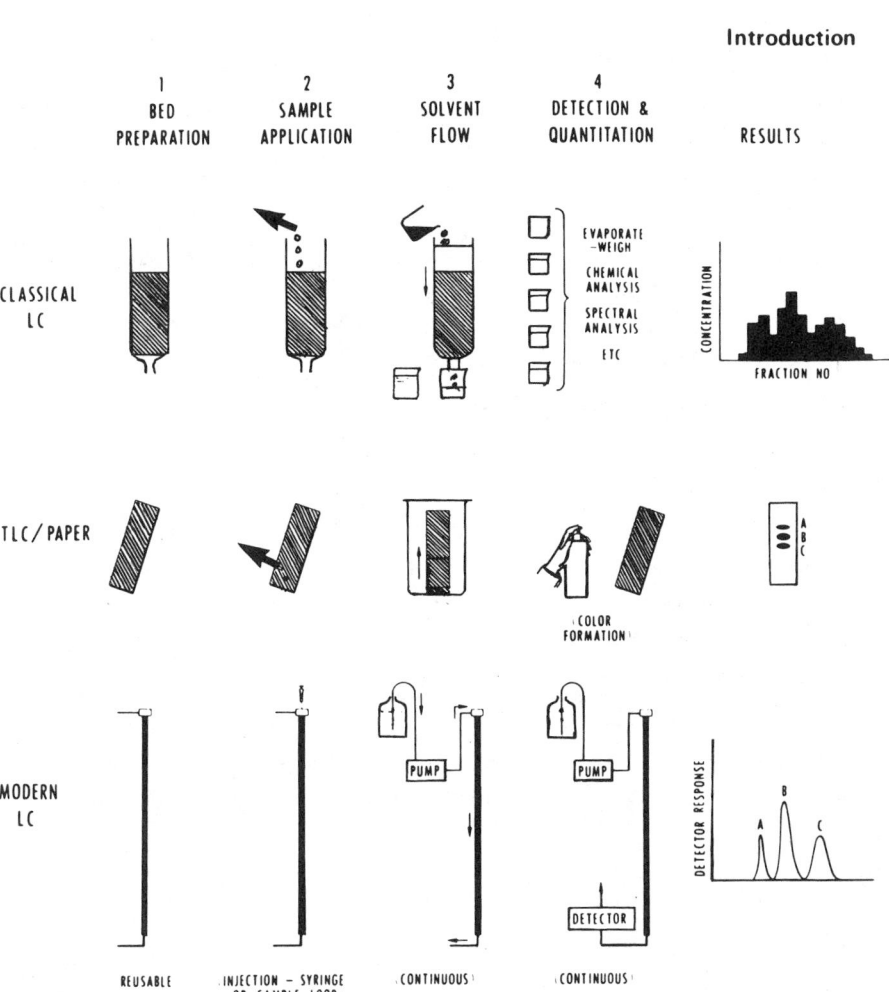

FIG. 1. Different forms of liquid chromatography. (From Ref. 10.)

A. Liquid-Solid Chromatography

In this separation mode the stationary phase is a solid that retains
solutes by adsorption. Usually, silica or alumina is utilized as the
adsorbent, with relatively nonpolar solvents such as hexane or
methylene chloride used as the mobile phase. This type of chromatography is known as normal-phase adsorption. Nonpolar polymer beads
can be used as a stationary phase, with relatively polar solvents
such as water, acetonitrile, or methanol as the mobile phase. In
the latter instance the separation mode can be referred to as reversed-phase adsorption.

B. Liquid-Liquid Chromatography

In this chromatographic mode the stationary phase is a liquid and
the separation occurs as a result of solutes partitioning between
two liquid phases. If the stationary phase is polar and the mobile
phase is nonpolar, the chromatographic mode is known as normal-phase
partition chromatography. In the convex case of reversed-phase
partition chromatography the stationary phase is a nonpolar liquid
and the mobile phase is polar. Liquid-liquid chromatography suffers
from the disadvantage that the stationary phase always has some solubility in the mobile phase and thus certain precautions must be
taken to limit the dissolution of the stationary phase. Normally,
a precolumn is employed where the mobile phase is presaturated with
stationary phase.

C. Bonded-Phase Chromatography

The salient feature of this chromatographic mode is the bonding of
an organic substrate to a silica-based support material. The resultant stationary phase is stable, and therefore the chromatographic
mode does not suffer from the disadvantage inherent in liquid-liquid
chromatography. If the bonded phase is polar and a relatively nonpolar mobile phase is employed, the technique is known as normal-
phase bonded-phase chromatography. On the other hand, if the bonded
phase is nonpolar and the mobile phase is relatively polar, the

chromatographic mode is reversed-phase bonded-phase chromatography.
Bonded-phase chromatography is by far the most commonly employed
liquid chromatographic mode employed today.

D. Ion-Exchange Chromatography

In ion-exchange chromatography, a charged stationary phase is employed containing oppositely charged counterions which are available
to exchange with solute ions of the same charge in the mobile phase.
The technique is known as cation-exchange or anion-exchange chromatography, depending on whether the solutes to be exchanged are positively or negatively charged. Ion-exchange chromatography requires
careful control of mobile-phase pH and ionic strength.

E. Size-Exclusion Chromatography

In this chromatographic mode the stationary phase contains polymer-
or silica-based support material of controlled pore sizes where the
predominant mechanism is separation based on the effective size of
solutes in the mobile phase.

IV. BASIC REFERENCES

Several excellent textbooks dealing with the technique of high-
performance liquid chromatography are listed in Table 1.

V. THEORETICAL CONSIDERATIONS

Of fundamental importance to any analyst using high-performance
liquid chromatography is the separation. Resolution (R) is a quantitative description of the separation that is obtained between two
peaks. This term, which is defined by the following relationship,
describes how good the separation is:

$$R = \frac{t_2 - t_1}{\frac{1}{2}(w_1 + w_2)} = \frac{2\Delta t}{w_2 - w_1} \tag{1}$$

In Eq. (1), t_1 and t_2 represent retention times of peaks 1 and 2
while w_1 and w_2 represent peak widths of peaks 1 and 2, respectively.

TABLE 1. Basic Textbooks on Liquid Chromatography

B. L. Karger, L. R. Snyder, and C. Horvath, *An Introduction to Separation Science.* Wiley-Interscience, New York, 1973.
R. P. W. Scott, *Contemporary Liquid Chromatography.* Wiley-Interscience, New York, 1976.
P. A. Bristow, *LC in Practice.* HETP Publishers, Handforth, Wilmslow, Cheshire, 1976.
C. F. Simpson, *Practical High Performance Liquid Chromatography.* Heyden & Son, London, 1976.
R. J. Hamilton and P. A. Sewell, *Introduction to High Performance Liquid Chromatography.* Chapman & Hall, London, 1978.
J. H. Knox, J. N. Done, A. T. Fell, M. T. Gilbert, A. Pryde, and R. A. Wall, *High Performance Liquid Chromatography.* Edinburgh University Press, Edinburgh, 1978.
H. Engelhardt, *High Performance Liquid Chromatography.* Springer-Verlag, Berlin, 1979.
L. R. Snyder and J. J. Kirkland, *Introduction to Modern Liquid Chromatography.* Wiley-Interscience, New York, 1979.
C. Horvath, *High Performance Liquid Chromatography -- Advances and Perspectives,* vols. 1 and 2. Academic Press, New York, 1980.

Source: R. W. Yost, L. S. Ettre, and R. D. Conlon, *Practical Liquid Chromatography: An Introduction,* Perkin-Elmer Corporation, 1980, p. 31.

The separation between two peaks that are assumed to be Gaussian is taken as the distance between the band centers divided by the average peak widths. Thus the greater the separation in retention times and the narrower the peaks, the higher the resolution, as shown in Fig. 2. In general, for quantitative work our aim would be for a minimum resolution of 1, while for qualitative work a smaller resolution could be tolerated. Snyder gives an excellent account of how much resolution is needed for both qualitative and quantitative work [2]. It will become apparent that the difference in retention time is related to thermodynamic considerations while the peak widths are determined by kinetic relationships.

Alternatively, assuming equal bandwidths, the following relationship equates resolution with fundamental chromatographic parameters k', α, and N.

$$R = \tfrac{1}{4}\left[\frac{k'}{k' + 1} (\alpha - 1)N^{\frac{1}{2}}\right] \tag{2}$$

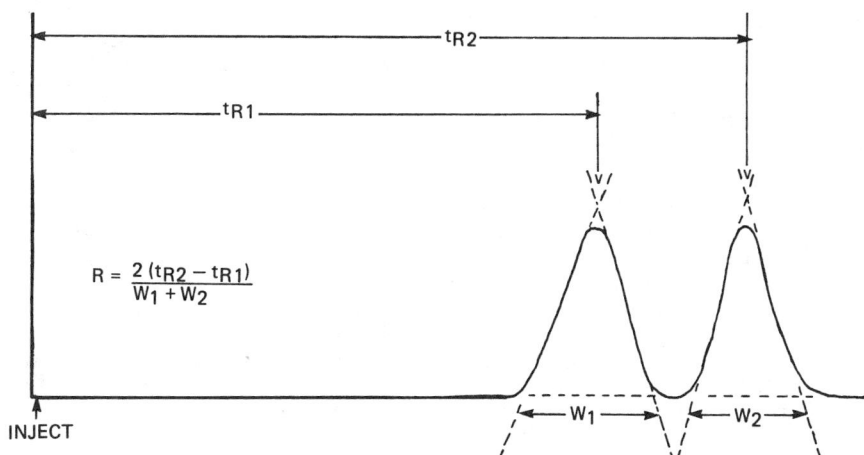

FIG. 2. Definition of resolution.

where k' is the capacity factor, α is the selectivity factor, and N is the efficiency term.

The capacity factor k' can be expressed as

$$k' = \frac{t_1 - t_0}{t_0} = \frac{V_1 - V_0}{V_0} \tag{3}$$

where V, the retention volume, is defined by

$$V = ft \tag{4}$$

where f is the flow rate and t_0 and V_0 are retention time and retention volume, respectively, of an unretained peak. The capacity factor k' is related to a fundamental thermodynamic parameter K, the equilibrium distribution coefficient, by the equation

$$k' = \frac{KV_s}{V_m} = \frac{\text{moles of solute in stationary phase}}{\text{moles of solute in mobile phase}} \tag{5}$$

where

$$K = \frac{\text{concentration of sample in stationary phase}}{\text{concentration of sample in mobile phase}} \tag{6}$$

V_s = volume of stationary phase

V_m = mobile phase interstitial volume or the volume occupied by the mobile phase in the voids of the column plus the volume occupied in the pores of the column

It is clear from the relationships above that various factors can effect k':

1. The relative distribution of a solute between stationary phase and mobile phase (K), which depends on the nature of solute, stationary phase, and mobile phase and on temperature. These interactions are thermodynamic in nature and the choice of which columns and mobile phase to be used for a given solute is discussed in subsequent chapters. Unlike gas chromatography, in liquid chromatography an increase in temperature will have only a small effect on reducing K. The relationship between k' and temperature is given by the van't Hoff equation,

$$\frac{d \ln k'}{dT} = \frac{-\Delta H_{s \rightarrow m}}{RT^2} \tag{7}$$

For liquid chromatography the $H_{s \rightarrow m}$ value normally will be positive and approximately one-fourth of the value found for gas chromatography.

2. The amount of stationary phase on a given column (V_s). This parameter is influenced by factors such as surface area of an adsorbent or bonded phase, loading of a liquid phase, and pore volume of a size separation column.

3. The void volume and pore volume occupied by the mobile phase (V_m) or void volume only in size-exclusion chromatography. This term is influenced by the nature of the column packing used and the way the column is packed.

If we examine the k'/(k' + 1) term in the resolution equation it becomes apparent that if k' is initially low, an increase in k' will give a substantial increase in resolution. However, for k' values greater than 5, k'/(k' + 1) approaches a constant and there is little to gain in resolution by increasing k'.

An unretained peak can be defined as

$$t_0 = \frac{L}{u} \tag{8}$$

where L is the length of the column in centimeters and u is the linear velocity of the mobile phase in centimeters per second. Equation (3) can be rearranged as follows:

$$t_1 = t_0(1 + k') \tag{9}$$

Substituting Eq. (8) into Eq. (9) gives

$$t_1 = \frac{L}{u(1 + k')} \tag{10}$$

Thus the longer the column, the greater the retention time; and the smaller the linear velocity of the mobile phase, the longer the retention of a component.

The selectivity factor α is defined as follows;

$$\alpha = \frac{k_2'}{k_1'} \tag{11}$$

Selectivity is a very important term to consider in optimizing resolution since as can be shown in Eq. (2), a small increase in α will affect resolution significantly. For example, suppose that $\alpha = 1$, $k' = 2$, and $N = 1600$. From Eq. (2), $R = 0$, which means no separation. This is to be expected since $k_2' = k_1'$. If we now increase α to 1.1 and keep the other conditions constant, according to Eq. (2), resolution will be approximately 0.7. If we increase α to 1.2 while k' and N are kept constant, resolution will be doubled. Since $\alpha = k_2'/k_1'$, the following factors that influence differences in the capacity factor will influence selectivity.

1. The nature of the mobile phase can influence the relative distribution of two solutes between stationary phase and mobile phase. For example, in reversed-phase ion-pairing bonded-phase chromatography, the selectivity can change with the amount of water, the size and concentration of the counterion, and the pH of the mobile phase.

2. The type of stationary phase employed can also influence the relative distribution of two solutes between stationary phase and mobile phase. In reversed-phase bonded-phase chromatography the size of the alkyl chain bonded to the silaceous packing can influence the selectivity factor.

3. The temperature utilized can influence selectivity. For example, at 53°C herion and acetylcodeine exhibit a lower selectivity factor in a reversed-phase bonded-phase ion-pairing chromatographic system than does the same system run at ambient temperature [3].

Unlike the first two terms in Eq. (2), which are primarily thermodynamic in nature, N, the efficiency term, depends on kinetic considerations. Efficiency is governed by the various rate processes that occur during a chromatographic separation.

In a typical liquid chromatographic separation a peak broadens with increased residence time on a column.[*] Efficiency or N is a quantitative measure of how a peak broadens with time, as shown in the equation

$$N = 16\left(\frac{t_r}{w}\right)^2 \tag{12}$$

N is approximately constant for all solutes in a chromatogram. Thus the smaller the peak width at a given time (t_r), the higher the efficiency. Since the measurement of N values via Eq. (12) becomes less accurate in the case of tailing peaks, other means for calculating N have been devised. Another method employs the formula [4]

$$N = 25\left(\frac{t_r}{w_{4.4}}\right)^2 \tag{13}$$

where $w_{4.4}$ is the width of the peak at 4.4% of the peak height.

Efficiency is also proportional to the length of a column L, as depicted in the equation

$$N = \frac{L}{H} \tag{14}$$

H is the height equivalent of a theoretical plate. Rearranging Eq. (14) gives the following relationship:

$$H = \frac{L}{N} \tag{14a}$$

H is a measure of the column efficiency per unit length. It is apparent from Eq. (14a) that at a constant value of 1, the larger the efficiency N, the smaller the value of H. The height equivalent to

---

[*] For gradient elution (covered later in this section), peak widths tend to be approximately constant. In size-exclusion chromatography, peak widths decrease with increased residence time on the column.

a theoretical plate can be related to the various rate processes that occur during the chromatographic separation. It is during these rate processes that band spreading occurs. Thus by minimizing values of H, band spreading can be kept to a minimum.

There are five processes of varying importance in liquid chromatography that can lead to band spreading, which was studied in detail by Giddings [5].

The first process, known as eddy diffusion, arises from the various paths a molecule can take through a column. Since these paths have varying distances, eddy diffusion can lead to band spreading. This process can be described by the equation

$$H_e = 2\lambda\, d_p \tag{15}$$

where λ is a constant that depends on the packing and d_p is the diameter of a particle of packing material. We can reduce the magnitude of λ by using particles of packing material that are uniform in size and densely packed. Equation (15) also tells us that the use of small particles will reduce eddy diffusion. Eddy diffusion is a significant process in modern high-performance liquid chromatography and the present trend is to use small particles (5-10 μm) that are nonsegregated and uniformly packed. This process is illustrated in Fig. 3.

The second process, known as longitudinal diffusion in the mobile phase, is given by the equation

$$H_d = \frac{2\gamma D_m}{u} \tag{16}$$

where γ is an obstructive factor that is a measure of how diffusion is restricted by the column packing. The magnitude of γ is reduced by particles of packing material that are small, uniform, and densely packed. D_m is the diffusion coefficient of a solute in the mobile phase, which can be approximated by the Wilke-Chang equation [6]:

$$D_m = \frac{7.4 \times 10^{-8}(\nu_2 M_2)^5 T}{n V_1^{0.6}} \tag{17}$$

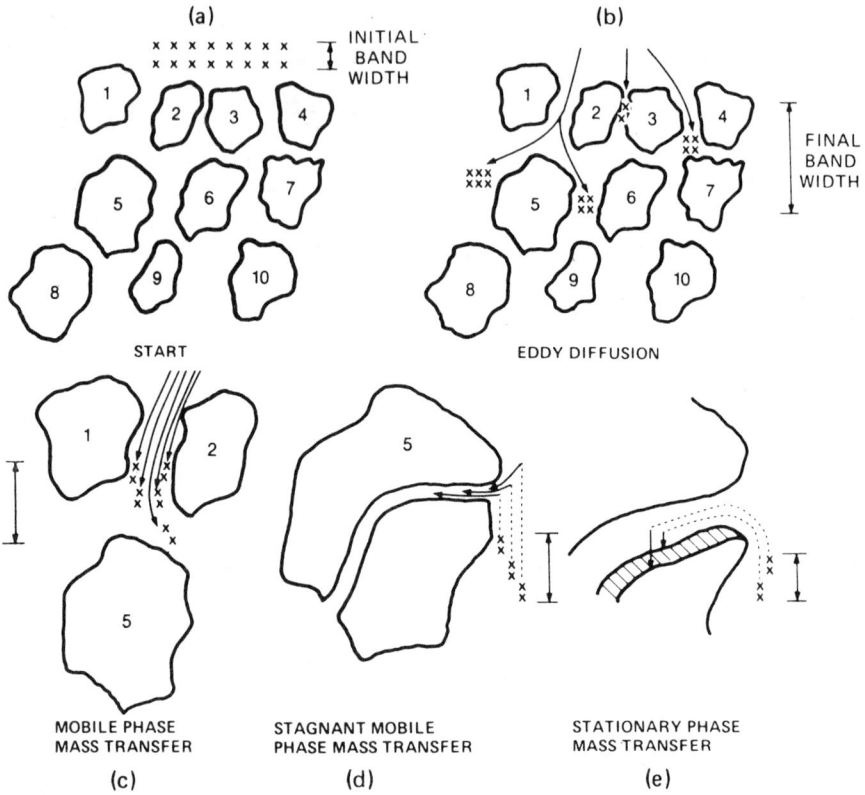

FIG. 3. Contributions to molecular spreading in liquid chromatography. (From Ref. 10.)

The association factor ν_2 is in general 1 except for strongly hydrogen bonding solvents, where it is greater than 1. M_2 is the molecular weight of the solvent, T is the temperature in kelvin, n represents the solvent viscosity, and V_1 is the molecular volume in milliliters. It is apparent from Eq. (17) that the diffusion co-efficient decreases with more viscous solvents and high-molecular-weight solutes and high linear velocities. Normally, longitudinal diffusion can be neglected at the linear velocities employed in high-performance liquid chromatography. This process is illustrated in Fig. 3.

In the third process, referred to as mobile-phase mass transfer band spreading arises due to flow inequalities along a given channel (Fig. 3). Flow tends to be greater in the center of a channel than along its walls. Mobile-phase mass transfer can be described by the equation

$$H_m = \frac{wd_p^2}{D_m} \qquad (18)$$

where w is a constant that will depend on how well the column is packed, d_p is the diameter of a particle, u is the linear velocity of the mobile phase, and D_m is the diffusion coefficient of a solute in the mobile phase. A well-packed column is one in which the particle distribution is narrow and the particles are densely packed. According to Eq. (18), H_m will decrease with smaller linear velocities, smaller particle sizes, less viscous solvents, lower molecular weight solutes, and good column-packing techniques.

The fourth process is known as stagnant mobile-phase mass transfer and is given by the equation

$$H_{sm} = \frac{C_{sm}d_p^2u}{D_m} \qquad (19)$$

H_{sm} arises from pools of mobile phase that become stagnant in the pores of the column-packing material, as depicted in Fig. 3. C_{sm} is a constant that is dependent on the fraction of mobile phase occupying interparticle space and the capacity factor [7]. The use of small particle sizes for packing material leads to tinier pores for mobile phase to become trapped in. Alternatively, pellicular packing, which consists of a solid core surrounded by a thin layer of porous material, reduces stagnant mobile phase.

The last factor influencing band broadening is stationary-phase mass transfer. This process, which results from slow desorption from the stationary phase, is given by the equation

$$H_s = \frac{C_sd^2u}{D_s} \qquad (20)$$

where C_s is a constant that depends on the shape of the pool of stationary phase and the capacity factor [7], d is the thickness of the
stationary phase, u is the linear velocity of the mobile phase, and
D_s is the diffusion coefficient of a solute in the stationary phase.
It is apparent from Eq. (20) that H_s will decrease with well-packed
columns, thinner stationary phases, smaller linear velocities, and
larger diffusion coefficients of solutes in the stationary phase.
Since the small particle columns (d_p = 10 μm) and the pellicular
packings that are presently employed have small d values, H_s can be
neglected for these columns. Stationary mass transfer is shown in
Fig. 3.

The process of eddy diffusion and mobile-phase mass transfer
can couple to give a band-broadening effect that is less than the
contribution of eddy diffusion and mass transfer individually, which
can be represented as follows:

$$H = \frac{1}{1/2\ \lambda d_p} + \frac{1}{D_m/wd_p^2 u} \tag{21}$$

The band-broadening processes described above can be combined
into a single equation,

$$H = H_e + H_m + H_d + H_{sm} + H_s \tag{22}$$

$$H = \frac{1}{1/2\ \lambda d_p + 1/D_m/wd_p^2 u} + \frac{2\gamma D_m}{u} + \frac{C_{sm}d_p^2 u}{D_m} + \frac{C_s d^2 u}{D_s} \tag{23}$$

By use of a set of constants A, C_m, B, C, and D, which depend for
the most part on the column employed, Eq. (23) can be written as
follows:

$$H = \frac{1}{1/A + 1/C_m u} + \frac{B}{u} + Cu + Du \tag{24}$$

For presently employed columns Du = 0, and thus Eq. (24) can
be written as

$$H = \frac{1}{1/A + 1/C_m u} + \frac{B}{u} + Cu \tag{25}$$

In general, the B/u term can be neglected except when employing low linear velocities with small particle packings. The Cu term is small when using microparticulate columns (particle sizes of 10 μm or less) and can be neglected when using pellicular packings.

To compare different columns operating at various linear velocities, we can use reduced parameters, defined by Giddings as follows [5]:

$$h = \frac{H}{d_p} \qquad (26)$$

$$v = \frac{ud_p}{D_m} \qquad (27)$$

If we substitute Eqs. (26) and (27) into Eq. (25), we obtain

$$h = \frac{1}{1/A + 1/C_m v} + \frac{B}{v} + Cv \qquad (28)$$

Pellicular and porous columns that are well packed can be represented by the following Knox equations [8]:

$$h = v^{0.33} + \frac{2}{v} + 0.003v \qquad (29)$$

$$h = v^{0.33} + \frac{2}{v} + 0.05v \qquad (30)$$

It is apparent from Eq. (2) that resolution increases with the square root of the efficiency term. Snyder presented a simple scheme for increasing resolution based on three ways of changing N [9].

1. If we decrease column pressure P (and flow rate) holding column length L constant, N will increase. This change in efficiency is based on Eqs. (14), (26), (27), (29), and (30) and the following relationship:

$$u = \frac{KP}{L} \qquad (31)$$

The column permeability K can be expressed by

$$K = \frac{K'd_p^2}{n} \qquad (32)$$

where K' is a constant, d_p is the particle diameter, and n is the solvent viscosity.

2. If we increase column length L while holding P constant, N will increase. This relationship follows from Eqs. (14), (26), (27), and (29) to (31).

3. If we increase both L and P while holding separation time t constant, N will increase. This relationship follows from Eqs. (9), (14), (26), (27), and (29) to (31).

Snyder has shown that the changes in chromatographic conditions required to increase resolution can be accomplished by the use of simple tables. All that is required is an approximate knowledge of the initial reduced velocity, which can also be obtained by looking at a table. These tables and the preceding relationships show that we will have to pay a price for an increase in N. If we increase N by decreasing pressure while L is held constant, the separation time will increase by a factor of 1/P. On the other hand, if N is increased by increasing L at constant P, we will be required to use a longer column, use two columns in series, or employ recycling. In addition, separation time will increase. Finally, if we increase N by increasing both L and P while holding t constant, the pressure required may exceed the limit of the column and/or instrument.

Suppose that in Eq. (2), which describes the effect of k', α, and N on resolution, α is kept constant; an expression can be written that gives the maximum resolution per unit time for a two-component mixture.

$$R = \tfrac{1}{4}(\alpha - 1)N_{\text{effective}}^{1/2} \tag{33}$$

where

$$N_{\text{eff}} = N\left(\frac{k'}{k' + 1}\right)^2 \tag{34}$$

If we combine Eqs. (10), (14), and (34), we obtain

$$N_{\text{eff}} = \frac{tu}{H}\ \frac{k'^2}{(1 + k')^3} \tag{35}$$

At constant t and u, N_{eff} has a maximum value at $k' = 2$ [10]. If u is allowed to vary, the optimum value of k' is between 3 and 5 [10].

For a given resolution, the greater the selectivity, the smaller the number of effective plates required, as shown by Eq. (33).

Most separations in liquid chromatography are carried out iso-cratically; that is, the mobile phase is kept constant throughout the chromatographic run. In all modes of separation except size exclusion, peaks will broaden as the k' values increase, which could result in later eluting peaks which are difficult to detect and in peaks that are not in the optimum k' range of $2 \leq k' \leq 5$. Conversely, early-eluting peaks could have k' values that are smaller than opti-mum and thus have poor resolution. If we employ gradient elution, a technique where the solvent strength is continuously increased during the chromatographic run, approximately equal bandwidths will result, with k' values approaching the optimum range. In practice, a weaker solvent is first employed to resolve the earlier-eluting peaks and the solvent strength is continuously increased to reduce the k' values of the later-eluting peaks. An example of an iso-cratic run versus a separation employing gradient elution is shown in Fig. 4.

Band spreading in liquid chromatography can result from ele-ments in the liquid chromatograph other than the column. Injectors, connecting tubing, end fittings, and detectors may all contribute to the final bandwidth depicted on the recorder. The relationship of this bandwidth to those of the various elements is given by

$$V_{c'}^2 = V_c^2 + V_i^2 + V_t^2 + V_f^2 + V_d^2 + \cdots \tag{36}$$

where $V_{c'}$ is the width of the final band in volume units and V_c, V_i, V_t, V_f, and V_d refer to the width of the band resulting from the column, tubing, end fittings, and detector, respectively. In order to keep band spreading from extra column effects smaller than 10%, V_i, V_t, V_f, and V_d should be less than one-third of V_c [11].

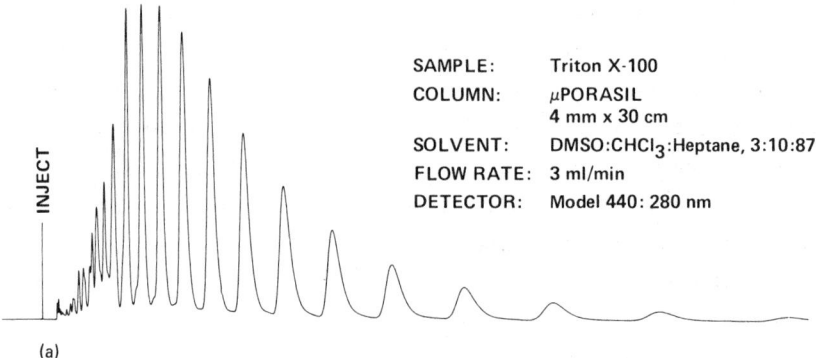

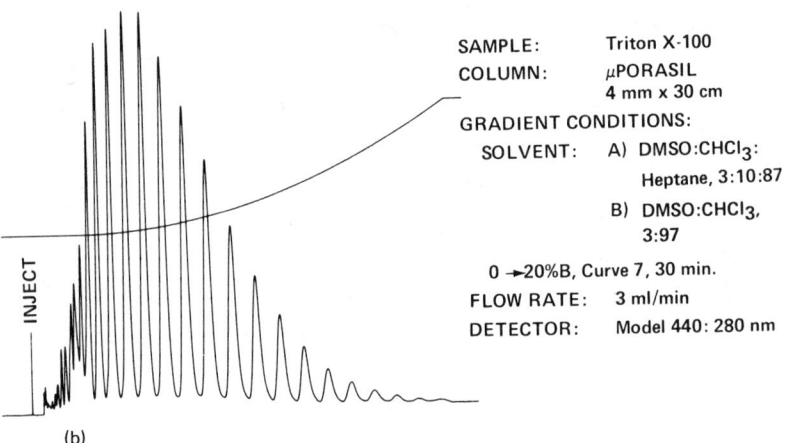

FIG. 4.   (a)   Initial isocratic run versus (b) gradient elution for Triton X-100.   (From *Waters Training Manual,* Waters Associates, Milford, Mass.)

REFERENCES

1. L. S. Ettre and A. Zlatkis (Eds.), *75 Years of Chromatography - A Historical Dialogue.* Elsevier, Amsterdam, 1979.
2. L. R. Snyder, *J. Chromatogr. Sci.* 10:200-212 (1972).
3. I. S. Lurie, Drug Enforcement Administration, unpublished studies, 1980.
4. L. R. Snyder and J. J. Kirkland, *Introduction to Modern Liquid Chromatography,* 2nd ed. Wiley-Interscience, New York, 1979, Chap. 5.

5. J. C. Giddings, *Dynamics of Chromatography*. Marcel Dekker, New York, 1965, Chap. 2.
6. C. R. Wilke and P. Chang, *Am. Inst. Chem. Eng. J.* 1:264 (1955).
7. B. L. Karger, in *Modern Practice of Liquid Chromatography,* J. J. Kirkland (Ed.). Wiley-Interscience, New York, 1971, Chap. 1.
8. J. N. Done, G. Kennedy, and J. H. Knox, in *Gas Chromatography 1972*, S. G. Perry (Ed.). Applied Science, London, 1973, p. 145.
9. L. R. Snyder, *J. Chromatogr. Sci.* 15:441-449 (1977).
10. L. R. Snyder and J. J. Kirkland, *Introduction to Modern Liquid Chromatography*, 2nd ed. Wiley-Interscience, New York, 1979, Chap. 2.
11. L. R. Snyder and J. J. Kirkland, *Introduction to Modern Liquid Chromatography,* 2nd ed. Wiley-Interscience, New York, 1979, Chap. 3.

2 Hardware for High-Performance Liquid Chromatography

John D. Wittwer, Jr.

South Central Laboratory
Drug Enforcement Administration
Dallas, Texas

I. INTRODUCTION

A liquid chromatograph is composed of the following hardware:

1. A pump or solvent delivery system
2. An injector of some type to introduce the sample onto the column head
3. A column packed with a suitable adsorbent
4. A detector to detect eluted solutes
5. A recorder or other means of displaying and utilizing the data generated in the chromatographic process.

In the following discussion, advantages and disadvantages of the various types of hardware available are described, together with their operating principles.

II. SOLVENT DELIVERY SYSTEMS

A pump or solvent delivery system should have the following characteristics:

1. It should provide a uniform, pulse-free flow.
2. The flow rate should be constant and reproducible.
3. It should provide a range of flow rates.

4. It should be capable of high-pressure operation.

5. It should have an essentially unlimited solvent reservoir.

6. It should be rapidly cleaned to facilitate mobile-phase changes.

7. With the proper ancillary equipment it should be capable of gradient formation.

8. It should be relatively maintenance free.

There are no solvent delivery systems that meet all the criteria listed above, but there are several types of pumps available, each type having its own characteristics. There are two basic classes of pump: constant pressure and constant flow. As the name implies, constant-pressure pumps operate at constant pressure and any changes in the column permeability result in variations in the flow rate. This obviously is not desirable in either qualitative or quantitative work that requires a high degree of precision. Constant-flow pumps, on the other hand, do deliver a constant flow rate even if the column permeability changes.

A. Constant-Pressure Pumps

Pneumatic. Holding coil pneumatic systems are the least expensive pumping system. In this system, gas (N_2) under pressure is applied directly to the mobile phase in a coiled tube or a cylinder. This system has several deficiencies: (1) it has a limited solvent reservoir; (2) the reservoir has to be flushed with relatively large volumes of solvents when changing from one mobile phase to another; (3) it operates at rather low pressures, 1500 psi or less; and (4) most important, the gas (N_2) dissolves in the mobile phase, which can result in baseline noise in ultraviolet (UV) detectors.

Pneumatic Amplifier. Pneumatic amplifier pumps are modified versions of holding coil pneumatic pumps. In these systems, as shown in Fig. 1, a large piston is driven by the N_2 gas at relatively low pressures, which in turn drives a smaller piston against the mobile phase in the cylindrical reservoir. The problem of dissolved N_2 gas is much less in this system than in the holding coil system.

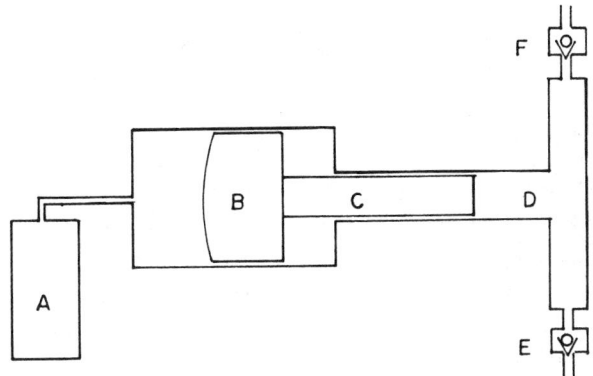

FIG. 1. Diagram of a pneumatic amplifier pump: A, low-pressure gas supply; B, large piston; C, small piston; D, liquid chamber; E, solvent inlet check valve; F, solvent outlet check valve.

There are two major drawbacks to systems of this type: (1) solvent capacity is limited to that of the smaller cylinder, which when emptied is automatically filled from a larger solvent reservoir, with a momentary stoppage of flow; and (2) when changing mobile phases, both the larger reservoir and the pressurized cylinder must be flushed with at least three volumes of the new mobile phase, which is rather time consuming and inconvenient. It should be pointed out that pneumatic amplifier pumps can be designed to operate as either a constant-pressure or a constant-flow device. The former is more common.

B. Constant-Flow Pumps

Positive-Displacement Pumps. In the positive-displacement syringe pump shown in Fig. 2, the piston is driven through the cylinder (solvent reservoir) by a motor-driven screw gearing device. Pumps of this type produce a pulse-free, constant flow of solvent, resulting in the lowest noise level of all pumping systems in those detectors that are flow sensitive.

The major drawbacks of these systems are: (1) they have a limited solvent capacity; (2) solvent changeover requires tedious

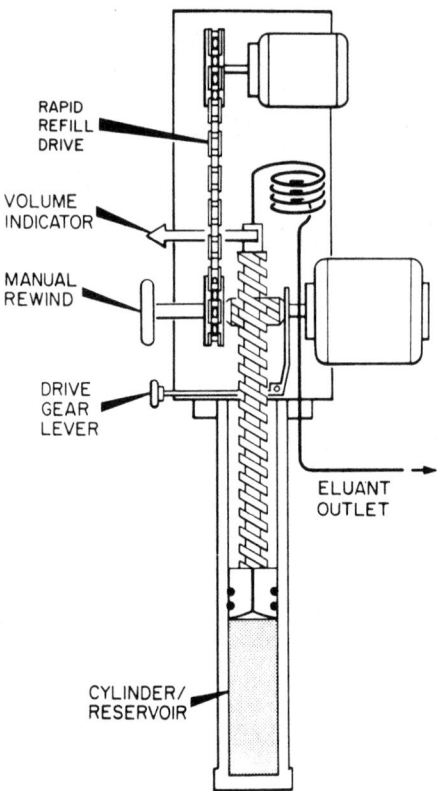

RAPID
REFILL
DRIVE

VOLUME
INDICATOR

MANUAL
REWIND

DRIVE
GEAR
LEVER

ELUANT
OUTLET

CYLINDER/
RESERVOIR

FIG. 2.   Schematic of a positive-displacement syringe pump.   (From Ref. 3.)

and time-consuming rinsing of the solvent reservoir (cylinder); (3) these pumps are more expensive than the other types available; and (4) two pumps must be used for gradient formation.

Reciprocating Pumps.   The pumping system currently used in most liquid chromatographs is the reciprocating pump.   In general, these systems use a gear-driven cam to drive a piston through a low-volume pump chamber.   Single-head, dual-head, and triple-head pumps are available.   Figure 3 is a schematic of a dual-head pump.   All produce a constant flow, the magnitude of pulsation being dependent on pump design.   The flow rate in these systems can be adjusted by increasing or decreasing the pump stroke rate, or by increasing or

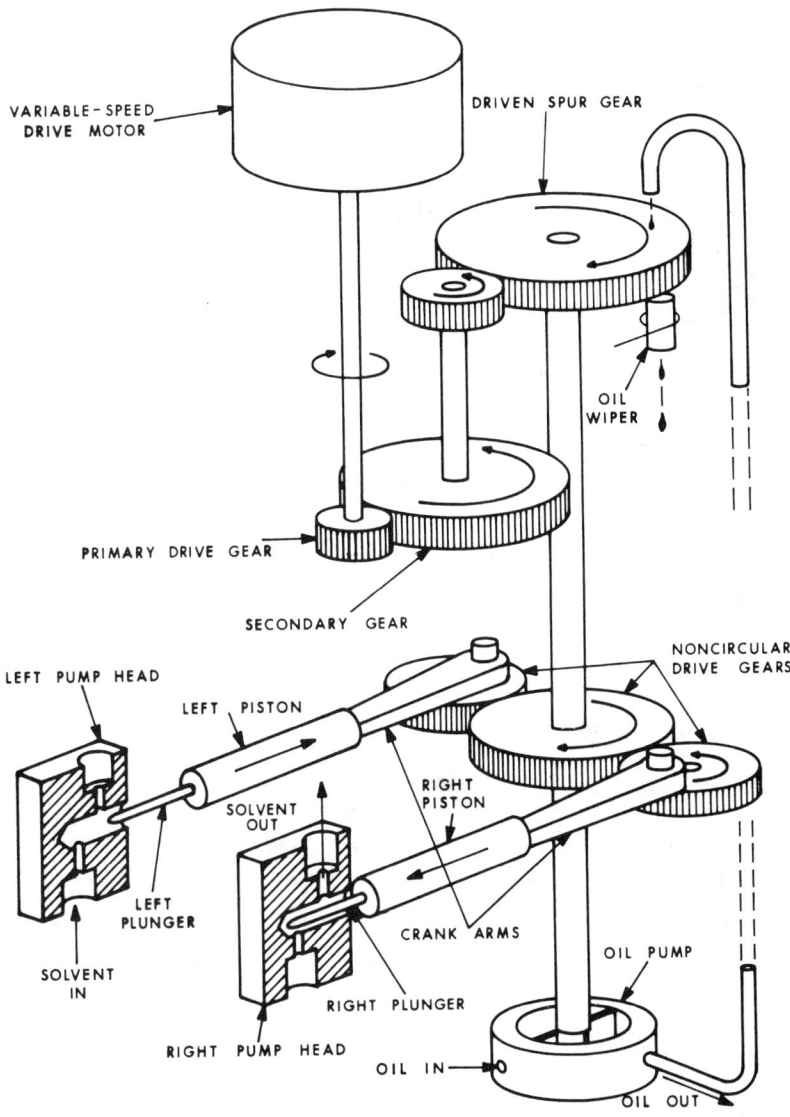

FIG. 3. Schematic of the Waters M-6000A, a dual-head reciprocating pump. (From *Manual IM-25745*, Sept. 1976, Rev. A. c 1975, Waters Associates, Milford Mass.)

decreasing the length of the pump stroke and thus the pump volume. Inlet and outlet check valves are used to control the direction of solvent flow through the pump. Failure of a check valve results in low solvent delivery and noise. Most check valves are actuated by pressure changes, but mechanical activation is also used.

The major difficulty with these pumps is that of pulsation, which with proper design or judicial use of pulse dampers can be nearly eliminated. An additional problem is that of cavitation, observed in some high-volatility solvents, such as methylene chloride and ethyl ether. This results when the pump looses prime or becomes starved. These pumping systems also have more moving parts (i.e., pistons, balls in the check valves, etc.), so that pump seals must be periodically replaced and check valves cleaned by pumping various cleansing solutions. These maintenance procedures are easily done by the user and present no serious problem.

III. SAMPLE INTRODUCTION

The highest chromatographic efficiencies are obtained by injection directly onto the top of the column. This can be achieved through the use of septum injectors or by stop-flow injectors.

A. Septum Injectors

The type of injector used in the earliest liquid chromatographs was that of the septum injector. This was a simple modification of the type used in gas-liquid chromatography. They are cheap and relatively easy to use at low pressures but they have many disadvantages.

1. Septum life is relatively short.
2. The mobile phase must be compatible with the composition of the septum. Often two or three types of septa were needed to cover the range of solvents used in most mobile phases.
3. Special syringes are required to avoid blowout and loss of sample.
4. Depending on injector design, septum material can clog the column end fitting, causing an increased pressure drop, usually

accompanied by a leaky septum.

5.   These injectors are limited to about 2000 psi.

This type of injector is seldom used in modern liquid chromatographs.

B. Stop-Flow Injectors

This injector has a valve or a cap seal, allowing the septum to withstand pressures up to 6000 psi. In use the flow is stopped; the valve is positioned or the cap seal is removed, giving access to the septum; the injection is made with a syringe; the cap is replaced or the valve repositioned in the run position; and the flow is resumed.

This technique is useful at pressures in excess of 2000 psi, but it has the distinct disadvantage that the precision of the technique depends on the skill of the operator.

C. Valve and Loop Injectors

Direct on-column injectors have generally been replaced by various types of valve and loop injectors. In all valve and loop injectors, the sample is introduced at atmospheric pressure, either as a measured volume by a microsyringe, or by a larger syringe into a fixed-volume sample loop, and then through the switching valve into the mobile-phase flow to the column.

This type of injector has several advantages over on-column injectors: (1) they operate at very high pressures, 6000 to 10,000 psi; (2) in certain instances they offer high precision since they depend only slightly on operator skill; (3) they can be used for analytical or preparative injections; and (4) through the use of accessory pneumatic actuator valves they can be adapted to unattended automatic injection.

There are also some disadvantages in the valve and loop injector type: (1) a small decrease in efficiency due to band spreading caused by the laminar flow pattern of the solvent in the injector loop system, and (2) this type of injector is relatively more expensive than on-column injectors.

There are basically four subtypes or variations of valve and loop injectors.

Type 1. Type 1 injectors were introduced in the early 1960s. These are six-port rotary valves, simple in design and operation, utilizing fixed-volume sample loops. With proper filling (i.e., 5-10 loop volumes [1]) extremely precise injections can be achieved. The major disadvantage of this system is that to change the sample injection volume, the sample loop must be changed. An additional disadvantage is that, as noted previously, in proper use 5-10 loop volumes are needed to fill the loop, resulting in sample waste. Another disadvantage is that the sample loop inlet line must be rinsed or flushed between injections to avoid cross contamination. Figure 4 shows the flow diagram of the Rheodyne Model 7010, a type 1 injector.

Type 2. The main differences between type 1 and type 2 injectors are as follows:

1. The sample loop in general has a larger volume than the sample injection volume, so that the volume of the injection is controlled or determined by a microsyringe for analytical work or a larger syringe for preparative work.

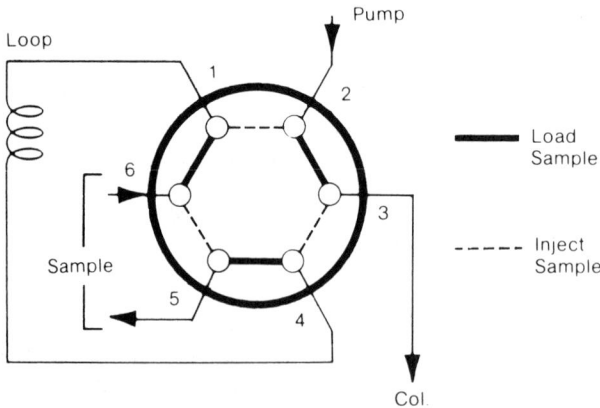

FIG. 4.   Flow diagram of the Rheodyne Model 7010, a type 1 injector. (From *Bull. 101,* Rheodyne Inc., Cotati, Calif.)

2. While the sample is being introduced into the injector (sample loading), the mobile-phase flow is switched through the bypass or restrictor coil. After injection, the injection port is sealed with the sample loading plug and the switching mechanism then changes the flow back to the sample loop, forcing the sample onto the column.

Figure 5 shows the operation of the Waters UK-6 injector, a type 2 injector.

The advantage of type 2 injectors are: (1) injection volumes are variable up to that of the sample loop; and (2) none of the sample is trapped in the injector, so that in normal usage it is not necessary to rinse the injector between injections.

Type 3. Type 3 injectors are a combination of types 1 and 2, in that the sample may either be introduced with a microsyringe (partial loop filling) or the sample loop may be totally filled using a larger syringe. One difference between type 3 and types 1 and 4 injectors is that the injection port is never exposed to high pressures, and therefore there is no need to plug it after the injection. Both six-port and four-port designs are available in type 3 injectors.

Type 3 injectors have the disadvantage that a small volume of sample, 0.5-2 µl depending on the design, remains trapped in connecting passages of the injector. The volume retained is constant and can be corrected either by injecting an extra amount of sample equivalent to that retained, or by the use of a specially calibrated syringe. This also means that to avoid contamination, the injector must be rinsed between injections.

Figure 6 shows the flow diagram of the Altex 210, a type 3 injector of four-port design.

Type 4. Type 4 injectors are like type 3 except that there are no connecting passages between the syringe needle tip and the sample loop. This means that no sample is retained, and therefore no flushing between injections is required in normal usage. Figure 7 shows the flow pattern of the Rheodyne Model 7125, a type 4 injector.

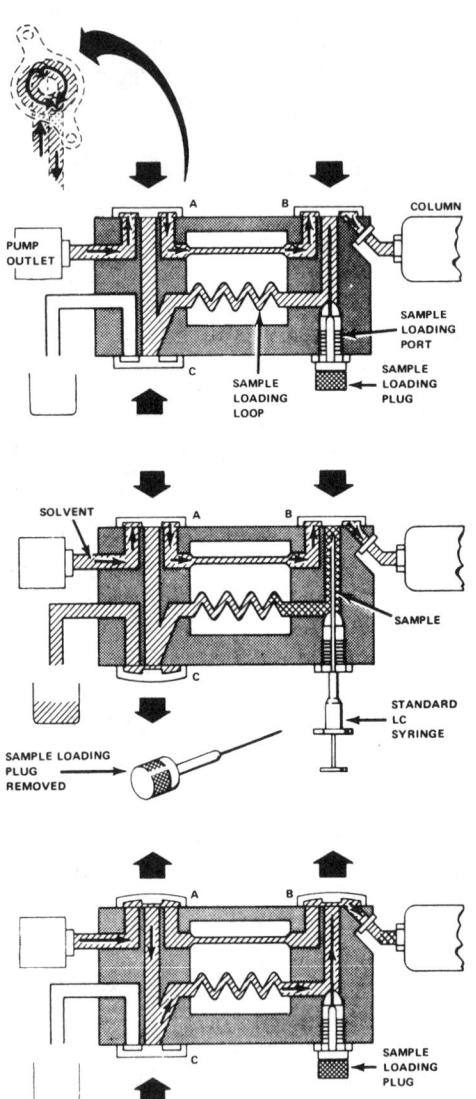

FIG. 5. Flow diagram and schematic showing the operation of the
Waters UK-6 injector, a type 2 injector. (From *Publ. An-153,* Jan.
1975, c 1975, Waters Associates, Milford, Mass.)

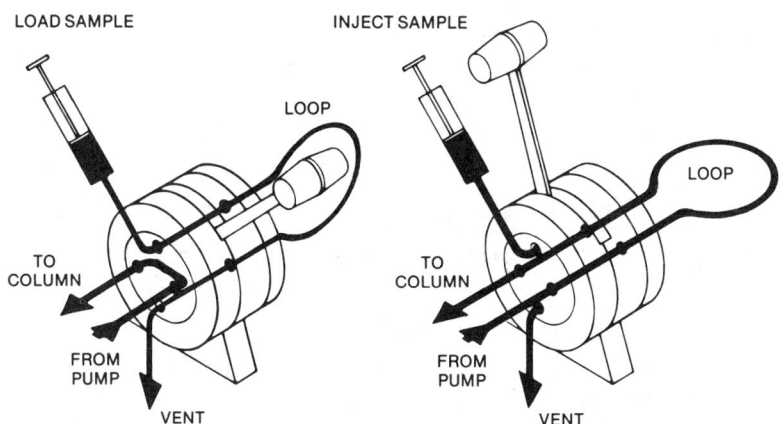

FIG. 6. Flow diagram and schematic showing the operation of the Altex 210, a type 3 injector. (From *Sample Injection Valves 130/279/10M,* Brochure-Series 210, Altex Scientific Inc., Berkeley, Calif.)

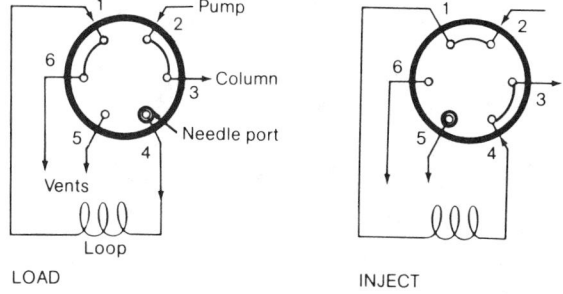

FIG. 7. Flow pattern of the Rheodyne Model 7125, a type 4 injector. (From *Bull. 106,* Rheodyne Inc., Cotati, Calif.)

IV. COLUMNS

This brief discussion will be concerned only with column design for analytical and preparative applications. Packing materials are discussed elsewhere in the book.

A. Conventional Columns

The column is formed from a stainless steel tube with various forms of terminators or end fittings. For optimum efficiency the inner walls of the column should have a crack-free, smooth, polished surface, but more important, the end fittings should contribute minimal dead volume. All columns employ a porous metal frit at both inlet and outlet: one to protect the column and the other to keep column packing fines from reaching the detector.

Figures 8-11 illustrate some of the different configurations to end fittings and frits that are used in analytical and preparative columns. The column tube shown in Figure 8 is drilled out to accept the frit. These are of a rather permanent nature, so if they become clogged, the entire column may need replacement. This type is not widely used in modern high-performance liquid

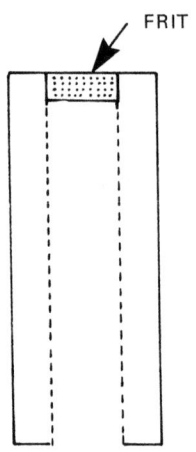

FIG. 8. Column end drilled out to accept the frit.

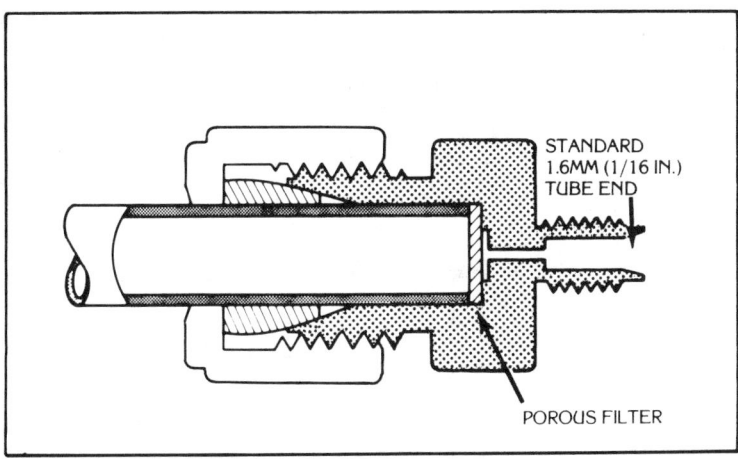

FIG. 9. Standard configuration end fitting. Accepts compression nut connectors. (From *HPLC Columns and Packings Product Guide - E37310,* DuPont Co., Wilmington, Del.)

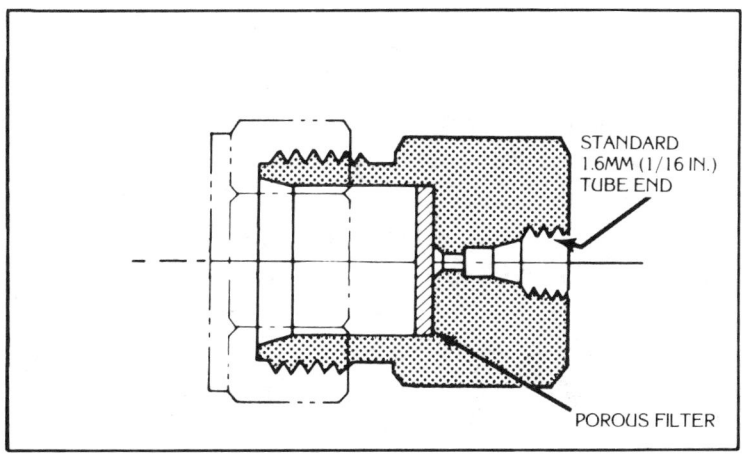

FIG. 10. Inverted configuration end fitting. Accepts compression screw connectors. (From *HPLC Columns and Packing Product Guide - E37310,* DuPont Co., Wilmington, Del.)

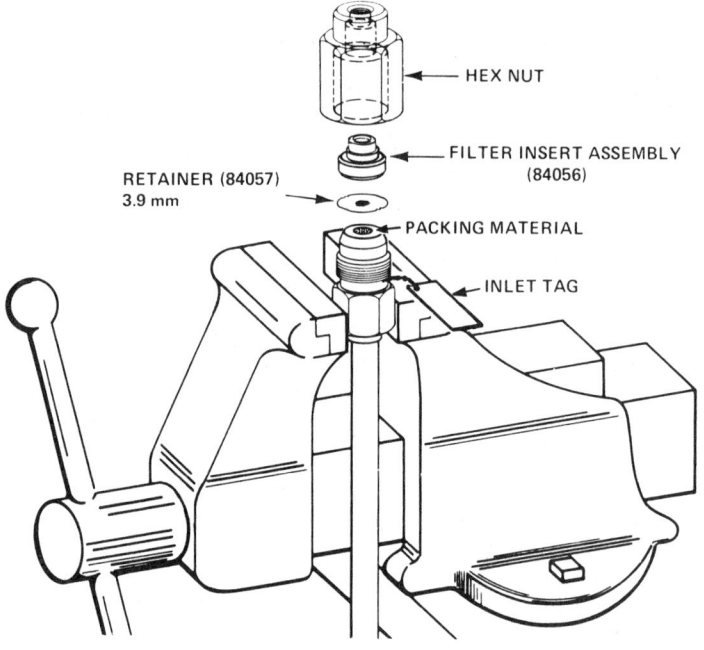

HEX NUT

FILTER INSERT ASSEMBLY
(84056)

RETAINER (84057)
3.9 mm

PACKING MATERIAL

INLET TAG

FIG. 11. Variation of the inverted configuration end fittings, which accepts compression screw connectors. (From *Publ. CU-27386,* Rev. B, c 1976, Waters Associates, Milford, Mass.)

chromatography (HPLC) columns. Figures 9 and 10 show standard and inverted configurations of the type widely used. Connecting tubing is fitted with compression nut or compression screw fittings, respectively. In general, the frits are replaceable in these end fittings. Figure 11 shows a variation of the inverted configuration end fitting. This end fitting also uses a replaceable frit.

Most analytical columns fall into two internal diameter ranges: 2-2.6 mm or 4.6-5 mm and are approximately 25 cm in length. Preparative or semipreparative columns fall in the range 7-35 mm ID and are usually 25 or 50 cm in length. Figure 12 is a diagram of the Whatman Magnum 9 preparative column. Note that the outlet end frit is of a smaller diameter than the inlet frit. It is claimed that this lets solutes be eluted from the large-diameter column (9.4 mm)

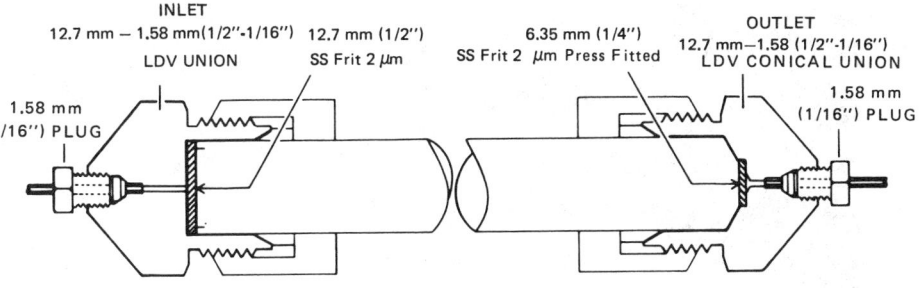

FIG. 12. A diagram of the Whatman Magnum 9 preparative column.
(From *Publ. 126*, Whatman Inc., Clifton, N.J.)

into the fine-bore tubing leading to the detector without tailing
or band spreading.

The larger-diameter columns, simply because they hold more adsorbent, have a greater sample capacity than the analytical columns.
They have the disadvantage of higher initial cost and greater operating cost due to increased solvent consumption.

B. Radial Compression

The concept of radial compression columns was first introduced in
1976 by Waters Associates as the column in the Waters Prep LC and
has been applied to a scaled-down analytical version. Columns for
the analytical version are either 5 mm ID or 8 mm ID, are 10 cm
long, and are made of a virgin, high-density polyethylene tube.
The radial columns fit into a special, mechanical device that applies hydraulic pressure to the column, thus creating the radially
compressed column. Figure 13 is a diagram of the end of a Radial-
Pak column, and Fig. 14 is a diagram of the column installed in the
radial compression module. These columns are claimed to (1) be
highly efficient, (2) eliminate channeling and voiding, (3) operate
at high flow rates at low back pressures, and (4) be convenient and
economical to use [2]. It would appear that the major disadvantage
of this system is the initial high cost of the radial compression
module, one module being required for each operating column.

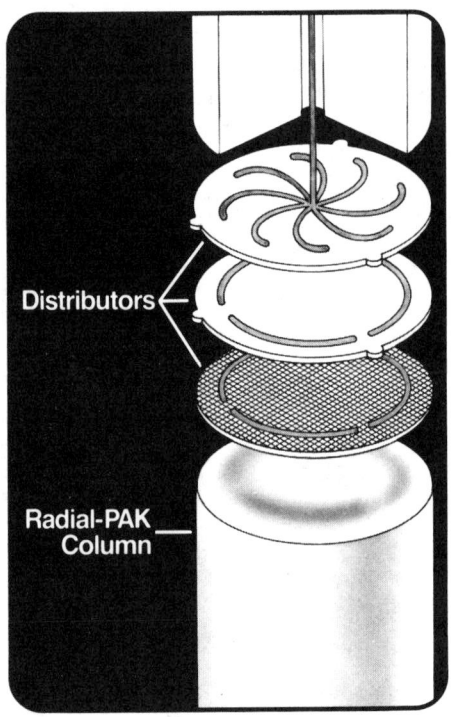

FIG. 13.   Diagram of the Radial-Pak column showing design features
of the column head.   (From *The Waters RCSS,* Brochure B30/Mar. 1981,
© 1981, Waters Associates, Milford, Mass.)

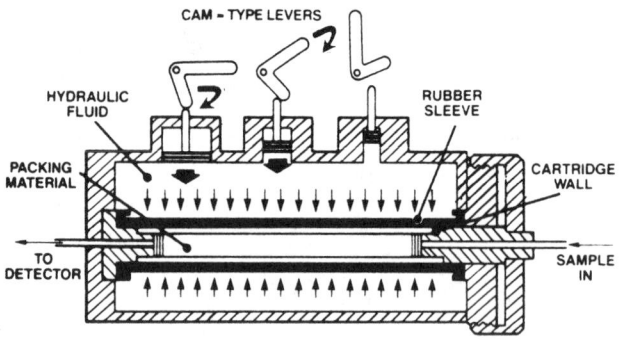

FIG. 14.   Diagram of the radial compression module with a Radial-Pak
column installed.   (From *The Waters RCSS,* Brochure B30/Mar. 1981,
© 1981, Waters Associates, Milford, Mass.)

V. DETECTORS

It has been suggested [3] that the introduction of the flow-through detector by Tiselius in 1940 was the single most important breakthrough in the advancement of liquid chromatography as it exists today. From this early use of the refractometer have evolved a number of highly developed detectors designed specifically as HPLC detectors.

Detectors are of two very broad types: (1) those that measure by difference a bulk property of the solute and mobile phase, the most common example being the refractive index; and (2) those that measure a specific property of the solute which is not possessed, or at least to a much smaller degree, by the mobile phase, the most common example being ultraviolet (UV) adsorption. In addition, some detectors measure these properties, generally bulk properties, after elimination of the mobile phase.

A. Properties of Detectors

The essential properties of HPLC detectors have been variously described [3-6], and are as follows:

1. Should respond to all solutes (universal) or have a predictable selectivity (specific)
2. Should not respond to mobile phase
3. Should produce predictable, highly sensitive response
4. Should be relatively independent of temperature and flow variations
5. Should have a wide range of linear response
6. Should not contribute significantly to extra column band broadening
7. Should have a short response time
8. Should provide some degree of qualitative information
9. Should be nondestructive, permitting further detection and identification
10. Should be convenient, easy to use, reliable, and inexpensive
11. Should be compatible with gradient elution

It is immediately obvious that no detector could possibly satisfy
all these characteristics. There are many detectors available, each
having some of these characteristics, and it is possible through the
use of a series of detectors to obtain not only a high degree of
specificity, but also a wide range of solute-type detectability
within a given sample. Snyder and Kirkland [5] contend that although all of the aforementioned characteristics are important, detector sensitivity may determine the detectors applicability for
any given sample.

In trace-level analysis detector noise becomes very important
where the detection of very small peaks is required. Detector noise
is that noise which is due only to the detector, under static conditions, and appears either as a rather uniform "fuzz," widening the
baseline, or as random peaks or spikes. The former is known as
high-frequency noise and the latter as *short-term noise*. Baseline
drift (up or down) with time also makes detection of small peaks
difficult. The detection limits of a detector are generally taken
as twice that of the signal-to-noise level.

In Sec. V.B several of the different types of HPLC detectors
are described briefly and discussed relative to desirable characteristics defined above.

B. Refractometers

As previously noted, the refractometer was the first flow-though
HPLC detector and is probably second only to UV detectors in current
usage. Since all solutes have a refractive index, the refractive
index (RI) detector comes closest to being a universal detector.
Solvents also have refractive indices, so that mobile-phase composition is critical. The ultimate sensitivity is determined by the
difference in the refractive index of the solute and the solute plus
mobile phase, and at the same time the mobile phase must satisfy the
chromatographic requirements of the separation. Sensitivity for
this detector is typically no better than 10^{-6} g/ml, which makes it

unsuitable for trace analysis. The lack of sensitivity makes this
detector useful in preparative work.

RI detectors are very sensitive to flow changes and temperature
variations, so consideration needs to be given to the type of pumping system used and to the temperature controls.

In addition to its use in preparative work, the RI detector is
widely used in gel permeation chromatography (GPC), where the mobile-
phase composition, usually a pure solvent, does not have "chromatographic" requirements imposed by other chromatographic modes. RI
detectors have the advantage of being nondestructive detectors, as
witnessed by their use in preparative work.

There are two basic types of refractometers in common use as
HPLC detectors: the Fresnel (reflective) and the deflection type,
as shown in Figs. 15 and 16, respectively. Both types measure the
change or difference in the refractive index of the pure mobile
phase via the reference cell and that of the solute plus the mobile
phase via the sample cell. Fresnel (reflective) RI detectors measure this difference via a change in the intensity of the reflected
light beam reaching the photocell. Deflection RI detectors measure
the change via the angle of deflection, which changes the focus of

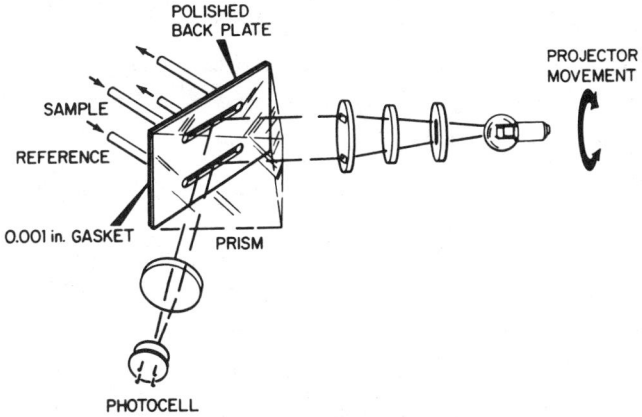

FIG. 15.   Diagram of a typical Fresnel refractometer.   (From Ref. 3.)

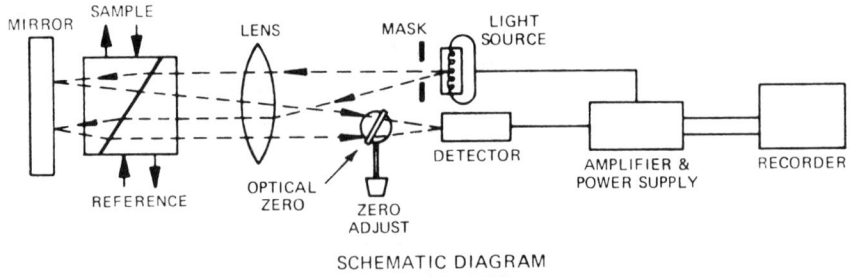

SCHEMATIC DIAGRAM
R–400 SERIES

FIG. 16. Schematic diagram of the Waters R-400 series, a deflection
refractometer. (From *Manual 48597*, Rev. B, c 1976, Waters Associates, Milford, Mass.)

the light beam on the detector as the refractive index of the liquid
in the sample cell changes.

Fresnel RI detectors have a greater potential sensitivity than
do deflection refractometers, but since they measure the intensity
of light changes, the cells must be kept very clean. Additional
disadvantages of this type of refractometer are (1) a limited linearity range and (2) they require two prisms to cover the RI range
normally covered, 1.33-1.63. Deflection refractometers, on the
other hand, have a wide linearity range, and since they do not
measure light intensity changes, are less sensitive to particles of
dirt, bubbles, and so on, that may accumulate in the cell. In addition, they require only one prism to cover the RI range 1.33-1.63.
The main disadvantage of deflection refractometers is that they are
more expensive than the Fresnel type.

C. UV Absorbance Detectors

A very high percentage of organic compounds possess some degree of
conjugation or functional groups that absorb UV and or visible light,
so that the UV absorbance detector is the most widely used detector.
It does possess some selectivity, in that only UV-absorbing compounds can be detected. Sensitivities for substances possessing
good UV properties can be as great as 10^{-10} g/ml for fixed-wavelength

detectors, so that UV detectors are applicable to trace analysis. UV detectors have a good linearity range, making them very useful for quantitative analysis. Depending on the operating wavelength, many solvents are compatible with UV detection. UV detectors are also compatible with gradient elution, again depending on the operating wavelength and UV properties of the solvents. These detectors are easy to use, reliable, relatively inexpensive, and are nondestructive. The main disadvantage in these detectors is not with the detectors, but the fact that the UV properties of the organic substances to be detected vary so widely.

There are basically three types of UV absorbance detectors: (1) fixed wavelength, usually 254 nm; (2) multiple wavelength, which through the use of filters can operate at several wavelengths, but in practice operate as a fixed-wavelength detector; and (3) variable wavelength, which can be set at any wavelength, UV to visible, usually from 190 to 700 nm.

The first two types commonly used low-pressure mercury vapor lamps, which produce intense radiation at 254 nm, as the radiation source. The second most commonly used wavelength in fixed-wavelength detectors is 280 nm, which is obtained through the use of a phosphor converter and a filter. Converter kits are available which produce other wavelengths, such as 313, 340, 365, 405, 436, and 546 nm. These converter kits give these detectors an extra degree of selectivity and in some cases greater sensitivity by measuring the absorbance at or nearer the wavelength of the solute's maximum absorptivity. This extra sensitivity may be lost in many cases due to increased baseline noise and reduced source strength.

In fixed-wavelength detectors the radiant energy is passed though the cell, which includes the flowing sample cell and an air reference cell, through an appropriate filter, and onto a dual photocell. The current generated by the sample and reference photocells is then converted, by log amplifiers, into an output signal that is linear with solute concentration. The block diagram of a dual-cell fixed-wavelength absorbance detector is shown in Fig. 17.

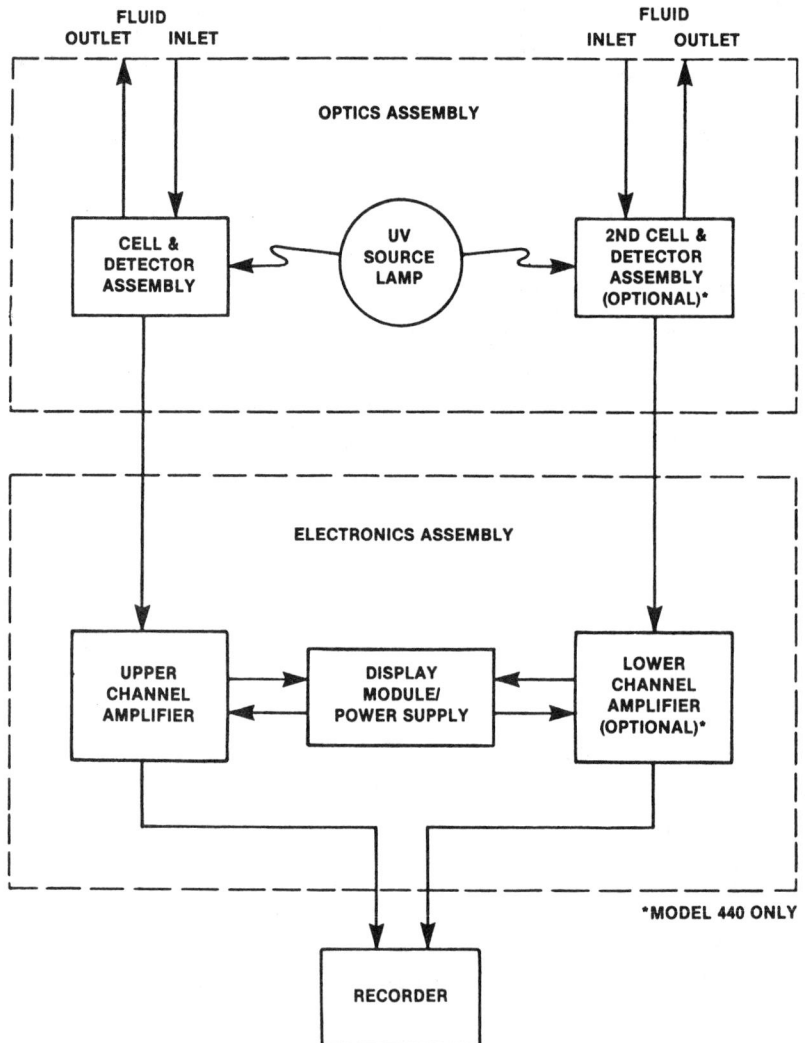

FIG. 17. Block diagram of the Waters Model 440 absorbance detector, a multiple, fixed-wavelength detector. (From *Series 440: Absorbance Detectors Operators Manual*, c 1981, Waters Associates, Milford, Mass.)

The optics path for one cell is shown in Fig. 18. This particular
absorbance detector can function either as two independent detectors,
that is, can monitor column effluent from two separate columns, or
the cells can be connected in series and one column effluent can be
monitored at two different wavelengths, thus adding greatly to the
qualitative information obtained for the detected compound via absorbance ratioing.

Variable-wavelength detectors allow the chromatographer to
select any wavelength, usually 190-700 nm, in the UV or visible
range. As in conventional UV-visible spectrophotometers, deuterium
and tungsten lamps serve as radiation sources for the UV and visible
regions, respectively. Thus compounds that are considered to be
nonabsorbing in the UV (e.g., sugars) can be detected (195 nm),
greatly expanding the utility of the UV absorvance detector for
HPLC. Some variable-wavelength absorbance detectors permit a peak
to be trapped in the cell and rapidly scanned, giving a spectrum
like that obtained on a conventional UV-visible spectrophotometer.
This provides additional qualitative information about the trapped
substance as well as its purity. The detector limits for highly
absorbing compounds using the variable-wavelength detector is about
10^{-11} g/ml. The major disadvantage of the variable-wavelength

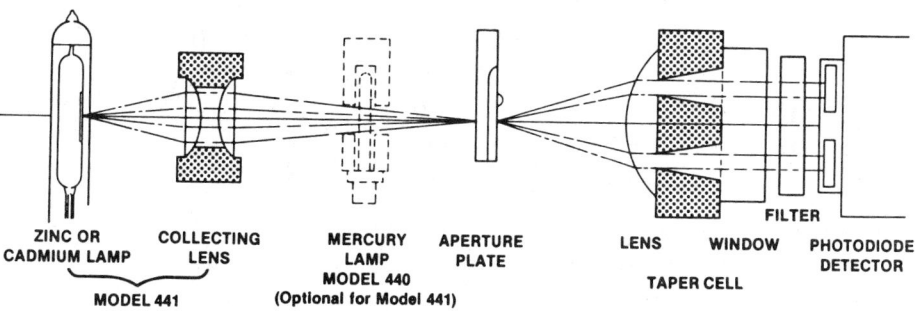

ZINC OR   COLLECTING   MERCURY   APERTURE   LENS   WINDOW   PHOTODIODE
CADMIUM LAMP   LENS   LAMP   PLATE   DETECTOR
                                   MODEL 440   TAPER CELL
        MODEL 441   (Optional for Model 441)

FIG. 18. Optics path of the Waters Model 440 absorbance detector,
a multiple, fixed-wavelength detector. (From *Series 440: Absorbance Detectors Operators Manual,* c 1981, Waters Associates,
Milford, Mass.)

detector is cost, about twice that of fixed wavelength, and in addition the deuterium lamps have a much shorter life expectancy than
do the cheaper mercury vapor lamps.

D. Fluorescence Detectors

A relatively small number of organic substances possess the ability
to absorb energy at one wavelength and almost instantaneously emit
energy at a different, higher wavelength. This phenomenon is known
as fluorescence. Because the number of fluorescing substances is
relatively small, fluorescence detectors have a relatively high
degree of selectivity. In addition, the detector limits can extend
down to 10^{-12} g/ml, making the fluorescence detector one of the most
sensitive HPLC detectors and very useful in trace analysis applications. The linearity range may be rather small, so that in quantitative analysis this range should be determined for each solute in
the concentration ranges of interest. This detector is compatible
with most solvents and can be used in gradient elution applications,
but it should be noted that chloroform and methylene chloride can
quench and lessen fluorescence.

The fact that both the excitation wavelength and the emmision
wavelength detected can be varied adds additional selectivity to
this detector.

There are two basic types of fluorescence detectors, the primary difference being whether they use filters or grating monochro-
meters as the mechanism to control wavelength selectivity. The
filter instruments are cheaper but are limited in selectivity,
whereas the more expensive monochrometer instruments are variable
throughout their range, typically 200-800 nm.

In the fluorescence detector, the energy from the lamp, after
passing through the excitation filter or through the optical system
of the grating monochrometer, passes through the sample in a quartz
cell. The emitted fluorescent energy is measured at right angles
to the incident excitation energy. The emitted energy is then
passed through an emission filter or through the optical system of

the emission grating monochrometer and is eventually focused on a
photomultiplier detector. Figure 19 is the schematic of a filter-
type fluorescence detector.

E. Electrochemical Detectors

Electrochemical detectors, although not widely used, can be very
useful detectors for some samples. The electrochemical detector
is based on measurement of the current produced by the oxidation
or reduction of the eluted solute at a suitable electrode by the
application of a controlled potential. Platinum, gold, mercury,
carbon paste, and glassy carbon have all been used as electrodes in
electrochemical detectors.

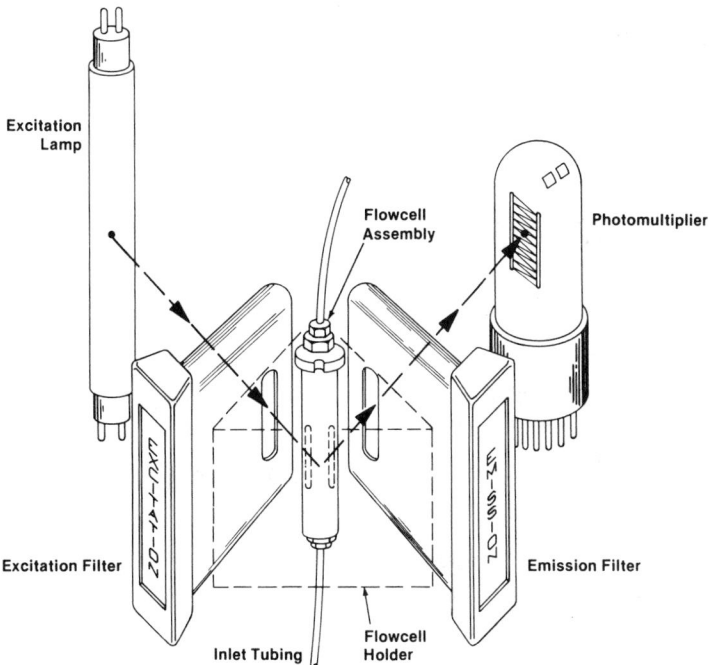

FIG. 19. Schematic of the Kratos Model FS950 Fluoromat Fluorometer.
(From brochure entitled *FS 950: Fluoromat Fluorometer*, SP5/78-102
R1-10-780. Courtesy of Kratos Analytical Instruments, Westwood,
N.J.)

For this detector to be useful, the sample to be detected must
be electroactive, that is, electrooxidizable or electroreducible.
Figure 20 shows the voltage ranges for some electroactive functional
groups.

Electrochemical detectors, primarily amperometric [7], offer
high sensitivity (10^{-12} g/ml), selectivity, and relatively low cost.
In addition to the limitations previously mentioned, severe restrictions are placed on the mobile phase. It must be electrically con-
ductive, which in most instances limits the detector to applications
utilizing reversed-phase, reversed-phase ion-pairing, and ion-
exchange systems, where suitable salts can be added. It should be
noted that in some cases the addition of a salt may change the
chromatography. Additionally, the mobile phase must be free of
halides, oxygen, and metal contaminants. If present, these contaminants add to the background current, producing noisy, drifting
baselines.

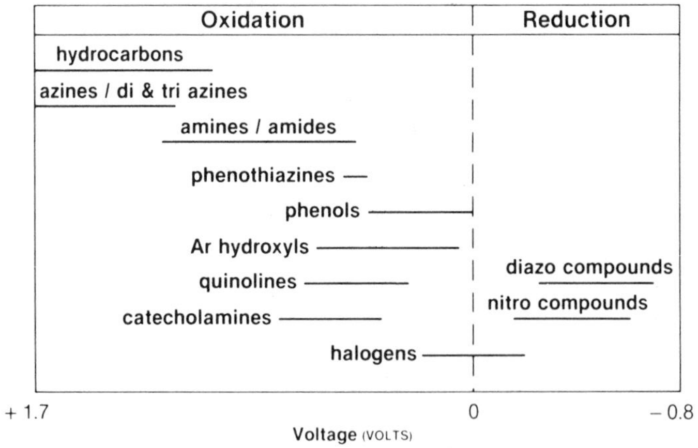

Oxidation	Reduction
hydrocarbons	
azines / di & tri azines	
amines / amides	
phenothiazines —	
phenols —————	
Ar hydroxyls —————	
quinolines ————	diazo compounds
catecholamines ————	nitro compounds
halogens ————	

+ 1.7   0   − 0.8
Voltage (VOLTS)

FIG. 20. Diagram showing electroactivity range of some electroactive
functional groups. (From brochure describing the CMX-20 Amperometric
HPLC detector, and accompanying literature. Courtesy of Chromatix,
Inc., Sunnyvalue, Calif.)

F. Infrared Absorbance Detectors

Although infrared detectors have been used in some gel permeation
chromatography applications [8] and recently in nonaqueous reversed-
phase gradient elution separation of saturated triglycerides [9],
this detector would appear to have little general utility as an
HPLC detector.

Severe limitations are placed on mobile-phase selection because
of spectral considerations and also because of detector cell incom-
patability. Infrared detectors are also relatively insensitive,
the minimum detection limits being about 10^{-6} g/ml.

G. Liquid Chromatography-Mass Spectrometry

Depending upon their selectivity, all the detectors discussed previously provide some degree of qualitative information in addition
to the retention time of eluted solutes. The mass spectrometer as
an HPLC detector not only permits detection of the eluted solute,
but can also provide absolute identification. It is therefore a
very powerful tool.

The primary difficulty with liquid chromatography-mass spectrometry (LC-MS) involves the interface, that is, the introduction of
the eluted solute into the mass spectrometer. There are basically
two types of interfaces used in commercially available systems:
the direct liquid interface (DLI), used by Hewlett-Packard and
Ribermag (France); and (2) the moving-belt interface, available
from Finnigan Instruments and V. G. Micromass (England).

In the DLI of Hewlett-Packard (Fig. 21), the total column
eluate passes into a water-cooled probe. Approximately 0.3-3% of
the eluate (at a flow rate of about 1 ml/min) flows from a pinhole
orifice directly into the chemical ionization (CI) source of the
mass spectrometer, while the major portion of the eluate flows back
out of the probe. The solvent itself acts as the CI reagent. The
most obvious drawback of this system is that the very low amount of
eluted sample utilized for detection severely limits the sensitivity
of this detector.

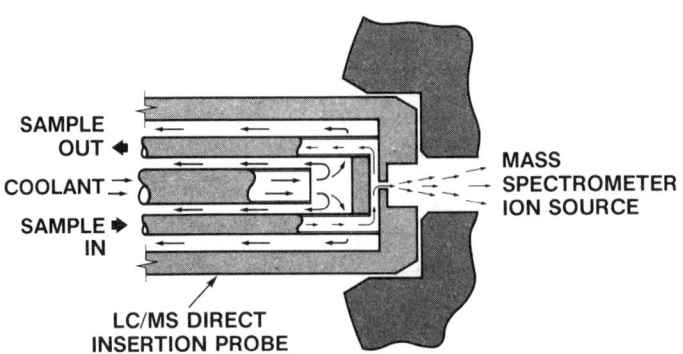

FIG. 21.   Schematic of the Hewlitt-Packard direct liquid interface.
(From *HP Publ. 23-5952-5935* (6/79).   Courtesy of Hewlett-Packard,
Palo Alto, Calif.)

A diagram of the Finnigan Instruments moving-belt interface is
shown in Fig. 22.   In this system the column eluate flows either
through a controllable splitter or directly onto the belt, which is
made of stainless steel or the polyimide material Kapton.   The belt
then transports the column eluate under an infrared lamp, which
evaporates most of the solvent, and then the solute and residual
solvent pass through two vacuum locks into a flash vaporizer.   The
vaporized sample then passes into the ion chamber.   This system can
operate either in the CI mode or in the electron impact (EI) mode.
After vaporization of the sample the belt moves through a cleanup
heater, where any residual sample is removed.

For favorable mobile phases (i.e., volatile nonpolar solvents),
the belt can handle flow rates up to 1.5 ml/min.   For unfavorable
mobile phases -- those containing large quantities of water, such
as those used in reversed-phase and reversed-phase ion-pairing
chromatography -- this may fall to as low as 0.05 ml/min [10].

Although the moving-belt interface introduces a larger amount
of sample for detection, at least for volatile nonpolar mobile
phases, the mobile phase still imposes restrictions on the potential
sensitivity of this detector.

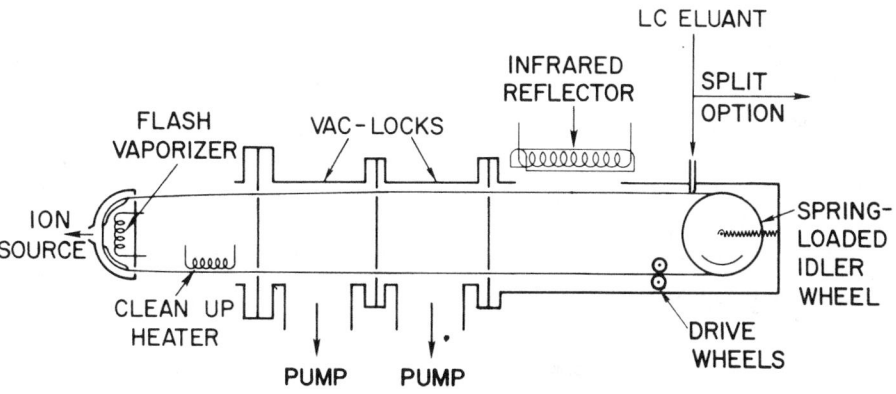

FIG. 22. Diagram of the Finnigan moving-belt interface. (From brochure entitled *LC/MS*. Courtesy of Finnigan Instruments, Sunnyvale, Calif.)

Henion and Maylin [11] have recently described a micro-DLI for use with a micro-LC system, and Henion [12] has demonstrated its usefulness in low-nanogram detectability in forensic drug applications. The entire effluent from the micro-LC is introduced into the mass spectrometer, where the mobile-phase solvents act as the ion source in the CI mode. It would appear that the use of micro-LC - micro-DLI may be a solution to many of the problems associated with conventional HPLC-DLI, particularly in trace analysis applications.

H. Other Detectors

Many additional detectors -- flame ionization, electron capture, radioactivity, thermal energy analyzer, atomic absorption, polarimetry, photoionization, and others -- have been investigated as HPLC detectors and for various reasons have not been widely used. These are not discussed in this chapter, but the use and applications of electron capture and thermal energy analysis detectors, as well as UV, MS, and electrochemical detectors, are discussed in Chap. 7.

It should be noted in passing that through the use of suitable pre- or postcolumn reactions or derivatization reactions, many solutes can be rendered detectable, or their detectability enhanced. These techniques are especially useful with UV [13-18] and in fluorescence [17, 19] detection.

VI. DATA HANDLING

In the simplest of chromatographic systems, the signal coming from the detector is fed into a strip chart recorder and a trace or chromatogram is produced. Manual qualitative and quantitative measurements of the peaks can be made using a ruler or a vernier caliper device. For quantitative measurements, either peak height or peak area measurements can be made. Peak height measurements are less dependent on flow rate variations and are well suited in those cases where flow rate control is poor [20] (i.e., a constant-pressure pump is used). Peak height measurements provide suitable accuracy and precision, provided that the chromatography is good (i.e., well-resolved, symmetrical peaks). Best results are obtained if the sample and standard solutions are approximately the same concentration. Detector sensitivity should be adjusted to give about one-half scale deflection at the desired concentration.

Manual peak area measurements using a ruler or a venier caliper device can be made in a variety of ways. The simplest, most widely used, and most accurate manual peak area measurement [3] is that of the peak height times the peak width at one-half peak height.

Two triangulation methods may be used, but these are less precise because of the problem of drawing tangents. In one triangulation procedure the peak height is measured from the baseline to the intersection of the tangents. In the other triangulation procedure the peak height is measured at the apex of the peak, and the tangents are drawn from this point to the baseline. In both cases area equals one-half peak width at the base times the peak height. These techniques require more time than the simpler peak height times peak width at one-half peak height.

An additional manual technique that can be used is that of

"cut and weigh," where the peak, or more often an xerographic copy
of the peak, is cut out and weighted. The success of this technique
depends on the care taken in cutting out the peak and on the uni-
formity of the paper thickness and moisture content. Precision and
accuracy can be improved by using a faster chart speed, thereby
making the peaks wider, heavier, and easier to cut out. Cutting out
and weighing several copies of one peak will also improve the precision of the procedure. Although quite time consuming, better results are obtained by cut and weigh than by triangulation for asy-
mmetrical peaks.

Planimetry can also be used, but its accuracy and precision
depend totally on the skill of the user and it is quite time consuming.

The more sophisticated HPLC systems employ electronic integra-
tors of varying capabilities, ranging from simple area and retention
time measurements for each eluting peak to those with computer
capabilities. These systems, employing printer-plotters, can be
programmed to give a very detailed analytical report, including not
only component concentrations and names, but chromatographic conditions, sample identification, date, and operator identification.
Greater accuracy and precision are obtained through the proper use
of these computer-integrators and the results are obtained almost
instantaneously. In one study [21] Scott and Reese observed that
the precision of computer measurements generally were better than
those of manual measurements. However, for peaks eluting at k'
values of about 5, the precision of manual measurements to that of
computer measurements was essentially equal.

The most highly sophisticated of the microprocessor-based com-
puters, in addition to data collection and reduction, may be interfaced with the HPLC to control the mobile-phase composition (isocratic or gradient operation), flow rate, detector wavelength,
sensitivity, and so on, and with ancillary equipment such as automated sample processors and fraction collectors.

REFERENCES

1. *Rheodyne Tech. Notes* 1, Sept. 1979.
2. G. W. Fallick and G. W. Rausch, *Am. Lab.,* p. 87 (Nov. 1979).
3. R. W. Yost, L. S. Ettre, and R. D. Conlon, *Practical Liquid Chromatography: An Introduction.* Perkin-Elmer Corp., Norwalk, Conn., 1980.
4. W. A. McKinley, D. J. Popovich, and T. Layne, *Am. Lab.,* p. 37 (Aug. 1980).
5. L. R. Snyder and J. J. Kirkland, *Introduction to Modern Liquid Chromatography,* 2nd ed. Wiley-Interscience, New York, 1979.
6. E. L. Johnson and R. Stevenson, *Basic Liquid Chromatography.* Varian Associates, Inc., Palo Alto, Calif., 1978.
7. P. T. Kissinger, *Anal. Chem.* 49:447A (1977).
8. G. Dallas and S. D. Abbot, *Ind. Res.* 19:58 (1977).
9. N. A. Parris, *J. Chromatogr. Sci.* 17:541 (1979).
10. W. H. McFadden, *J. Chromatogr. Sci.* 18:97 (1980).
11. J. D. Henion and G. A. Maylin, *Biomed. Mass Spectrom.* 7:115 (1980).
12. J. D. Henion, *J. Chromatogr. Sci.* 19:57 (1981).
13. T. H. Jupille, *Am. Lab.,* p. 85 (1976).
14. C. R. Clark, J. D. Teague, M. M. Wells, and J. H. Ellis, *Anal. Chem.* 49:912 (1977).
15. C. R. Clark and J. L. Chan, *Anal. Chem.* 50:635 (1978).
16. C. R. Clark and M. M. Wells, *J. Chromatogr. Sci.* 16:332 (1978).
17. S. Ahuja, *J. Chromatogr. Sci.* 17:168 (1979).
18. T. Jupille, *J. Chromatogr. Sci.* 17:160 (1979).
19. J. F. Lawrence, *J. Chromatogr. Sci.* 17:147 (1979).
20. R. W. Ross, *J. Pharm. Sci.* 61:1979 (1972).
21. R. P. W. Scott and C. E. Reese, *J. Chromatogr.* 138:283 (1977).

3 Adsorption Chromatography and the Use of Silica

John D. Wittwer, Jr.

South Central Laboratory
Drug Enforcement Administration
Dallas, Texas

I. INTRODUCTION

Chromatographic techniques, regardless of the degree of sophistication of the instrumentation, are separation techniques. Therefore, the purpose of a liquid chromatographic procedure is the separation of the components of a mixture placed on the head of a column by elution with a suitable solvent.

In adsorption chromatography the columns are usually packed with either silica or alumina. Silica is the most widely used adsorbent. Some of the reasons for the predominance of silica are as follows:

1. Silica columns are more efficient than corresponding alumina columns.

2. In general, silica is less reactive than alumina.

3. Most silicas have a relatively high surface area, which means that they will separate samples in amounts sufficient for additional analysis by infrared spectrophotometry, mass spectrometry, and so on.

4. Properly prepared analytical silica columns show pressure drops of 2000 psi or less when used at flow rates of 1-2 ml/min.

5. A wide range of types of compounds can be separated using silica.

6. A large number of solvents and solvent mixtures can be used, affording almost unlimited separational selectivities.
7. Silica is not subject to degradation by solvents normally used in adsorption chromatography.
8. Silica columns showing poor performance with usage can be regenerated to nearly their original efficiency.
9. Silica is available in many forms from many commercial sources in bulk and in packed columns.

II. THEORY

As previously stated, the objective of the chromatographic process is the separation of at least two components in a sample mixture. The chromatographic process then involves placing the sample mixture, dissolved in a suitable solvent, onto the top or head of a bed of adsorbent packed in a column and then by passing the solvent or solvent mixture through the column, causing the sample components to separate into bands moving down or through the column at different rates, thus effecting separation of the components. The solvent or solvent mixture is more correctly described as the mobile phase. Each of the participants in this process -- the sample, the adsorbent, and the mobile phase -- contribute to the relative success or failure of the chromatographic separation.

A. Mobile Phase

In liquid-solid chromatography (LSC) the ability to predict how a given solvent will effect the separation of sample solutes is a matter of great practical importance. Toward this end several theoretical models of adsorption in LSC have been proposed. These models are discussed and compared in this section.

Snyder Model. The original "competition" model for LSC was developed by Snyder in the 1960s. Snyder's prolific works during this period formed the nucleus for a monograph [1] on this topic published in 1968. The majority of Snyder's work was done using alumina as the adsorbent.

In the adsorption model as developed by Snyder [1], four basic assumptions are made:

1. The active sites at the surface of the adsorbent are totally covered by a monolayer of either solvent molecules, S, or by solvent plus solute molecules, S + X. Solute molecules X and solvent molecules S compete with each other for active sites on the surface of the adsorbent. For every solute molecule X that is adsorbed, one or more solvent molecules S must be desorbed.

2. The preferred orientation for the adsorption of solute molecules is flat. Other possible orientations are edgewise and verticle. Flat orientation takes up the greatest amount of adsorbent surface of the three possible orientations; that is, the solute molecular area is the greatest and thus would displace the most solvent molecules.

3. The active-site strength or surface is effectively homogeneous, leading to constant adsorption energies of solvent molecules at all locations on the adsorbent surface.

4. Solution energy terms arising from mobile-phase interactions of solute X and solvent S molecules are canceled by adsorbed-phase interactions of a similar nature.

Langmuir Isotherm. An adsorption isotherm results if we plot the concentration of the sample X in the adsorbed phase against the concentration of X in the liquid phase. The Langmuir isotherm, whose general shape is shown in Fig. 1, is the simplest of the adsorption isotherms. The Langmuir isotherm assumes monolayer adsorption, adsorption sites of equal strength, and that no interactions between neighboring adsorbed molecules take place. Thus Snyder's [1] theoretical model for LSC is approximately represented by the Langmuir isotherm.

The Langmuir isotherm for liquid-solid adsorption [1] takes the form

$$\theta = \frac{K_{th} N_x}{1 + K_{th} N_x} \tag{1}$$

where θ is the fraction of available adsorption sites occupied by adsorbed molecules of sample X, K_{th} is the thermodynamic equilibrium constant for the adsorption-desorption of sample molecule X, and N_x is the mole fraction of sample molecule X in the liquid phase.

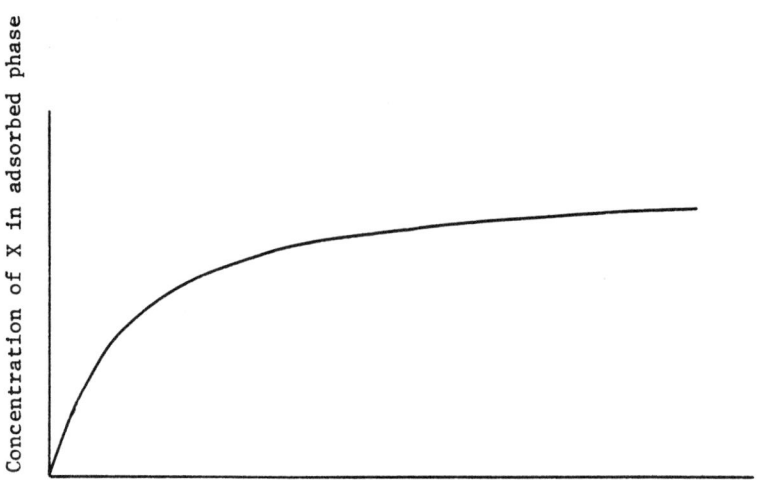

Concentration of X in liquid phase

FIG. 1.   Langmuir adsorption isotherm.

Adsorption Equilibrium. If the molecular area of the solute
is A_s and the molecular area of the solvent is A_e, the number of
solvent molecules displaced by each adsorbed solute molecule is
A_s/A_e, or n. The adsorption equilibrium can then be stated as
follows:

$$X_m + nS_a \rightleftharpoons X_a + nS_m \tag{2}$$

where the subscripts a and m refer to adsorbed phase and mobile
phase, respectively. The position of the equilibrium is determined
by the thermodynamics (free energies) of the adsorbed and nonad-
sorbed phases. The chance of a molecule of X being adsorbed is
dependent on the relative amount of the adsorbed phase and on the
net energy of adsorption of X. The larger the adsorbed phase and
the greater the net energy of adsorption of X, the greater are the
odds of the adsorption of a molecule of X.

The net free energy of reaction E results when the sum of the
energies on the left side of Eq. (2) are subtracted from the sum
of the energies on the right side:

$$\Delta E = E_{xa} + nE_{sm} - E_{xm} - nE_{sa} \tag{3}$$

According to assumption 4 above, solution energy terms resulting from dispersion interactions between mobile phase and solute molecules in solution are canceled by similar adsorbed-phase interactings, resulting in

$$\Delta E \approx E_{xa} - nE_{sa} \tag{4}$$

Solute Retention. An expression relating solute retention, or capacity, k' to mobile-phase composition was developed [1] from Eq. (4):

$$\log k' = \frac{\log V_a W}{V_m} + \alpha(S^\circ - \varepsilon^\circ A_s) + \Delta_{eas} \tag{5}$$

in which V_a is the volume of adsorbed monolayer per gram of adsorbent, W the weight in grams of adsorbent in the column, V_m the column void volume, α the adsorbent activity parameter, S° a relationship of the interaction energy of the solute molecules with the adsorbent surface [E_{xa} of Eq. (4)], ε° the solvent-strength parameter which relates the interaction energy of the mobile-phase molecules with the adsorbent surface [E_{sa} of Eq. (4)], A_s the relative area of the solute molecule when adsorbed (proportional to n), and Δea_s a second-order term correcting for secondary solvent effects (localization and hydrogen bonding) which were ignored in Eq. (4) but are observed to occur when mobile phases employing certain polar solvents are used. If wide-pore silica is used, the surface is considered to be essentially homogeneous [2], so that Eq. (5) reduces to

$$\log k' = \log\left(\frac{V_a W}{V_m}\right) + S^\circ - \varepsilon^\circ A_s + \Delta_{eas} \tag{6}$$

Solvent Strength. The following expression [1] describing the variation of solute retention with solvent strength for pure solvents has been derived from Eq. (5):

$$\log\left(\frac{k_2}{k_1}\right) = \alpha'A_s(\varepsilon_1 - \varepsilon_2) \tag{7}$$

where k_1 and k_2 are the capacity factors of solute X in pure solvents 1 and 2; α' is an adsorbent activity parameter, and as before, can be ignored for wide-pore silicas; A_s is the solute molecular area; and ε_1 and ε_2 are the strengths of pure solvents 1 and 2. The quantity $(\varepsilon_1 - \varepsilon_2)$ is equivalent to ε°. The solvent strength parameter [1] is equivalent to the adsorption energy of the solvent per unit area or

$$\varepsilon^\circ = \frac{E_{sa}}{A_e} \tag{8}$$

where A_e is the solvent's molecular area. Using Eq. (8), it is possible to calculate values of ε° for pure solvents relative to pentane for which ε° has been defined as zero [1].

It has been shown [3] that when solvent 1 in Eq. (7) is pure B, Eq. (7) may be rearranged to

$$\log k_2 = \log k_1 - n \log N_b \tag{9}$$
or
$$\log k_2 = \text{constant} - n \log N_b \tag{10}$$

Binary-solvent mobile phases are used to a much greater extent in high-performance liquid chromatography (HPLC) than are pure solvents. An additional expression has been developed [1] that facilitates calculation of the solvent strength of a binary mixture composed of a weak solvent A and a strong solvent B:

$$\varepsilon_{ab} = \varepsilon_a + \log \frac{N_b 10^{n_b(\varepsilon_b-\varepsilon_a)} + 1 - N_b}{\alpha'n_b} \tag{11}$$

where ε_a and ε_b are the solvent strengths ε° for pure solvents A and B; N_b the mole fraction of solvent B in the binary; n_b the molecular area, A_e, of solvent B; and α' the adsorbent activity parameter, which can be ignored for wide-pore solicas.

When using nonpolar solvents such as hexane, heptane, isooctane, and so on, and moderately polar solvents such as butyl chloride, benzene, methylene chloride, chloroform, and so on, Eqs. (7) and (11) will accurately determine the solvent strength for pure solvents and binary mixtures, respectively. In addition, Eqs. (6) and (7) will predict solute retention as a function of mobile-phase composition. When using nonpolar and moderately polar solvents, the Δ_{eas} term of Eq. (6) reduces to zero.

Strong Polar Solvents -- Secondary Solvent Effects. The Snyder model breaks down when strongly polar solvents of class P [2], those that cannot self-hydrogen bond, such as methyl ethyl ketone, tetrahydrofuran, ethyl acetate, and acetonitrile, and class AB [2], those that can self-hydrogen bond, primarily alcohols such as methanol, propanol, and water, are used in mobile phases to elute strongly retained polar solutes. This is due largely because of the oversimplification of assumption 3, concerning the homogeneity of the adsorbent surface, and assumption 4, the cancellation of solute-solvent interactions in the mobile phase by similar adsorbed-phase interactions. The solvent effects observed when mobile phases utilizing class P or AB solvents are due largely to either localized adsorption of solute and/or solvent on strong sites or solute-solvent interactions, largely hydrogen bonding in nature. These effects, referred to as secondary solvent effects, are recognized in the Δ_{eas} term in Eq. (6).

The surface of silica is not homogeneous with respect to adsorption-site strength for strongly polar solutes and/or solvents [1], with the result that adsorption energies of some solvent and solute molecules are different at various adsorption sites. Thus selective localized adsorption of some solvent or solute molecules occur at the stronger adsorption sites [1]. Equations (7) and (11) are valid, however, if the values used for A_s and n_b are larger than the actual molecular areas of the solute and solvent molecules involved [1]. Recall that for localized adsorption the adsorption energy of the solvent, E_{as}, is larger so that if ϵ° is to remain

constant, the molecular area, $A_e(n_b)$, must be larger as per Eq. (8).
Values for n_b for many solvents are available in Ref. 1 as well
as data used to calculate solute molecular areas A_s, so that solvent
strengths of binary solvent mixtures and solute retention can be
predicted with a high degree of confidence in many LSC systems
utilizing silica as the adsorbent.

In the case of hydrogen bonding, the polar component B of the
A-B binary largely covers the adsorbent surface, which therefore
has a greater concentration of B than the mobile phase. The result
is that the adsorbed-phase energy term is much larger than the
mobile-phase energy term, meaning that the effect on k' of solute-
solvent hydrogen bonding is determined by the adsorbed phase.

The practical result of secondary solvent effects is solvent
selectivity. Solvent selectivity has been defined as "the ability
of a change in solvent to create a difference in the *relative* retention of two similar solutes, particularly when the separation
factor for the two solutes is initially close to one" [1].

Soczewinski Model. Soczewinski [4] and Soczewinski and Golkiewicz
[5,6] have developed complementary competition models for adsorption
onto silica. The primary differences in the Snyder [1] and
Soczewinski [4-6] models are as follows [3]: Soczewinski assumes
discrete adsorption sites of equal energy. The strong component B,
in a solvent binary A-B, is assumed to cover the active sites completely when the molar concentration N_B is 0.25 or greater. In addition, Soczewinski assumes that the n of Eq. (2) is equal to the
number of adsorbable functional groups that comprise the solute
molecule and is equivalent to the number of solvent molecules displaced [6].

Working with thin-layer chromatography, Soczewinski and
Golkiewicz [6] related solute retention and solvent composition in
the expression

$$R_M = \text{constant} - n' \log N_B \tag{12}$$

where N_B is the molar concentration of the stronger component of a

binary mixture and n' is as previously defined. $R_M = \log k'$. It
has been shown [2] that Eq. (12) can be written in a slightly different form:

$$\log k' = \log k_B' - n \log k'N_B \tag{13}$$

where k_B' is the k' value for the solute in pure solvent B. Note
that Eq. (13) is of the same form as Eq. (9) in the Snyder model.
Soczewinski [7] has noted that Eq. (12) can be expressed in yet another form:

$$\frac{1}{V'} = \text{constant''} \cdot C_s^n \tag{14}$$

in which V' is the corrected retention volume of the solute, C_s^n is
the molar concentration of the polar solvent S, and "the constant is
a function of the ratio of solvent volume to adsorbent weight ratio,
the specific surface area of the adsorbent and the properties of the
solute and solvent" [7]. Logarithmic plots of V' versus %B show a
linear relationship, so solvent strength or solute retention can be
predicted.

This model, then, is capable of predicting solute retention as
a function of mobile-phase composition using strong solvents of
class N or P when the concentration of the stronger solvent is about
10% or greater and complements the Snyder model. Secondary solvent
effects are ignored in this model, so it has little utility in predicting solvent selectivities [3].

Scott and Kucera Model. Scott and Kucera [8-12] offered an alternative model for adsorption onto silica based on solute-solvent interactions in the mobile phase. This model has become known as the
sorption model.

In their first paper concerning solvent effects, Scott and
Kucera [8] advocated the following exceptions to the competition
theory: (1) They assumed complete coverage of the adsorption sites
by the strong component B in a binary mixture A-B when the concentration of B exceeded 2%. (2) They proposed that a continual increase in B to a final concentration of 100% caused a more gradual

increase in solvent strength due to mobile-phase interactions.
(3) They also proposed that if nonpolar forces were holding the
solute onto the adsorbent surface, dispersion forces would be important in determining solvent strength. Dispersion forces were stated
to be approximately proportional to the solvent's molecular weight.

In a later work, Scott and Kucera [9] and Scott [10] expanded
their model to include the following expression defining the role of
the solvent

$$\frac{1}{V'} = A + Bc_p \tag{15}$$

in which V' is the corrected retention volume; A and B are constants
incorporating dispersive and polar interactions, respectively; and
c_p is the concentration (w/v) of the polar solvent in the dispersive
solvent. The concentration (w/v) of the polar solvent in the phase
concerned was stated to govern the probability of solute interactions with that phase.

The similarity of Eq. (15) with Eq. (14) in the Soczewinski
model has been noted [2, 7, 21].

The relationship of $1/V'$ versus c_p is linear for concentrations
of the polar solvent greater than 3% and becomes nonlinear for concentrations less than 3%. For solutes weakly retained at 2% (w/v)
of polar solvent, where all silanol groups are assumed to be covered
by the polar solvent, decreasing the concentration of polar solvent
causes a rapid increase in retention and a rapid decrease in $1/V'$,
reportedly due to the increased number of silanol groups available
for interaction directly with the solute. For solutes that are
strongly retained at 2% (w/v) of polar solvent, decreasing the concentration of the polar solvent caused only a moderate increase in
retention and thus only a moderate decrease in $1/V'$. It was proposed
that because the solute was already interacting directly with the
silanol groups, freeing additional silanol groups has only a moderate effect. The linear relationship of $1/V'$ versus concentration of
polar solvent is demonstrated in Fig. 2. Three solutes are chromatographed on a Partisil 10 column using three polar solvents at varying concentrations in n-heptane.

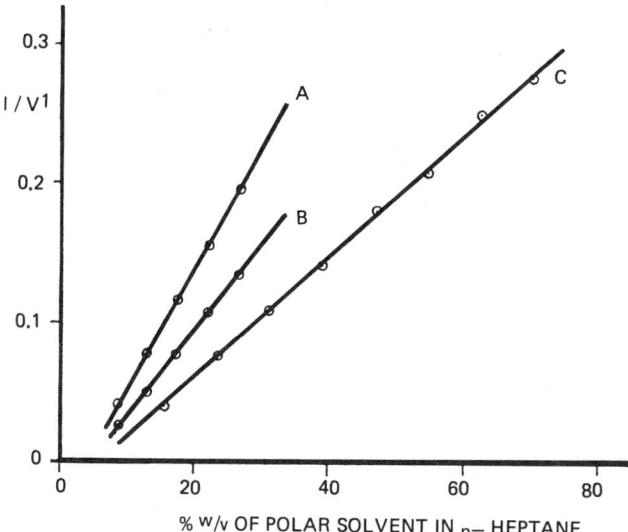

FIG. 2. Graphs relating the reciprocal of the corrected retention volume V; to the % w/v of polar solvent in n-heptane for different solutes. Column: 25 cm × 4.6 mm ID, packed with Partisil 10; solutes: A, benzyl alcohol; B, 3-phenyl-1-propanol; C, deoxycorticosterone alcohol; polar solvents: A and B, tetrahydrofuran; C, isopropanol. (From Ref. 10.)

The magnitude of polar interactions of the solute and solvent, monofunctional substances only, were claimed [9, 10] to be proportional to the exponent of the polarizibility per milliliter, and thus the dielectric constants, of the solute and solvent involved.

Scott and Kucera [9] proposed that by holding polar forces constant the dispersive interactions could be examined. Equation (15) then becomes

$$\frac{1}{V'} = A + Bd \tag{16}$$

where d is the density of the dispersion solvent and A represents the constant polar forces. If solutes of varying polarity are chromatographed in a series of solvents composed of a constant concentration (w/v) of polar modifier and dispersion solvents of differing densities, the plot of 1/V' versus density of the dispersion

solvent, according to the premise discussed above, should demonstrate
the effect of dispersive interactions on solute retention. Figure 3
shows the results of such plots. When polar forces are weak, a plot
of 1/V' versus the density of the dispersion solvent is linear, increasing with the density of the dispersion solvent. If the polar
forces are strong, the plot of 1/V' versus the density of the dispersion solvent is a straight line parallel to the density axis, indicating no significant dispersive interactions. For dispersive
interactions to be important, they must be approximately equal to or
greater than the polar interactions. The net force of solute-solvent
versus solute-adsorbent interactions was stated to determine the
degree of retention. Solute-adsorbent forces can be reduced by

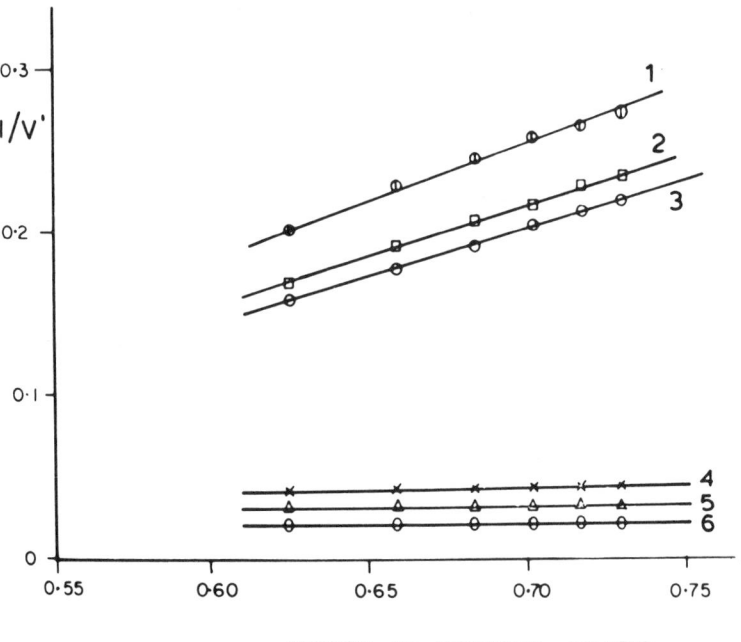

DENSITY OF DISPERSION SOLVENT

FIG. 3.   Graphs of the reciprocal of the corrected retention volume
against the density of the dispersion solvent. 1, 2-ethyl anthraquinone; 2, 2-methyl anthraquinone; 3, anthraquinone; 4, phenyl-
methyl carbinol; 5, benzyl alcohol; 6, 3-phenyl-1-propanol. (From
Ref. 10.)

using a more polar solvent, while solute-solvent polar forces can be reduced by using a lower concentration of the polar solvent [10].

More recently, Scott and Kucera [11, 12] and Scott [13] have studied the nature of the solute-solvent interactions with the silica surface. Utilizing both stoichiometric and chromatographic experiments, they proposed that depending on the nature and concentration of the stronger component of a binary solvent mixture, both monolayer and bilayer adsorption occurs.

Nonpolar solvents such as butyl chloride, chloroform, and benzene were found to have adsorption isotherms that conformed to the Langmuir isotherm for monolayer adsorption which was described by the expression

$$N_A = \frac{c}{A + Bc} \tag{17}$$

where N_A is the surface area covered by the monolayer of solvent, c is the concentration of the stronger component in the binary solvent, and A and B are constants.

Polar solvents that can hydrogen bond with the silica surface, such as ethyl acetate, methyl ethyl ketone, and tetrahydrofuran, were found not to fit the Langmuir isotherm for monolayer coverage, but did fit the following expression that was developed [11] for bilayer adsorption:

$$M = A - \frac{A + ABc/2}{1 + Bc + Dc^2} \tag{18}$$

where M is the total mass of solvent of a bilayer, c is the concentration of stronger solvent in the binary, and B and D are constants.

It was determined from the isotherm data for ethyl acetate on silica [11] that the first layer of ethyl acetate was held to the silica two orders of magnitude more strongly than the second layer of ethyl acetate was held to the first layer.

From chromatographic data it was proposed that plots of $1/k'$ versus c are linear at all concentrations for solvents where only monolayer coverage is demonstrated [11]. For solvents deemed capable of double-layer formation, two regions of linearity were

observed [11]. The first region of linearity of 1/k' versus c is
observed at low concentrations of the polar component of the binary,
where, as proposed, the monolayer is being formed and the polar
solvent is interacting with the silica surface. The second region
of linearity occurs after the initial layer has been formed, usually
1-2% of the polar solvent, where according to Scott and Kucera [11]
the polar solvent is interacting with the initial layer of solvent.

From stoichiometric data it was concluded [12] that nonpolar
solvents, butyl chloride, chloroform, and benzene, gave a monolayer
coverage that was calculated to contain an average of 6.4×10^{20}
molecules per gram of silica. Calculations also showed that the
first layer of bilayer coverage of the polar solvents, ethyl acetate, methyl ethyl ketone, and tetrahydrofuran, contained an average
of 6.9×10^{20} molecules per gram of silica, or essentially the same
number of molecules as found for the nonpolar, monolayer adsorbing
solvents. A completely formed second layer would also contain
6.9×10^{20} molecules per gram of silica.

Scott and Kucera [12] and Scott [13] have indicated that they
now believe that under certain conditions, both the sorption and the
competition processes occur and have proposed the following adsorption process to explain the mechanism of solute adsorption over a
wide range of adsorption systems.

For polar solvents deemed capable of bilayer formation and at
concentrations of the polar modifier where the monolayer is near
completion and the second layer is minimal [i.e., 0.35% (w/v) ethyl
acetate in n-heptane], it was shown stoichiometrically that solutes
which would be eluted from a column (employing the same conditions)
at k' values through 10 did not displace any ethyl acetate from the
silica. According to Scott and Kucera [11, 12], the solute was re-
acting with the primary layer of solvent. This is an example of
the sorption process where the interaction is by association, as
shown in Fig. 4B. Solutes that would elute from a column with a k'
value of 20 or greater displace ethyl acetate from both layers and
interact directly with the silica surface. The latter is an example
of the competition process.

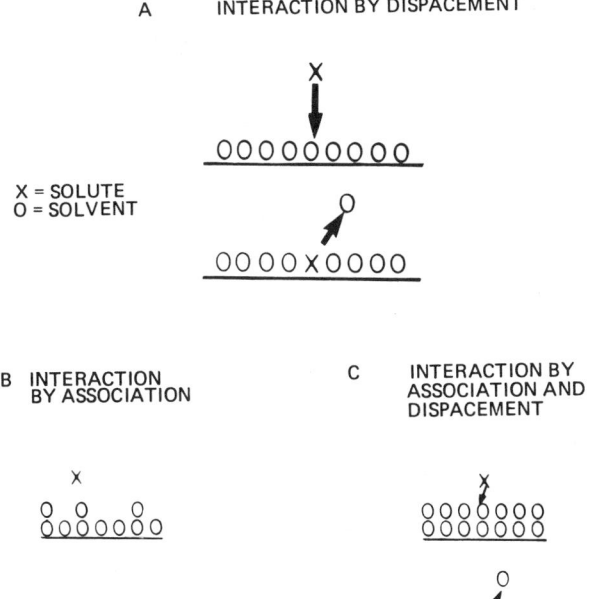

A INTERACTION BY DISPACEMENT

X = SOLUTE
O = SOLVENT

B INTERACTION BY ASSOCIATION

C INTERACTION BY ASSOCIATION AND DISPACEMENT

FIG. 4. Different types of interactions of a solute with a silica surface. (From Ref. 13. Reproduced from the *Journal of Chromatographic Science*, by permission of Preston Publications, Inc.)

Utilizing a higher concentration of polar solvent (i.e., 13% ethyl acetate in n-heptane), where both the primary and second layers were assumed well formed, it was shown by frontal analysis chromatography [11] that solutes eluting with k' values up to 9.5 displaced the second layer but interacted only with the primary layer of ethyl acetate. This is an example of the sorption process, where both displacement and association occur, and is shown in Fig. 4C. Once again it was shown that a very polar, strongly retained solute, with a k' value of 20 or greater, displaced both layers of solvent and interacted directly with the silica surface. For this to occur, the solute must have a polarity of the same order or even greater than that of the solvent.

For the case of weakly held nonpolar, non-hydrogen bonding solvents that form only monolayer coverage, Scott [13] has stated that the solute competes with and displaces the solvent molecules and interacts directly with the silica surface. This is shown in Fig. 4A.

Other Workers. The complementary Snyder-Soczewinski competition model and the Scott-Kucera "sorption" model currently represent the opposing theoretical treatments of adsorption onto silica. Many other workers have contributed to the understanding of the adsorption process. Some of these will be presented, but since their works are largely confirmatory in nature, they are not discussed in great detail.

Jandera and Churacek. Jandera and Churacek [14-18] and Jandera et al. [19], working in the area of gradient elution, developed and tested a model of adsorption chromatography similar to that of Soczewinski and Golkiewicz [6].

Slatts et al. Slatts et al. [20] have proposed a model combining aspects of both the competition and the sorption models. Utilizing activity coefficient measurement of solutes and moderator (B solvent) in the mobile phase and the capacity ratios, they found that it was possible to evaluate individually the dependence of the competition and solvent interaction (sorption) terms on the mobile-phase composition.

Interestingly, they propose just the opposite conditions to be favorable for the existence of the competition and sorption processes as those proposed by Scott and Kucera [12]. Slatts et al. [20] proposed that at low and intermediate concentration ranges of the polar modifier the competition process dominates, and that only for high concentrations of the polar modifier does the sorption process dominate. They also suggest the possibility of the formation of multiple layers on the silica surface.

Jaroniec et al. Jaroniec et al. [21] developed a very general expression relating solute capacity k' and binary mobile-phase composition for a system assuming nonideality of both the mobile and adsorbed phases. It was shown that the expressions of Snyder [Eq. (10)], Soczewinski [Eq. (12)], and Scott and Kucera [Eq. (15)] were all special cases of this general equation.

Thomas et al. Using isohydric solvents, a linear relationship between solute capacity factors k' and the reciprocal of the molar fraction of water in the isohydric solvent or solvent mixtures was observed [22]. This relationship may be expressed as

$$k' = f \frac{1}{N_{H_2O}^{isoh}} \tag{19}$$

or in the form of Snyder's equation (9) with n equal to 1,

$$\log k' = \log k'_0 - n \log N_{H_2O}^{isoh} \tag{20}$$

Isohydric solvents [22] are solvents whose water content corresponds to the water content or activity α of the adsorbent. When changing from one isohydric solvent to another, the activity of the adsorbent is unchanged. In a series of isohydric solvents, the more polar solvents will have a greater water content.

Some advantages of the use of $1/N_{H_2O}^{isoh}$ as a measure of mobile-phase polarity or relative strength compared to the use of the other parameters, such as solvent strength $\epsilon°$ or ϵ_{AB}, were suggested [23]. (1) The use of the inverse of $N_{H_2O}^{isoh}$ instead of logarithmic scales means that k' values near zero can be plotted, whereas with logarithmic scales they cannot. (2) Equation (19) is a simpler relationship than Eq. (20), for example, and gives a direct linear relationship. (3) The water content of the solvent and the adsorbent, both factors affecting retention, are taken into account.

Hara et al. Hara [24] monitoring reactions in organic synthesis, Hara et al. [25] studying retention of mono- and difunctional steroids relative to binary mobile-phase composition, Hara et al. [26] reporting on the optimized separation of some indole alkaloids on silica, and Hara et al. [27] in the deisgn of binary mobile phases for the separation of protected oligopeptides have further demonstrated the validity of the Snyder-Soczewinski model.

Overview. Snyder [3], and more recently Snyder and Poppe [2], have
been highly critical of the Scott-Kucera model of adsorption on
silica in the case of solutes and solvents of class P.

In very brief and very general terms, some of these criticisms
[2] of the Scott-Kucera model are as follows: (1) The model of
Scott-Kucera does not adequately describe solute-solvent interaction
effects in the mobile phase or solute retention. It was shown [2]
that the experimental results validating the Scott-Kucera model
could also be explained using the Snyder hypothesis of a displacement mechanism (competition) and approximate cancellation of solute-
solvent mobile- and stationary-phase interactions. (2) The polar
components of solvent-solute interactions cannot accurately be described by one term (i.e., solute-solvent polarizability) [9, 10].
The degree of dispersive interactions between solute and solvent
can better be related to their refractive indices rather than their
densities [9, 10]. (3) Scott and Kucera [11-13] assume that solutes
less polar than the B solvent component cannot displace the B solvent from the adsorbed monolayer. Snyder and Poppe [2] have proposed that if sorption occurs at all, it occurs when the B concentration is less than 5%. They then suggest that in fact in this
situation the mechanism is competitive with solvent B localization
and displacement of the nonpolar A solvent instead of the more polar
B solvent. (4) Snyder and Poppe [2] point out that one of the
theoretical requirements for Langmuir adsorption is constant interactions between solute and solvent in the mobile phase as solvent composition changes (i.e., the concentration of B changes).
The Scott and Kucera model, assuming changing interactions between
solute and solvent in mobile phase as B changes, is in direct op-
position to the requirements for Langmuir adsorption, and therefore
any use of Langmuir adsorption in support of postulates of the sorption process is not logical. In addition, Snyder and Poppe [2] note
that because of strong polar interactions and localized adsorption,
class P solvents will in general not meet Langmuir adsorption requirements, so that the use of such isotherm data to support bilayer

formation is questionable. (5) Scott and Kucera [11, 12] use isotherm shape to support bilayer adsorption for class P solvents.
Snyder and Poppe [2] point out that it is very easy to misinterpret
isotherm shape relative to bilayer formation. This was demonstrated
by the isotherm of the aromatic hydrocarbon dibenzyl, a class N substance giving only monolayer adsorption, but which gave a discontinu-
ous isotherm plot similar in shape to those of the class P solvents
that Scott and Kucera [11, 12] claim exhibit bilayer adsorption.
Snyder and Poppe [2] indicate that because of localized adsorption,
class P solvents would be expected to show discontinuity in the
monolayer isotherm.

In spite of the differences in the theoretical approaches of
the Snyder-Soczewinski competition model and the Scott-Kucera sorption model, it has been noted [7, 28] that in practice both yield
essentially equivalent predictions of solute retention and solvent
strength or composition.

B. Adsorbent

Adsorbent Surface. The surface of chromatographic silica is described by Snyder [1] as being composed of free, reactive, and bound
silanol groups, as shown in Fig. 5A, B, and C, respectively. It is
generally accepted that for polyfunctional adsorbate molecules the
site strength increases in the order: bound silanol, free solanol,
and reactive silanol groups. For monofunctional adsorbate molecules, free and reactive silanol groups are equivalent. In a more
recent study, Bather and Gray [29] have concluded that free hydroxyls are much stronger sites than hydrogen-bonded (reactive) silanols.

The silanol composition of wide- and narrow-pore silicas are
different, in that free silanols predominate for wide-pore silicas,
while reactive and bound silanols comprise the majority of active
sites for narrow-pore silica. In a study of narrow-pore silica,
Bather and Gray [29] found 3.4 hydrogen-bonded silanols per 100 $\overset{\circ}{A}^2$
versus 1.2 free silanols per 100 $\overset{\circ}{A}^2$. It is well known [1] that the

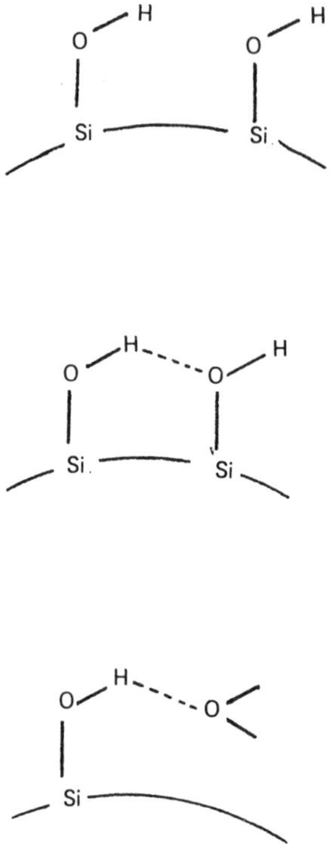

A. Free Silanol

B. Reactive Silanol

C. Bound Silanol

FIG. 5. Surface silanols.

total silanol concentration is 4-5 per 100 $\overset{\circ}{A}^2$ of surface area, or about 8 $\mu mol/m^2$. In the study previously cited, Bather and Gray [29] found a total silanol concentration of 4.6 per 100 $\overset{\circ}{A}^2$.

Silica is characterized by Snyder [1] as having two types of water present: molecular or adsorbed water and bulk water which is not associated with the surface. The surface of silica, with respect to the water concent as proposed by Snyder [1], could be represented as shown in Fig. 6. Note that the term "surface" includes all the surface associated with the pores.

The layer of molecular or adsorbed water may be removed either by heating between 150 and 200°C or by extraction with dry solvents. This would leave the surface covered with exposed silanol groups

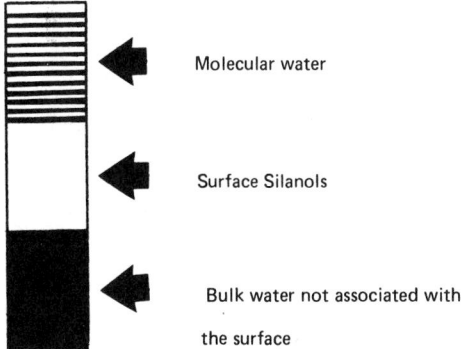

Molecular water

Surface Silanols

Bulk water not associated with
the surface

FIG. 6. Water content of silica according to Snyder.

and is known as *activated silica*. As described by Snyder [1],
further heating between 200 and 400°C causes free silanol groups to
move about the surface, form temporary reactive silanol groups, and
then condense to form siloxane groups and water. There can also be
a permanent loss of surface due to particle-particle fusion. On
heating at temperatures of 800 to 1000°C, a large amount of the bulk
water is lost.

Water that is added to activated silica to deactivate it is
preferentially adsorbed by the bound and reactive hydroxyls [1],
purportedly because these groups might form multiple hydrogen bonds
with the water molecules.

Scott [13] and Scott and Traiman [30] have recently made observations concerning the nature of the water content of the silica
surface that differ considerably from those of Snyder [1] and many
others. From the results of studies utilizing thermogravimetric
analysis (TGA), infrared spectrophotometry, and chemical reaction
with dimethyl octylsilylchloride (DMOSC), they have proposed that
the surface of silica is covered with three layers of water, as
shown in Fig. 7. The first layer of strongly held constitutional
water is hydrogen bonded to the silanol groups of the silica surface, on a one-to-one basis which was calculated to be 8.5×10^{20}
silanol groups and 8.5×10^{20} water molecules. Two additional

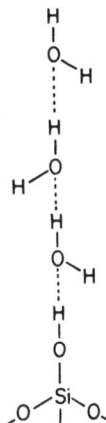

3rd LAYER OF WEAKLY ADSORBED WATER, LOSS BETWEEN ROOM TEMPERATURE AND 70 °C, MAXIMUM LOSS AT 40 °C, REVERSIBLE, REMOVED BY DRY SOLVENTS

2nd LAYER OF WEAKLY ADSORBED WATER, LOSS COMPLETE AT 120 °C, MAXIMUM LOSS AT 100 °C, REVERSIBLE, REMOVED BY DRY SOLVENTS

1st LAYER OF STRONGLY HYDROGEN BONDED WATER, LOSS COMMENCES AT 200 °C AND APPEARS COMPLETE AT 650 °C, REVERSIBLE, NOT REMOVED BY SOLVENTS

SILANOL GROUPS LOSE WATER TO PRODUCE SILOXYL GROUPS. COMMENCES AT 450°C, COMPLETE AT 1100 °C. LOSS IS IRREVERSIBLE

FIG. 7. Schematic impression of multilayer formation of water on silica gel. (From Ref. 30.)

layers of physisorbed water, 16.1×10^{20} molecules of water, complete the surface of silica as described by Scott [13] and Scott and Traiman [30]. The two layers of physisorbed water are removed either by heating at approximately 110°C or by extraction with dry solvents. The layer of constitutional, hydrogen-bonded water is removed only by heating between 400 and 800°C. Above 400°C some condensation of surface silanol groups occurs; however, it was reported [13] that even at 600°C only about 10% of the silanol groups were destroyed, and 25% at 750°C [30]. Additional heating between 800 and 1000°C causes the remaining silanol groups to condense into siloxyl groups, with the release of additional water.

Activated silica is then pictured as having a layer of constitutional or hydrogen-bonded water, so that adsorption, according to Scott and Traiman [30], takes place on the hydrated silanol group and not on free silanol groups as proposed by Snyder [1].

Snyder and Poppe [2] have been very critical of the technical aspects of the studies of Scott [13] and Scott and Traiman [30] and the conclusions drawn from these studies concerning the nature of the silica surface.

Surface Area and Adsorbent Activity. According to Snyder [1] the
relative adsorption of an adsorbate molecule is related to two properties of the adsorbent. These are the surface volume V_a, which
is directly related to the surface area, and the adsorbent activity
α, which is the relative tendency of the adsorbent to adsorb different absorbates and is related to the water content of the adsorbent.
Adsorbent activity is greatest for adsorbents that have had the
molecular water removed either by heating between 150 and 200°C, or
by dry solvent extraction. The adsorbent activity and the surface
volume both decrease with the addition of water to an initially
activated silica, depending on the amount of added water. These two
adsorbent properties, V_a and α, are independent of the nature of
the sample and the solvent.

For silicas thermally deactivated at 600°C and above, Scott
and Kucera [31] have reported that solute retention is a linear
function of the water content of the silica, indicating a homogeneity of surface hydroxyl groups. They concluded that surface
area does not determine retention characteristics, but that in the
case of normal chromatographic silicas (i.e., those not thermally
deactivated), surface area appears to do so because the surface is
saturated with hydroxyl groups, which relates the surface area directly with the hydroxyl and water content.

Adsorption Forces. There are basically two types of interactions
that bind adsorbates to adsorbents. These are dispersion and selective or nondispersion interactions. Among the selective interactions are (1) induction forces, (2) electrostatic forces, (3) hydrogen bonding, (4) charge transfer, and (5) ion exchange or chemisor-
ption. As discussed previously, Snyder [1] contends that dispersion
forces are unimportant in the adsorption of polar compounds onto
silica because of cancellation of solution- and adsorbed-phase
energy terms, and that hydrogen bonding due to solute-solvent interactions is the major interaction affecting the adsorption of polar
adsorbates onto silica. It was previously noted that Scott and
Kucera [11, 12] advocate the importance of both dispersion and

hydrogen-bonding interactions, depending on the chromatographic
conditions.

Silica and Selectivity. Silica and LSC in general show little selectivity for homologs, whereas selectivity is good for compounds
differing in functional groups.

LSC and silica show unique selectivities in the separation of
isomers, which is apparently due to the fact that one of the two
compounds functional groups match up with the adsorption sites
better than the others [28].

c. Role of the Sample

The number and nature of the functional groups a compound possesses
largely determines how strongly it is adsorbed. As the polarity of
the functional groups increases, the relative adsorption or k' value
increases. The relative adsorption of a compound is then the total
or summation of all the interactions of the functional groups of
the sample molecule with the geometrically fixed surface silanol
groups of the adsorbent.

It has been shown by Snyder [1] that group adsorption energies
can be determined experimentally and these values can be used to
calculate sample adsorption energies for a given adsorbate sample
and then these values used to estimate the samples k' value in a
given chromatographic system.

There are several factors that complete the use of group adsorption energies in the prediction of a sample adsorbate adsorption
energies. One such complication occurs when one group in an adsorbate molecule interacts by localization on a strong adsorption site,
thereby interfering with a second molecular group by restricting
its interaction with the surface. Additional complications can
occur by interactions of the groups within the adsorbate molecule,
thus changing their individual contributions to the total adsorbate
molecule's adsorption energy, and thus the total molecular adsorption energy. Such interactions may be steric, electronic, or
chemical in nature.

These complications can cause predictions of an adsorbate's k'
value to be considerably in error and thus of questionable practical value. Readers desiring a comprehensive discussion of sample
structure and its effects on k' value should consult Chapters 10
and 11 of Ref. 1.

Of a more practical nature are general elution orders of adsorbates based on functionality. One example of such an elution
order on silica [32] from the least to the most strongly retained
is shown in Table 1. Those classes of compounds having about the
same retention are shown on the same line.

One aspect of the sample that should be discussed is that of
sample size as related to concentration, or amount of sample that
is placed on the column. The Langmuir adsorption isotherm as described by Eq. (1) is useful to show sample-size effects. At low
sample concentrations the isotherm is linear. Snyder [1] has defined the term *linear capacity* "as the sample size $\theta 0.1$ (grams of
sample per grams of adsorbent) which is just sufficient to cause a
10% change in K, relative to the linear isotherm value K°."

The practical implications of chromatography within the linear
isotherm are constant, reproducible retention volumes and optimum
resolution between adjacent peaks. To assure operation within the
linear isotherm the linear capacity of the column adsorbent bed
must not be exceeded. At higher sample concentrations where the
linear capacity is exceeded, retention volumes are variable and resolution and peak shape are poor.

TABLE 1.   Elution Order on Silica from Least to Most Strongly
Retained
_____
Saturated hydrocarbons
Aromatic hydrocarbons-organic halides
Sulfides
Ethers
Nitro compounds
Esters-aldehydes-ketones
Alcohols-amines
Sulfones
Sulfoxides
Amides
Carboxylic acids
_____

III. SILICA PACKINGS

Chromatographic silicas are usually classified according to many of
their physical properties. Included in these physical properties
are the following: average pore diameter, particle shape (spherical or irregular), specific surface area (usually determined by the
BET procedure), and average particle size.

The silicas that are produced by gelation have an amorphous
surface composed of an agglomerate mass of spherical particles.
The pH of the gelation solution determines the surface area of the
resultant silica. The amount of surface area of a silica is analogous to the percent of liquid loading in partition chromatography
and is an important parameter in describing the silica. Surface
areas of silicas range from 100 to about 900 m^2/g [33], with 400-
600 m^2/g being the most common. The surface area of silica particles
includes all of the internal surface of the pores, so that for a
given silica particle, the smaller the pore diameter, the greater
the surface area.

There are two generally recognized categories of silicas based
on the size of the average pore diameter. Pore diameters of commercially produced silicas for HPLC use range from a low of 20 Å to a
high of over 300 Å, with most falling in the range 40-60 Å. Those
silicas having an average pore diameter of 100 Å or greater are
labeled wide-pore silicas and those having an average pore diameter
of less than 100 Å are labeled narrow-pore silicas.

Both irregular-shaped and spherical silicas are produced commercially. Spherical silicas are more expensive to manufacture and
usually have a tighter size distribution than irregular-shaped
silicas. In general, most analytical and preparative columns are
now packed with 5- or 10-μm particles of either irregular or spherical silica. Equivalent columns packed with the same size spherical
and irregular silicas give columns having essentially the same efficiency [34].

As the particle size gets smaller, the column permeability becomes smaller (i.e., the pressure drop gets larger. Columns packed

with spherical particles have a greater permeability than those
packed with an equivalent-size irregular silica [35]. This may be
an important consideration in the packing selection if solvents of
a relatively high viscosity will be used.

A column packed with spherical objects has inherently greater
stability than a column packed with irregular-shaped objects, which
tend to shift, and even fracture, if pressured or jostled. Chromatographic columns also exhibit these characteristics, so that columns
packed with spherical silica particles are less likely to be damaged
by physical rough handling or by operation at high pressures.
Columns packed with irregular silicas have a tendency to be more
easily damaged, resulting in loss of efficiency.

A. Packing Procedures

The major drawback of the microparticle silica adsorbents is that
they are not amenable to dry packing. It has been shown that irregular silicas smaller than about 50 μm [36, 37] and spherical
silicas smaller than 20 μm [38, 39] must be slurry packed. Many
slurry packed procedures have been reported [40-50] that will produce uniform, efficient columns. An excellent discussion of slurry
packing techniques is presented in Ref. 51.

All of these procedures, with the exception of Ref. 49, require expensive, additional equipment, so that many chromatographers,
depending on the frequency and numbers of columns to be replaced,
may find it more economical to buy prepacked columns.

B. Silicas for Other Uses

Silicas in the particle size range 15-40 μm find some use in preparative HPLC. Larger silicas are most widely used in solvent
purification and in the preparation of solvents of a constant water
content.

In addition to the silicas described previously, porous layer
beads (PLBs) have been produced. These consist of a glass bead core
and a thin porous outer layer of silica that is bonded to the glass

bead. Some PLBs have two layers of silica bonded to the glass
bead. The average particle size of the PLBs is 30-50 μm. The PLBs,
having a surface area of approximately 12-30 m^2/g, offer high efficiencies, but low sample capacities. Because of their larger particle
size, PLBs can be readily dry packed. Because of their very low
sample capacities, PLB packings are seldom used in modern analytical columns, but they perform admirably as packing for guard columns.

C. Commercial Availability

Tables 2 and 3 list many of the irregular and spherical silica adsorbents that are available, together with some of their properties.
Table 4 is a listing of some sources for these adsorbents and prepacked columns. In some cases, not all of the sources supply all
available forms of a given adsorbent.

Most analytical columns supplied are 15-20 cm in length and
have inside diameters of 2-5 mm. Those columns specifically designed for preparative chromatography usually are available 25-50
cm in length with about 10 mm inside diameter.

IV. MODERATORS

The usefulness of water as a moderator to reduce the activity of
silica, and thus the retention volumes of solutes, has been reported by many early workers [52-57]. Additional benefits are
greater column efficiencies [57, 58] and increased sample amounts
that can be separated (i.e., greater linear capacities) [1].

Moderators are usually strongly polar compounds that are added
to the mobile phase and are therefore added to the adsorbent in
situ. Water is the historical and still the most widely used
moderator.

A. Moderator Action

As previously stated, the surface of silica is inhomogeneous with
respect to the relative strength of the active silanol groups.
Approximately 5% of the wide-pore silica sites are reported to be

TABLE 2. Irregular Silicas

Product	Average Particle Size (µm)	Average Pore Diameter (Å)	Surface Area (m²/g)	Form Supplied[a]	Use[b]	Suppliers[c]
Biosill A	2-10	60	400	B	A	7, 35
Chromegasorb 60R	10		500	C	A	12
Chrom Sep SL	5, 10	60	400	B, C	A, A	36
Hi Eff Micro Part	5, 10		250	C	A, A	6
ICN Silica	3-7	60	500-600	B	A	16
ICN Silica	7-12	60	500-600	B	A	16
LiChrosorb Si 60	5, 10	60	500	B, C	A	1, 2, 11, 14, 15, 19, 21, 22, 27, 32, 35-38
LiChroprep Si 60	15-25	60	500	B, C	P	
LiChroprep Si 60	25-40	60	500	B, C	P	
LiChrosorb Si 100	5, 10	100		B, C	A	
Partisil	5, 10	50	400	B, C	A	1, 2, 5, 6, 8, 9, 14, 15, 17, 20, 22, 25-28, 30, 36
Polygosil	5, 7.5, 10	60	500	B, C	A	
Polygosil	15, 20, 30	60	500	B, C	P	
µ Porasil	10		300-350	C	A	1, 40
RSL Silica	5, 10	57	200	B, C	A	1, 30
Sil 60	5, 10, 20	60	500	B, C	A	9, 20, 30
Silica A	8-21		400	B	P	23
Ultra Sil	10			C	A	27, 2

[a]B, bulk; C, column.
[b]A, analytical; P, preparative.
[c]Names of suppliers are listed in Table 4.

TABLE 3. Spherical Silicas

Product	Average Particle Size (μm)	Average Pore Diameter (Å)	Surface Area (m²/g)	Form[a] Supplied	Use[b]	Suppliers[c]
Hypersil	5-7		200	B	A	31
LiChrospher Si 100	5, 10	100	370	B	A, A	1, 2, 11, 14, 15, 19, 21, 22, 27, 32, 35-37
Nucleosil 50	5, 10	50	500	B, C	A, A	1, 9, 27, 30
Nucleosil 100	5, 10, 30	100	300	B, C	A, A, P	1, 6, 9, 23, 26, 28, 37,
Spherisorb SW	3, 5, 10	80	220	B, C	A, A, A	
Spherosil XOA 600	5-8	83	550	B	A	9, 29, 30, 34
Spherosil XOA 1000	5-8	35	860	B	A	
Super Microbead Si	5, 10	95	380	B, C	A	13
Ultrasphere Si	5				A	2
Vydac TP		330	100	B, C	A	1, 2, 4-6, 15, 20, 22-24, 26, 30, 31
Zorbax Sil	6	70	350	B, C	A	10, 42
Chromasorb LC-6	5, 10			B, C	A	1, 18, 37

[a] B, bulk; C, column.
[b] A, analytical; P, preparative.
[c] Names of suppliers are listed in Table 4.

TABLE 4. Suppliers

1.	Alltech	22.	Micromeritics
2.	Altex Scientific (Beckman)	23.	Perkin-Elmer
3.	American Scientific Products	24.	Phase Separations
4.	Analabs	25.	Pierce Chemical Co.
5.	Applied Chromatography Systems	26.	Pye Unicam
6.	Applied Science Laboratories	27.	Rainin Instrument Co.
7.	Bio Rad Laboratories	28.	Regis Chemical Co.
8.	Bodman Chemicals	29.	Rhone-Poulenc (France)
9.	Chrompack (Holland)	30.	RSL (Belgium)
10.	E. I. du Pont de Nemours	31.	Separations Group
11.	E. Merck -- EM Labs	32.	Siemens
12.	E. S. Industries	33.	Spectra Physics
13.	Fuji-Davidson Ltd. (Japan)	34.	Supelco
14.	Glenco Scientific	35.	Touzart and Matignon
15.	Hewlett-Packard	36.	Tracor
16.	ICN Co.	37.	Universal Scientific
17.	J. A. Jobline (U.K.)	38.	Varian Associates
18.	Johns-Manville	39.	VWR
19.	Kipp and Zonen (Holland)	40.	Waters Associates
20.	Machery-Nagel and Co. (Germany)	41.	Fisher Scientific
21.	MCB Manufacturing Chemists		

strong sites [1]. The moderator is preferentially adsorbed from the eluant by the most active sites [1]. The distinction to be made between a solvent and a moderator is the degree of attraction to the active surface of the adsorbent. Thus solute molecules can compete with solvent molecules for active sites, but not with moderator molecules.

In the case of adsorption of solute molecules from solution using a highly activated wide-pore silica, the minority of the strong sites play the primary role. The strength of solvent needed to elute the sample molecules from the strong sites is of such a strength that they are not adsorbed to any degree by the weak sites. Since there are so few strong sites available for the adsorption process, a highly activated adsorbent is readily overloaded. As moderator is added it occupies most of the strong sites, and since the sample molecules do not compete with the modifier for these sites, the more predominant weak sites begin to play the dominant role in the adsorption process. Obviously, since there are more weak sites it follows that the amount of sample that can be separated

using a deactivated or moderated column is larger than what can be
separated using an activated column. What is accomplished by the
use of a moderator in reality is the conversion of an inhomogeneous
surface to a surface more nearly homogeneous with respect to site
strength, thereby increasing the linear capacity of the adsorbent.

It is important to add the proper amount of water. The addition of too much water results not only in the strong sites being
covered, but also some of the weak sites, with an accompanying decrease in sample capacity.

B. Sample Retention

An expression has been derived [59] that describes sample retention
in terms of sample and modifier concentrations on an adsorbent possessing two types, strong and weak, of active sites. This expression takes the form

$$K'_x = \frac{\theta_x}{N_x} = \frac{N_1 K_{1x}}{(1 + N_x K_{1x})(1 + N_y K_{1y})} + \frac{N_2 K_{2x}}{(1 + N_x K_{2x})(1 + N_y K_{2y})} \quad (21)$$

in which N_1 and N_2 are the mole fractions of the strong and weak
sites, respectively, and K_{1x} and K_{2x} are the thermodynamic equilibrium constants for the sample on the strong and weak sites, N_x is
the mole fraction of X in solution, N_y is the mole fraction of moderator in solution, and K_{1y} and K_{2y} are the thermodynamic equilibrium constants for the moderator on the strong and weak sites,
respectively. This expression describes the sample retention at
concentrations greater than the linear capacity of the adsorbent.
At these high concentrations sample retention is dependent on the
sample concentration. When the sample concentration is low and
within the linear isotherm region, sample retention is not dependent on sample concentration and Eq. (21) is simplified as follows
[59]:

$$K_x^\infty = \frac{N_1 K_{1x}}{1 + N_y K_{1y}} + \frac{N_2 K_{2x}}{1 + N_y K_{2y}} \quad (22)$$

where K_x^∞ is defined as the linear isotherm distribution coefficient.

Although the concentration and type of surface hydroxyls present on narrow-pore silicas are different from those of wide-pore silicas, the use of moderators gives the same results: increased linear capacity, smaller K' values, and greater column efficiencies.

C. Moderator Requirements

The moderator must be much more strongly retained than the sample; therefore, the nature of the sample dictates what may be used as a moderator. For example, if the sample is readily eluted with a nonpolar solvent such as hexane, a solvent of intermediate polarity might serve as a moderator. If, on the other hand, the sample is more strongly retained and requires a stronger, more polar solvent, then a more polar, more strongly retained moderator, such as water, is needed.

Moderators are generally used in the concentration range 0.01-1% by volume [35]. If a greater concentration is required to deactivate the strong sites, then the choice of moderator was probably wrong and a more polar moderator is needed.

D. Use of Water as a Moderator

In classical liquid-solid adsorption chromatography, the water was added directly to the silica, which after an equilibration period was then packed on a column. The columns were generally used only one time and were then emptied and filled with new silica. In modern liquid chromatography where the column is used many, many times it is not practical to add the water directly to the silica. The moderator is therefore added to the mobile phase and is adsorbed by the active sites as the mobile phase flows through the column.

There are situations where the use of water as a moderator presents problems of a practical nature. The optimum amount of water to be added to wide-pore silicas has been shown [60] to be from one-half to one monolayer. This is approximately 0.2-0.4 g per

100 m^2 surface area of silica. In the case of nonpolar hydrocarbon
solvents, the solubility of water in the solvent is very low and a
very large number of column volumes of solvent must be passed through
the column before equilibration is reached. It has been reported
[60] that with hexane as solvent, over 300 column volumes were needed
to reach equilibrium. An additional problem with water as the moderator in low-polarity solvents is maintaining a constant level of
water content in the solvent.

The water content of nonpolar solvents can change in the process of pouring from one container to another and can even be adsorbed by the glass [59] storage containers themselves. It is especially important to maintain a constant water content in low-
polarity solvents having poor water solubility. Small changes in
the water content result in greater changes in sample retention
than would result for small changes in water content for solvents
having a greater degree of water solubility [61].

There are several ways of preparing solvents saturated with
water for use as HPLC solvents. One method is that of shaking the
solvent in a separatory funnel with water. If the solvent has a
low solubility for water, this method is not very satisfactory. A
similar technique uses some type of mechanical stirring to mix the
solvent and water together over a period of time. A highly reliable
method uses an inexpensive grade of silica prepared to contain
20-30% water. This is packed in a conventional open chromatographic
column and then the solvent is passed through the column, or alternatively the prepared silica can be added to about 1000 ml of the
solvent in a flask followed by mechanical stirring for about 30 min.
The resulting water-saturated solvent can be used as is, or can be
mixed with dry solvent prior to use.

One procedure [62] is totally different in that the water is
added to the solvent and to the column in a one-step process. The
procedure employs a thermostated funnel which contains an inexpensive grade of silica containing a known, but variable, water content. This is placed on the liquic chromatography (LC) system

before the pump. The solvent is passed through the funnel and then
pumped through the column and is then recycled back through the
funnel. Equilibration times are still very long for nonpolar hydrocarbons. The main advantage of this system is the constancy of the
water content of the solvent resulting from the thermostated funnel.
By changing the water content of the silica in the funnel the water
content of the solvent can also be changed. Since the solvent is
continually being recycled, all substances eluted from the column
are also being recycled back through the column. Even though these
previously eluted substances are being diluted in a comparatively
large volume of solvent, their mere presence makes this technique
of questionable value for use in forensic applications.

The use of isohydric solvents, previously defined, to maintain
a constant level of adsorbent activation has been proposed [21].
It has also been suggested [28, 60, 63] that constant water saturated
solvent/dry solvent mixtures of 50% of each component be utilized,
the rationale being that if all solvents were prepared in this manner, they would all be in equilibrium with the water content of the
column and equilibration time when changing solvents would be minimal since the new solvent would neither add nor strip water from
the previously equilibrated column. Constant-ratio wet-dry solvents utilized in this manner approximate isohydric solvents [28].
A note of caution in the preparation of isohydric solvents composed
of organic solvent mixtures has been offered [28]. The organic
solvent mixture should be initially prepared and then divided into
two equal portions. One portion is saturated with water and then
mixed with the dry portion for the 50:50 wet/dry isohydric mixture.
Any other method of preparation, such as preparation of the individual 50% wet-dry solvents followed by mixing to the desired composition, does not result in an isohydric solvent. Because the
solubility of water in organic solvent mixtures is not linearly
related to the percent volume composition of the organic mixture,
it is possible that water could come out of solution.

E. Other Polar Solvents as Moderators

Because of the problems encountered when using water as a moderator, associated primarily with the nonpolar solvents, other polar
solvents have been investigated. Acetonitrile and methylene
chloride [59] were added as moderators to dry hexane with metadi-
phenoxybenzene as the sample. Acetonitrile worked well in this application, but methylene chloride was unsatisfactory as a moderator,
since about 10% volume was needed to reach the desired deactivation.
Alcohols have been investigated by several workers [60, 64-66].
Poor peak shape, slow column equilibration, and lower column efficiencies were observed.

F. Thermal Deactivation

It should be pointed out that the deactivation of silica can be
achieved by heating between ´200 and 900°C. This process is known
as thermal deactivation. In one study [67] the chromatographic
properties of silica thermally modified at 200, 400, 500, 600, 700,
800, and 900°C were evaluated by chromatographing several solutes
in dry n-heptane. Because of the constant nature of the surface
(i.e., homogeneous with respect to hydroxyl groups), it was concluded that silica deactivated at temperatures between 600 and
900°C acts as an ideal polar-bonded phase. The relative merits of
thermal deactivation and water deactivation of silica have been reported [68]. Utilizing a mobile phase of dichloromethane-isopropanol
(98;2), amobarbital, phenobarbital, and barbital were chromatographed
on untreated silica, thermally deactivated at 450 and at 500°C.
These and additional barbiturates were chromatographed on silica
using a mobile phase of dichloromethane-isopropanol (97.5:2.5),
saturated with water [68]. Although both modes of deactivation reduce the effects of strong site adsorption, water deactivation results in a greater modification of retention, while equilibrium
kinetics are slower. The more rapid equilibrium kinetics of thermally deactivated or moderated silica were seen to be advantageous
in preparative work and in gradient elution.

It would seem that even though there are problems in its use, water remains the moderator of choice for most applications.

V. MOBILE PHASE

The selection of the best mobile phase for the resolution of a given sample mixture is unquestionably the most difficult aspect of practical adsorption chromatography. There are several factors that must be considered in the selection of the solvent or solvents that make up the mobile phase.

A. General Considerations

1. The sample components must be soluble to some degree in the mobile phase. If the separation is to be a preparative separation the solubility has to be quite good.

2. The mobile phase has to be compatible with the detector to permit detection of the eluted sample components. For example, ultraviolet (UV) detector requires that the mobile phase have a UV cutoff below the operating wavelength.

3. The solvents must not cause the sample to degrade or be otherwise changed.

4. The mobile-phase properties should permit easy recovery of the resolved sample components. In most cases in preparative or semipreparative methods using adsorption chromatography, the solvent is merely evaporated, leaving a residue of the isolated component. The mobile-phase solvent should be volatile enough that it will evaporate in a reasonable time. It must also be free of nonvolatile residues.

5. Solvent purity is an important consideration in solvent selection. All solvents are going to contain some impurities, so it is very desirable that the nature and amounts of these impurities be the same from bottle to bottle, or from lot to lot. Obviously, this has to do with good manufacturing and quality control techniques employed in the manufacture and bottling of the solvent. Except for certain special circumstances (e.g., purification of a

solvent not commercially available as a high-purity solvent, etc.),
most laboratories find it more economical to purchase high-purity
solvents rather than to purify a lower-grade solvent. In such cases,
purification procedures for various solvents are readily available
[69, 70]. Of special concern are those solvents that can undergo
chemical changes during storage. Like the ethers that are subject
to potentially hazardous peroxide formation (e.g., diethyl ether,
tetrahydrofuran, dioxane, and isopropyl ether), most changes occur
once the solvent container has been opened and the manufacturer's
innert nitrogen atmosphere has escaped. To avoid problems these
solvents should be used on a timely basis. It is also a good idea
to purchase smaller quantities of these solvents so that they are
not sitting around in the solvent storeroom for extended periods.
Dating bottles of ethers upon receipt and also upon opening is a
good safety practice. If there is any question of the possibility
of the presence of peroxides in a bottle of one of the ethers, it
should be tested before use. An easy, convenient test for peroxides
is to mix 10 ml of the ether with 1 ml of water containing 100 mg
of potassium iodide (added just before use). The absence of yellow
color after 1 min is a negative test for peroxides [69]. If present, peroxides can be easily removed from these ethers by passing
them through a column of highly activated alumina [70]. Table 5
shows the volumes of peroxide-free ethers that can be produced by
passing the dry solvent through 25 g of Woelm basic alumina. Super
I Activity, in a column of such a diameter that the adsorbent bed
is 15-20 cm deep. The presence of water in the solvents and or the

TABLE 5. Volume of Peroxide-Free Ethers Obtained from a Column
Containing 25 g of Woelm Super I Activity Alumina

Solvent	Volume (ml)
Diethyl ether	500
Diisopropyl ether	500
Dioxane	400
Tetrahydrofuran	250

use of a lower-activity alumina will result in smaller volumes of peroxide-free ether per gram of alumina. The alumina containing the adsorbed peroxides should be wet with water before disposal. In no case should the alumina be heated!

One impurity present in practically every solvent at levels of 50-2000 ppm is water [71]. In most cases this is of minor importance. However, if the adsorbent is used in a highly activated form, it can be a problem, as the preferential adsorption of the water will result in the deactivation of the adsorbent. Water can be removed from solvents by passing the solvent through one of the less expensive grades of highly activated large-pore silica. The dried solvents must be handled very carefully, as they readily adsorb moisture from the air and it has been shown [71] that glass containers can contribute water to dried solvents. Thus glass containers intended for use as storage vessels for dried solvents must be carefully oven dried to remove all moisture.

In addition to the contaminants or impurities described previously, which are all unintentional, some solvents contain chemicals added to retard reactions. For example, antioxidants are added to ethers to retard oxidation, and chloroform, which is light sensitive, is stabilized with approximately 1% of ethanol or a nonpolar hydrocarbon. If the nature of the added chemical is known, it can readily be removed. For example, nonvolatile stabilizers can be removed by distillation of the solvent, whereas alcohols can be removed by aqueous extractions followed by drying the solvent as previously described.

The spectral purity of solvents and their UV cutoffs can be ascertained by examining the UV spectrum of the solvent in a 5- to 10-cm cell versus air as the reference. The UV cutoff should correspond with literature values. The amount and nature of nonvolatile impurities can be observed by evaporating 100-200 ml of the solvent. If a residue is observed, it can be subjected to thin-layer chromatography (TLC) and the plate can be examined under

short-wavelength UV light and by other visualization techniques,
such as iodine vapor or sulfuric acid charring.

  6. An additional, but somewhat less important consideration
is the viscosity of the solvent. Lower-viscosity solvents give
lower-pressure drop and greater column efficiencies.

  7. Most important of all, the mobile phase must sufficiently
resolve the sample components for qualitative and/or quantitative
analysis and accomplish the separation so that the k' values are
reasonable.

  Some of the important properties of many of the most commonly
used HPLC solvents are shown in Table 6.

  Snyder [72, 73] developed a scheme to classify solvents by
their "polarity" as well as their relative abilities to take part
in hydrogen-bonding or dipole interactions. Eight selectivity
groups were designated. The data in the last three columns of
Table 6 are taken from this work. Those solvents in groups I,
strong proton acceptors (ethers and amines); II, acceptor-donor
(alcohols); V, neutrals (methylene chloride); VIII, strong proton
donors (water and $CHCl_3$), are of particular importance in LSC.

B. Mobile-Phase Composition

Armed with the solvents listed in Table 6, nearly all separations
that are feasible using adsorption chromatography with silica as
the adsorbent should be possible provided that the proper binary
or ternary mixture has been selected. The initial step is finding
the solvent or solvent mixture that will provide a mobile phase of
such a strength that the desired component or components are eluted
with a practical k' value. Practical k' values are in the range
2-20, but in general k' should not exceed 10.

TLC Scouting. Thin-layer chromatography offers the capability of
rapidly examining the usefulness of several solvent systems for the
given separation. One way of using this technique would be to have
a series of chromatographic tanks, with a solvent series of increasing strength. The sample would be chromatographed in each system

TABLE 6. Properties of Some Useful LSC Solvents

Solvent	UV Cut off (nm)	Refractive Index at 20°C	Viscosity (cP) at 20°C	Boiling Point (°C)	Solvent-Strength Silica	Solvent Acidity X_d^a	Solvent Basicity X_e^b	Selectivity Group
n-Pentane	210	1.358	0.23	36	0.00			
Hexane	210	1.375	0.31	69	0.00			
Heptane	210	1.388	0.31	98	0.00			
Isooctane	210	1.391	0.51	99	0.01			
Cyclohexane	210	1.427	1.00	81	0.03			
Isopropyl ether	220	1.368	0.37	68	0.22	0.14	0.30	I
Isopropyl chloride	225	1.378	0.33	35	0.22			VI
Chloroform	245	1.446	0.57	61	0.26	0.41	0.25	VIII
Methylene chloride	245	1.424	0.44	40	0.32	0.18	0.29	V
Tetrahydrofuran	220	1.408	0.55	66	0.35	0.20	0.38	III
Ethyl ether	220	1.352	0.23	35	0.38	0.13	0.53	I
Ethyl acetate	260	1.372	0.45	77	0.38	0.23	0.34	VI
Ethylene chloride	230	1.444	0.79	84	0.38	0.21	0.30	V
Dioxane	220	1.422	1.54	101	0.49	0.24	0.36	VI
Acetonitrile	210	1.344	0.37	82	0.50	0.27	0.31	VI
Isopropanol	210	1.378	2.3	82	0.63	0.19	0.55	II
Ethanol	210	1.361	1.20	78	0.68	0.19	0.52	II
Methanol	210	1.329	0.60	65	0.73	0.22	0.48	II
Water	210	1.333	1.00	100	large	0.37	0.37	VIII

[a]Proton donor.

[b]Proton acceptor.

and by the use of a suitable visualization technique, each system's
usefulness could be determined. A variation of this technique makes
use of circular chromatography. The sample is spotted at several
locations across the TLC plate and then the various solvent mixtures
are applied using either a hypodermic needle with the tip squared
off or a serological pipette with a fire-polished tip, very nearly
closed. There is at least one commercially available system that
will spot and develop up to 16 sample applications on one 20 cm ×
20 cm standard plate or sheet.

TLC is very susceptible to "solvent demixing", which is the
change in solvent composition that occurs during the development of
the TLC plate. As the solvent migrates up the TLC plate, the
stronger component(s) of the binary or ternary mixture are adsorbed,
resulting in a progressively weaker solvent as development proceeds.
This can be a serious problem with low concentrations of the stronger
component(s) in the mobile phase and can largely be avoided if the
stronger component(s) are each present at 5% or greater concentrations. In addition to the concentration of the stronger component,
its strength ε° relative to the weaker solvent is also important.
None of the components in the solvent system should exceed the
strength ε° of the next weaker solvent by more than 0.4 unit.

Equilibrating the plate with the solvent vapors prior to developing the plate in conjunction with the concentration and solvent
strength considerations previously discussed will give TLC solvent
parameters that will usually be applicable to the same separation
using HPLC.

LC Scouting. The same scouting approach can be done directly with
LC using a series of solvents of increasing solvent strengths.
However, this is rather time consuming and rather expensive since
some of the solvents will probably be discarded. If the LC is
equipped with a system capable of generating gradients, this can
be used to find rapidly the composition of a binary, ternary, or
quaternary mixture, depending on the gradient-forming device, that
will elute the substance(s) at a suitable k' value. As with TLC,

solvent demixing can occur at low concentrations of the stronger solvent as the gradient is proceeding. The composition of the solvent at the time the sample component(s) was or were eluted can then be determined from the gradient profile. To convert this to isocratic operation a slightly lower concentration of the stronger solvent will be used. This can then be conveniently tested by using the solvent programmer to produce the desired composition of the stronger solvent in the solvent mixture, operating at a constant, isocratic composition.

Equieluotropic Mobile Phases. Quite often two or more components are not resolved by a binary solvent of a given strength. More often than not, merely changing the strength of the solvent by varying the percentage of the stronger solvent will change only the k' values of the sample components, not the resolution.

As described earlier, selectivity results largely from the adsorbed-phase interactions of the stronger component in the solvent mixture. If the sample components are not resolved by one binary, then a second binary mixture using the same weak, nonpolar solvent but a different strong solvent is tried. If the k' values are about right, it is convenient to prepare the second binary so that the solvent strength ε_{AB} is the same as the first binary. It is useful to draw a graph relating solvent strength with volume percent of the strong solvent in the binary mixture. This is known as an equieluotropic solvent series [1]. Figure 8 is one such series.

If, for example, a solvent strength of 0.35, ε_{AB}, elutes the sample components at a k' of 5, but does not resolve them, other binaries should be tried at the composition shown by dropping down a vertical line at the solvent strength ε_{AB}, 0.35. A change of $0.05\varepsilon_{AB}$ unit will produce a change in k' by a factor of 2-4 [71].

It is important to realize that the solvent strengths ε_{AB} of solvent mixtures such as those shown in Fig. 8 are a function of the adsorbent activity α, which is dependent on the amount of added water. This is in contrast to the solvent strengths, $\varepsilon°$, of pure solvents, which are independent of the adsorbent activity [1].

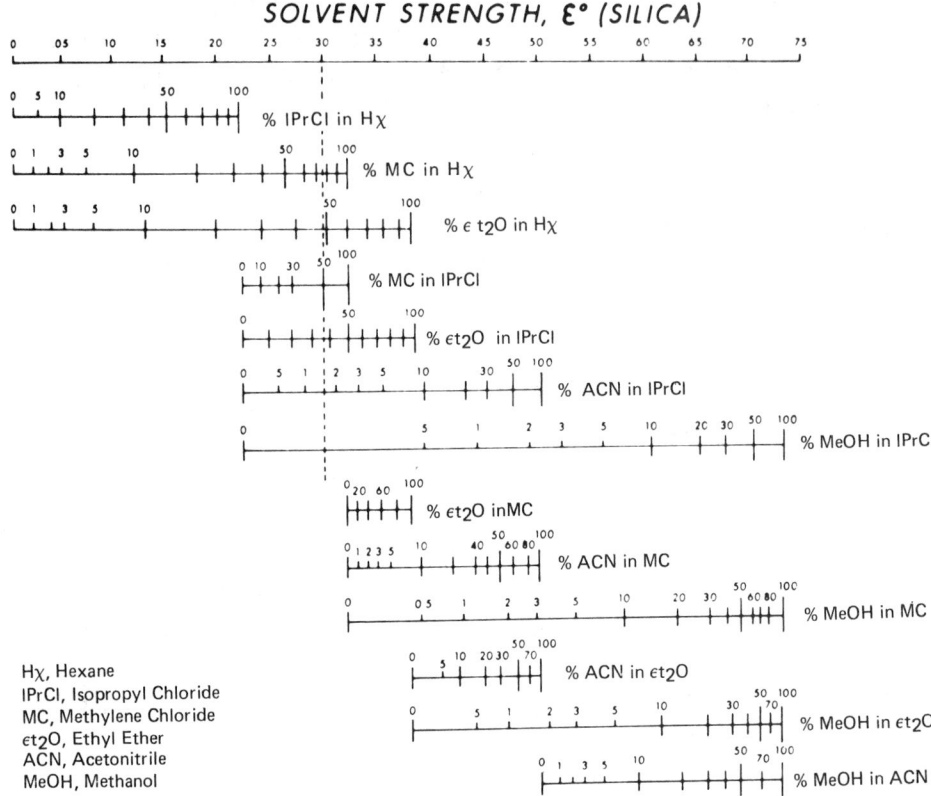

FIG. 8. Solvent strength of mixed solvents on silica. (From Ref. 71. Reproduced from the *Journal of Chromatographic Science*, by permission of Preston Publications, Inc.)

Table 7 shows how ε_{AB} for the binary solvent mixture 10% methanol-90% chloroform varies with the degree of deactivation beginning with fully activated (no added water) small-pore silica through 20% (w/w) added water.

Equieluotropic solvent composition can be obtained using Eq. (11), as developed by Snyder [1] or by the somewhat simplified modification of Snyder's calculations proposed by Saunders [74].

It is also possible to calculate the composition of equieluotropic binary mobile phases from chromatographic data using the

TABLE 7. Solvent Strength Versus Adsorbent Activity for the Binary
Solvent 10% Methanol in Chloroform

Percent Water	α^a	ε_{ab}
0	1.12	0.66
1	1.05	0.66
2	0.98	0.65
4	0.86	0.64
7	0.76	0.63
10	0.69	0.62
15	0.61	0.61
20	0.54	0.59

^{a}Taken from Table 6-1 of Ref. 1.

procedure of Hara et al. [25]. In practice the retention of (k')
of a solute is determined at two or more concentrations of a polar
solvent S_1 in a nonpolar solvent W_1. The solute is then chromatographed similarly in a mobile phase composed of $W + S_2$, where W is
the same nonpolar solvent and S_2 is a different polar solvent.
Utilizing the relationship log k' = c - n log X_s [25] and a plot of
log k' versus log X_s, the slope (n_1 and n_2) and axis intercepts
(C_1 and C_2) can be determined for the solute in mobile phases $W + S_1$
and $W + S_2$. To determine the concentration of polar solvent S_2 in
$W + S_2$ that will be equieluotropic composition with a given concentration of polar solvent S_1 in $W + S_1$, the following expression was
developed [25]:

$$\log X_{S(2)} = \frac{C_2 - C_1}{n_2} + \frac{n_1}{n_2} \log X_{S(1)} \tag{23}$$

The values previously obtained are plugged into Eq. (23), and the
concentration of $S_{(2)}$ to give the equieluotropic systems $W + S_2$
and $W + S_1$, with respect to the solute chromatographed, is calculated.

In reference to selectivity effects, those binaries having the
lowest concentration of the strong solvent show the largest selectivity effects. Selectivity effects can also be large if hydrogen
bonding occurs to a greater degree for one of the unresolved com-

ponents and the solvent. Samples containing amine, phenolic, and alcoholic moieties are among those showing hydrogen-bonding effects. To exploit these effects to the greatest degree, the stronger component of the binary should be one of the solvents from selectivity group I, II, V, or VIII [28].

If the sample contains both weak and strong proton donor (Lewis acid) components, the use of a strong solvent having either weak or strong proton acceptor (Lewis base) tendencies will result in maximum selectivities [28].

It has also been suggested [24] that the stronger solvent in the mobile phase binary should contain a functional group similar to the active functional group contained by the solute.

Ternary Mobile Phases. In some cases, ternary mixtures utilizing a nonpolar solvent and one solvent from two different selectivity groups (I, II, V, or VIII) can result in the selectivity needed to resolve complex mixtures [28].

Thomas et al. [23] evaluated solvents used in LSC on the basis of their properties as shown in Table 6, as well as their human toxicity. Shown in Table 8 are the six solvents that Thomas et al. deem to be most useful in LSC based on the criteria described above. In addition, they use acetic acid and diethylamine to adjust the pH of the mobile phases when necessary.

Six ternary combinations, each employing low-polarity isooctane plus one intermediate polarity solvent and one polar solvent, are possible. The mobile phases are prepared from isohydric solvents as in the following example. Isooctane-isopropyl ether (1 + 1) forms the nonpolar A mixture; isopropyl ether-methanol (1 + 1) forms

TABLE 8. Six Most Useful Solvents

Solvent	Polarity	Selectivity Class
Isooctane	Low	--
1,2-Dicholorethane	Intermediate	V
Diisopropyl ether	Intermediate	I
Ethyl acetate	Intermediate	VI
Acetonitrile	High	VI
Methanol	High	II

the polar B mixture. The polarity of the ternary mobile phase is determined by the concentration of the B mixture in the mobile phase. It is claimed [23] that more than 90% of the possible separations in LSC can be achieved using various ternary mobile phases, with appropriate pH adjustment, prepared from the A and B solvent mixtures just described.

As a result, procedures for accurate prediction of the optimum mobile-phase composition for the resolution of sample components have been proposed [23].

For relatively complex mixtures each solute is chromatographed, one at a time, in two mixtures of A + B. The straight lines representing T versus $1/N_{H_2O}^{isoh}$ are drawn, and then from this graph the retention times at other concentrations of A and B are estimated. $T_2 - T_1/T_2 + T_1$ for each pair of adjacent solutes are plotted versus $1/N_{H_2O}^{isoh}$. From the curved lines the optimum mobile-phase composition can be predicted. Using the following equation, it is possible to calculate the number of theoretical plates needed to give the desired resolution:

$$N = \left[\frac{2R_s(T_2 + T_1)}{T_2 - T_1}\right]^2 \tag{24}$$

Figure 9 shows the optimized separation of six barbiturates, which resulted from the use of the procedure described. Figure 10 shows the straight lines, T versus $1/N_{H_2O}^{isoh}$ for each barbiturate obtained from the retention data at 10% and 30% B. From this figure the estimated retention times at other concentrations of B were obtained and the values of $T_2 - T_1/T_2 + T_1$ for each adjacent pair of solutes were calculated and then plotted versus $1/N_{H_2O}^{isoh}$ resulting in Fig. 11.

As can be seen from Fig. 11, the poorest resolution was obtained for components b and c. The optimum mobile-phase composition was determined to be 10% B.

If the sample composition is simple, the optimum mobile-phase composition can be determined without calculations. Each component

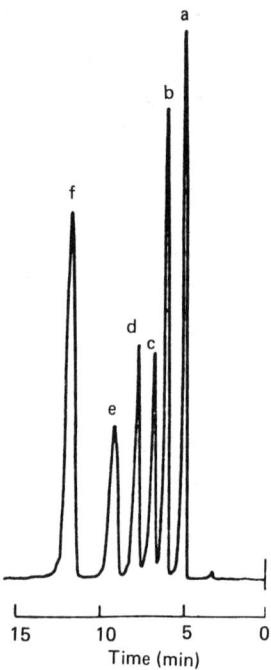

FIG. 9. Optimized separation of barbiturates. Column: 15 cm ×
0.6 cm Spherosil XOA, 6 μm; mobile phase: 10% of B, that is, isooctane-diisopropyl oxide-methanol (45:50:5), containing 0.2% of
acetic acid and 0.26% of water; flow rate: 1 ml/min; pressure: 25
bar; detector: UV at 254 nm; solutes: a, penthiobarbital;
b, methylphenobarbital; c, amobarbital; d, allobarbital; e, phenobarbital; f, mephobarbital. (From Ref. 23.)

is chromatographed, one at a time, in two mixtures of A and B. In
addition to the retention time, the starting and ending times of
the components are plotted versus $1/N_{H_2O}^{isoh}$ for each component, in
each mixture of A and B. As before, the straight lines are drawn,
but in addition the area between the starting and ending lines for
each component is shaded in, so that the optimum mobile-phase composition can be visually observed. The optimum mobile phase would
be the most polar combination of A and B that would, when observing
the graph, show an unshaded area between each of the components.

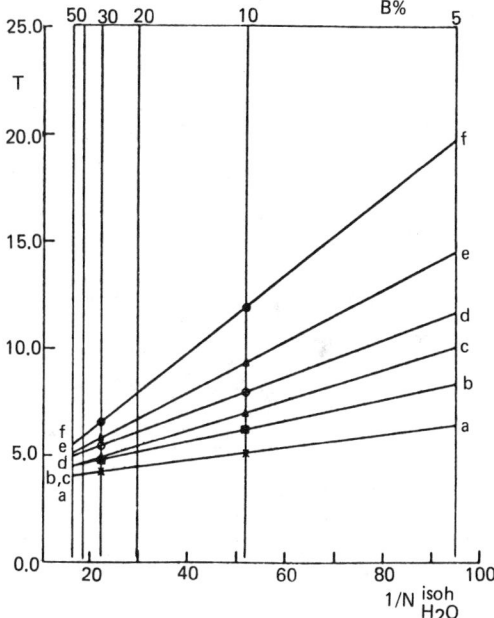

FIG. 10. Relationship between retentiom times of barbiturates and the mobile-phase polarity. Column: 15 × 0.6 cm Spherosil XOA 600, 6 μm; mobile phase A: isooctane-diisopropyl oxide (1:1); mobile phase B: diisopropyl oxide-methanol (1:1); acidity: 0.2% of acetic acid in A and B; adsorbent activation state: α = 1; flow rate: 1 ml/min; solutes: see the legend for Fig. 9. (From Ref. 23.)

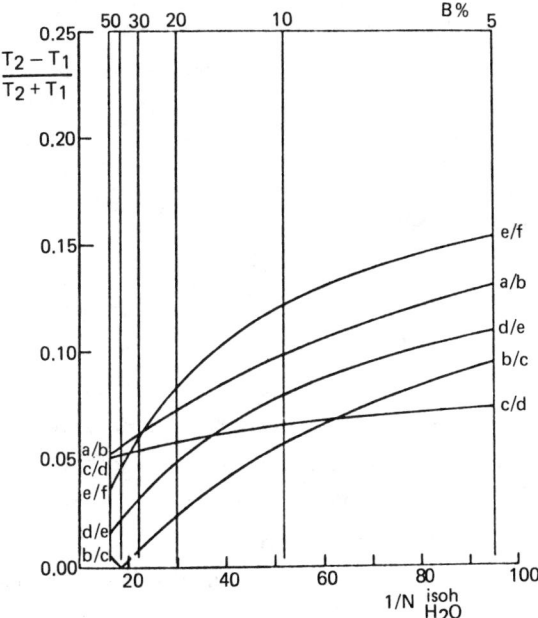

FIG. 11. Optimization of separation of barbiturates. Expression of $T_2 - T_1/T_2 + T_1$ versus the polarity, according to the values obtained from the straight lines plotted in Fig. 10. (From Ref. 23.)

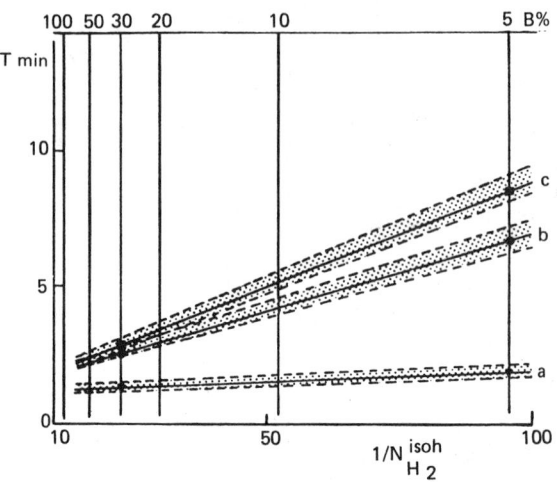

FIG. 12. Optimization of separation of the active constituents of
the injection solution "357 V." Column: 15 cm × 0.6 cm Spherosil
XOA 600, 6 μm; mobile phases: (a) isooctane-diisopropyl oxide
(1:1); (B) diisopropyl oxide-methanol (1:1); activation state:
α = 1 (water content of solvent B: 2.6%); acidity: 0.2% triethylamine; flow rate: 3 ml/min; ΔP: 70 bar. (From Ref. 23.)

Figure 12 shows the optimization of a two-component veterinary
injection: "357 V" and an internal standard. A mobile-phase composition of 10% B can be seen to resolve the mixture in the shortest
time.

VI. EVALUATION OF COLUMN PERFORMANCE

Chromatographic columns, especially those packed with silica adsorbent, generally give good service for relatively long periods of
time. The use or abuse of a column dictates how long it is usable,
and will be discussed in the next section. At any rate, as time
progresses the column's performance deteriorates. The deterioration is probably very slight, so that the subtle changes in the
chromatograms may go unnoticed over a period of time. Generally,
column deterioration is observed when resolution of neighboring
components diminishes and the peaks lose their sharpness and become

broader. Retention times for known components are observed to be
noticebly shorter and pressure drops increase.

At this time this particular column's performance has been
recognized as unsatisfactory and the column will be replaced or
cleaned. In the example of subtle deterioration described above,
a number of analyses of questionable value could have been performed
before the deterioration was recognized. This need not happen, as
it is a simple matter to monitor the column's performance on a
regular basis. Monitoring implies that such data will be compared
with some standard data that will lead to a quantitative assessment
of the column's current performance.

A. Use of Test Mixtures

The performance characteristics of commercially packed columns are
determined using a standard test mixture under a specified set of
operating conditions. This gives a standard chromatogram from which
a number of column parameters can be measured. The use of test
mixture and the parameters that can be calculated from them have
been discussed by several workers [71, 75-77].

Most, if not all, manufacturers of packed columns submit a test
chromatogram with the column as well as the calculated performance
parameters. It would therefore be possible to run the same test
mixture at time intervals determined somewhat by the amount of
column usage. It is highly unlikely that the mobile-phase system
used by the column manufacturer to run the test mixture would be
the mobile phase in routine use in your laboratory. This would
then mean a change in mobile phase in order to repeat the test conditions. There would be a period of equilibration time on both ends
of the mobile-phase change, the duration being dependent on the
degree of difference between the two mobile phases. Obviously,
this amounts to a big chunk of a day's work time being lost from
productive work. The possibility also exists that the test chromatogram could be run before equilibrium was reached, resulting in
erroneous test results.

A much better way to accomplish this is to derive a test mixture especially for the combination of column and solvent system
used in daily chromatographic separations. If one is fortunate
enough to have a large supply of columns and have columns dedicated
to specific mobile phases, a test mixture can be devised for each.
The test mixture is composed of three components, one of which is
unretained, one having a k' value of about 2-3, and one having a k'
value of about 5-6. Assuming that a UV detector is being used, the
three components should be fairly good UV absorbers at the wavelength
used, or at 254 nm if a fixed wavelength is used, so that low concentrations will give a good response. The combination of concentration of the test mixture components and detector sensitivity
should be such that a 5- to 10-μl injection will give approximately
a half-scale deflection at a moderately high detector sensitivity
setting having a noise level less than 1%. The test mixture is then
injected into the LC and the test chromatogram is produced. Remem-
ber that the measurements on the chromatogram will be made with a
good-quality ruler, so use a recorder chart speed fast enough to
give a rather broad peak for the last component. Alternatively,
retention times can be obtained from an electronic integrator, which
can then be converted to length in millimeters for the calculations.
The following operating conditions should be recorded for future
reference:

1. Composition of the test mixture
2. Mobile-phase composition
3. Flow rate
4. Column pressure drop
5. Recorder chart speed
6. Detector wavelength and sensitivity
7. Column temperature (e.g., ambient or thermostated temperature)

The best place to record these conditions is directly on the test
chromatogram, so that the complete record of the test can be filed
in a log book or notebook of a permanent nature. The column serial

number or other identifier should also be recorded together with
the date the column was placed in service.

From the test chromatogram, represented by Fig. 13, and the
operating conditions, the following parameters can be measured or
calculated from the measurements:

1. The retention of unretained component A (e.g., the distance
   measured from the injection point to the apex of peak A), designated Rt_A
2. The retention of component B (e.g., the distance measured from
   the injection point to the apex of peak B), designated Rt_B
3. The retention of component C (e.g., the distance measured from
   the injection point to the apex of peak C), designated Rt_C
4. k' value for component B: $k'_B = (Rt_B - Rt_A)/Rt_A$

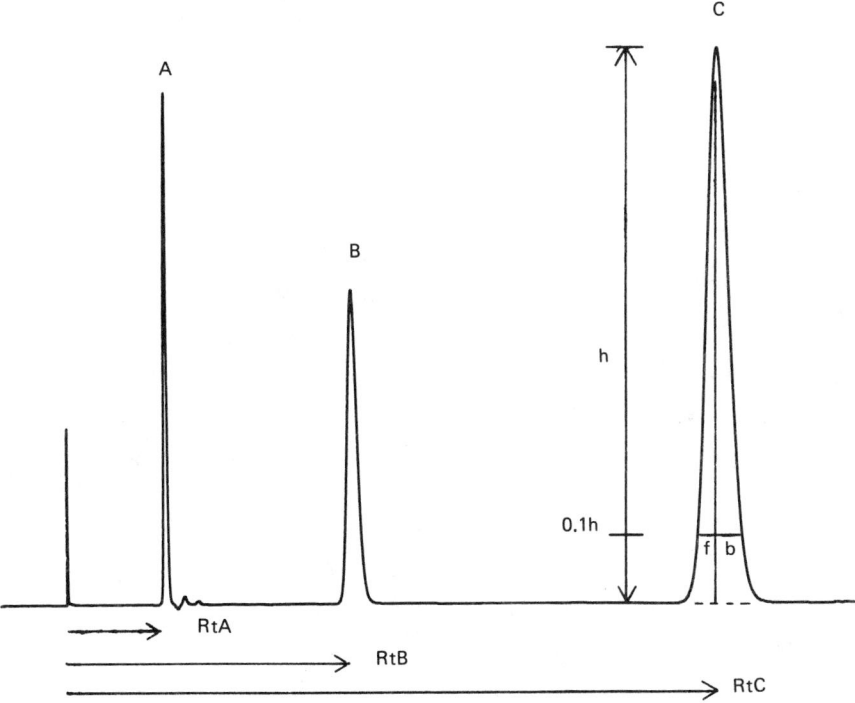

FIG. 13. Test sample chromatogram.

5. k' value for component C: $k_C' = (Rt_C - Rt_A)/Rt_A$
6. The relative retention of component C/component B $= k_C'/k_B'$
7. The degree of peak asymmetry for component C
8. The plate number, N, for component C
9. Void volume, $V_0 = Rt_A \times$ flow rate/recorder chart speed

B. Peak Asymmetry

The peak asymmetry factor measures the degree of tailing. Referring
to Fig. 13, the asymmetry of peak C is determined by initially
dropping a perpendicular line from the apex of the peak to the baseline and then drawing a line segment from the front edge of the
peak to the back edge, parallel to the baseline at a point on the
perpendicular equal to 10% of the peak height. The asymmetry of the
peak is then equal to the length of the segment b divided by the
length of the segment f. Asymmetry = b/f.

C. Plate Number

There are several methods of calculating the plate number N, each
of which gives different values. The magnitude of the variation is
most exaggerated for peaks showing a high degree of peak asymmetry.
Each method measures the value for peak width at different positions
relative to the peak height; thus each is affected to a different
degree by the peak asymmetry. Four methods of calculating N are
shown in Figs. 14-17. In each case Rt_C is the retention of component C, W is the peak width, and h is the peak height.

The asymmetry of the peak may be a factor in the selection of
the method used to calculate N. Ranking these four methods in the
order of increasing sensitivity to asymmetry gives the following:
first, method 4 (5α); second, method 1 (tangent); third, method 2
1/2 peak height); and fourth, method 3 (0.607 peak height). To obtain the most accurate plate count, and most meaningful evaluation
of the column, component C of the test mix should give a highly symmetrical peak. It has been reported [78] that plate count measurements on peaks with asymmetry factors greater than 1.2 will result

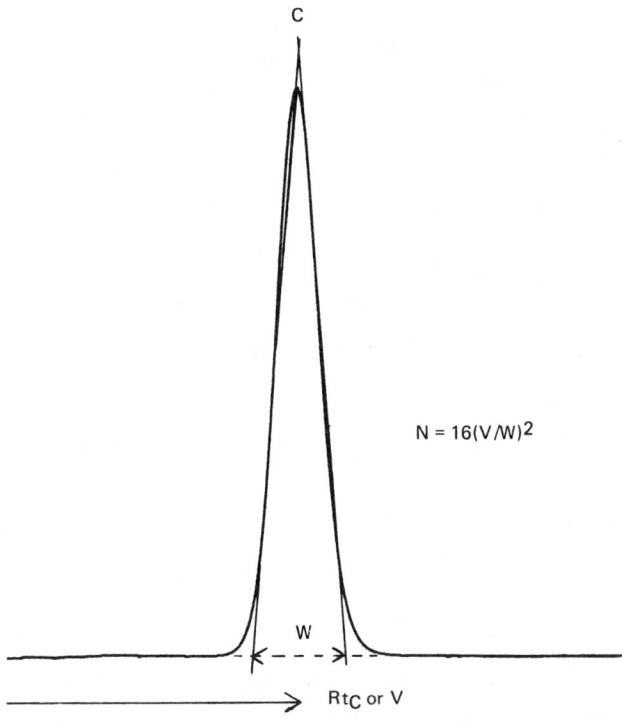

FIG. 14. Method 1, commonly known as the tangent method [78] for determining the number of theoretical plates, N.

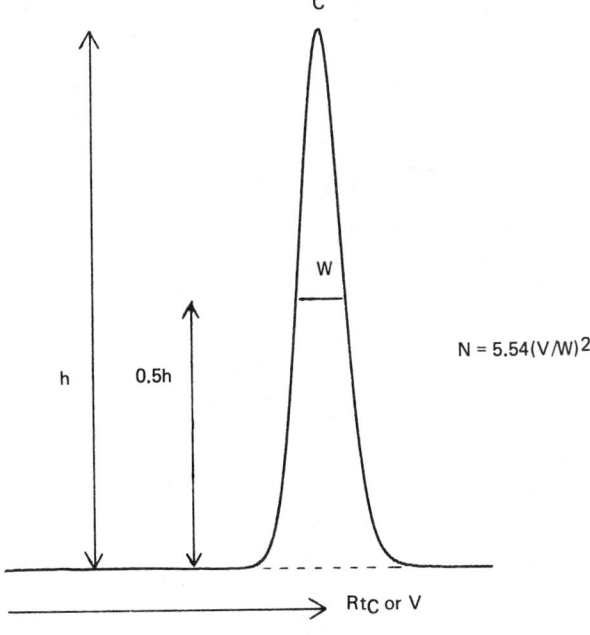

FIG. 15. Method 2, the 1/2-peak-height method [78] for determining the number of theoretical plates, N.

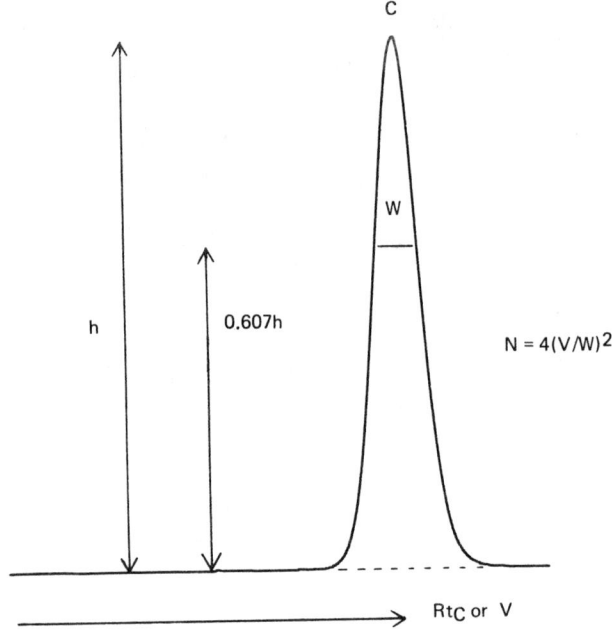

FIG. 16. Method 3, the 0.607-peak-height method [71] for determining the number of theoretical plates, N.

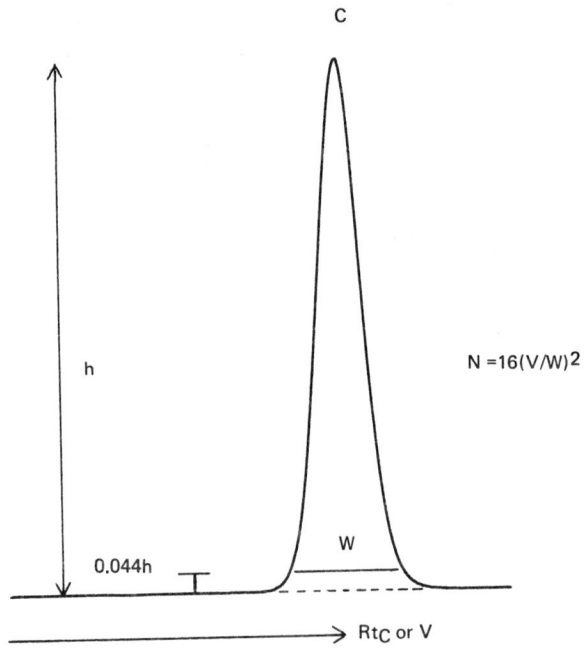

FIG. 17. Method 4, the 5α method [79] for determining the number of theoretical plates, N.

in significant positive error. It was suggested that not only
should these peaks be avoided in plate count measurements, but that
the chromatographic system be modified, or the column changed.

If the peak is highly symmetrical, all four would be suitable
and probably give nearly equivalent values for N. Assuming a peak
with some asymmetry, methods 2 and 3 give larger values for N because the peak width is measured at a point across the peak far
above the asymmetric portion of the peak. Method 4 measures the
peak width much lower and very close to the baseline, while method
1 measures the peak width on the baseline where the tangents of the
peak intersect the baseline. Methods 1 and 4 give a lower and probably more meaningful evaluation of N for asymmetric peaks.

For the purpose of evaluation by comparison with prior data,
it is obvious that the method used to determine the plate count
should be highly reproducible. The accuracy and reproducibility of
method 4 by manual calculations has been questioned [77]. It should
also be noted that drawing tangents reproducibly is difficult, and
for this reason, assuming a symmetrical peak, one or the other more
reproducible methods may be desirable.

Now that the performance parameters have been explained and
illustrated, a discussion describing how these parameters change as
the column deteriorates is in order. The retention times, or distances as measured with a ruler, get smaller. The $k'B$ and $k'C$
values get smaller, but probably not at the same rate. This means
that the relative retention, $k'C/k'B$, will probably change. Day-
to-day variations in retention are to be expected due to slight
differences in the mobile-phase composition and should not be interpreted as column deterioration. As previously noted, peaks will
lose sharpness and get broader, which means that the value of N will
get smaller. If the substance is subject to tailing, this will be
observed by a larger b/f ratio. A change in the void volume could
indicate the presence of trapped gas or immiscible solvent retained
in the packing pores, or even worse it could indicate channeling in
the adsorbent bed. A change in the pressure drop would probably

accompany these situations. If the column head itself or the protective filter used at the head of LC columns gets plugged, there
will be an obvious increase in pressure drop.

This entire discussion of column performance and evaluation
assumes that all the other components (e.g., pump, injector, detector) are in good working condition and that all plumbing is done
with the proper ID tubing and low-dead-volume fittings. In addition,
the strip chart recorder must also function properly, giving reproducible chromatograms.

It should be noted that although this material was presented
in this chapter, it is pertinent to the evaluation of all chromatographic columns.

VII. CARE AND CLEANING OF COLUMNS

Every chromatographer should, at least one time, carefully read the
information pamphlet that comes with a new chromatographic column.
These pamphlets contain many recommendations concerning the do's and
don'ts of column usage and handling, which if followed will help
extend the useful life of the column.

The old adage that cleanliness is next to godliness is certainly
true in the practice of liquid chromatography. If one does not
practice good "chromatographic hygeine," one should be prepared to
suffer the consequences of poor performance and plugged columns.
Fortunately, most "chromatographic hygeine" is based on common sense
and is neither difficult nor particularly expensive to practice.

A. Mobile Phase

The worst enemy of the chromatographic column is particulate material, either in the solvent or in the sample solution. It has been
the practice in the past to filter all solvents through a membrane
filter to remove particulate material. It is possible that this
practice, if not carefully done, may lead to "dirtier" solvents th
than those one started with because of the extra handling and containers that come in contact with the solvents. The quality control

measures currently employed by the manufacturers and "bottlers" of
solvents for HPLC use are resulting in solvents of greater purity
and cleanliness, so that filtration of HPLC or equivalent-grade solvents may no longer be necessary. In addition, all liquid chromatographs have at least two filters between the solvent supply and the
injector that will remove particulate material before it reaches the
column. There is one case, however, where filtration through a
membrane filter is an absolute necessity. This is when a dry chemical substance is dissolved either in water, to be mixed with methanol,
or in an organic solvent. The dry chemical may contain insoluble
particulate material that eventually could clog the in-line filters,
possibly causing pump or injector malfunction.

B. Sample Solutions

It is undoubtedly the sample solution that provides the greatest
potential for the introduction of particulate matter onto the head
of the column. Some type of clarification procedure is an absolute
necessity. Centrifugation and filtration using filter paper can be
used, but neither is totally satisfactory. The former can leave
particles in suspension and the latter small particles of cellulose
that have sloughed off the filter paper. A better procedure for
sample clarification uses a syringe with a Luer-Loc fitting that
will accept a Swinny filter holder. This is a two-piece screw-
together device with a wire screen which the filter pad rests. The
particulate material is removed by forcing the liquid through the
filter pad using the syringe. Filter pads for both aqueous and nonaqueous solvents are manufactured. Both types of filter pads and
the filter holders are readily available.

Other methods of sample cleanup which utilize some type of
disposable column containing a suitable adsorbent are available commercially. At least one of these is attached to a syringe and used
as described for the filter pad and syringe. Alternatively, homemade disposable minicolumns can be made from serological pipettes
by replacing a 3 to 4-cm layer of either silica or alumina over a
glass wool pledget. These are used as gravity flow columns.

In addition, classical wet chemistry utilizing separatory fun-
nels and conventional open column chromatography using Celite 545
mixed with a suitable aqueous phase and elution with an immiscible
organic solvent are still viable tools for sample cleanup prior to
HPLC. It should be noted that if the cleanup procedure involves
more than simple filtering, more than just insoluble materials may
be removed, depending on the nature of the substance and the solvent used in the cleanup procedure. Whether this is or is not desirable is dependent on the requirements of the analysis.

C. Guard Column

One protective device that has been used for some time is the guard
column. The guard column consists of a short column, 5-10 mm in
length, having the same internal dimensions [80] as the analytical
or preparative column with which it will be used. If for any reason there should still be particulate matter in the sample solution,
the guard column will keep it from getting to the analytical column.
    Ideally, the guard column should be packed with the same ad-
sorbent as the analytical column; however, due to the difficulties
in dry packing the microsilicas, most workers use one of the PLB
pellicular silica adsorbents, which are readily dry packed. Care
should be taken that the guard column is properly packed so that
there is no dead volume. An improperly packed guard column will
cause band spreading or broadening. An additional source of band
broadening arises from the tubing connecting the guard column to the
analytical column. This tubing should be the smallest possible
length of microbore tubing.
    Natural products or illicit drugs manufactured from natural
products are frequently encountered in forensic drug analysis. Because of the manufacturing techniques involved, these products may
contain varying amounts of natural pigments and other materials derived from the vegetable substance. These are very harmful to adsorption columns because a large amount of this material is either
strongly retained, or totally, irreversibly retained. This would

be especially harmful if the analytical column was being used as a semipreparative column and rather large amounts of this type of sample would be injected.

Irreversible adsorption can be readily observed by taking a chloroform extract of any colored vegetable substance and passing this extract through a silica column made from a disposable serological pipette. As chloroform is passed through the column, some of the colored substances elute readily, while others migrate very slowly and some not at all. If the eluant is changed to methanol, additional broad bands of color will elute. A large amount of a dark-brownish substance will be retained at the top of the silica column. This substance will not be eluted using water or acetic acid and is in fact totally, irreversibly adsorbed.

The guard column in this instance is of particular importance since it will retain these harmful substances and prevent them from entering the analytical column. As the guard columns hold a small amount of adsorbent, the expense of repacking it with new adsorbent is very minimal compared to the cost of a new column and therefore should be periodically repacked.

D. Silica Dissolution

Most manufacturers recommend that silica columns and silica-based bonded-phase columns should not be used above a pH of 8. It has been shown [81-83] that silica has some solubility in basic solutions. Silica has the greatest solubility in aqueous solutions containing tetraalkylammonium salts [82] commonly used in ion-pairing chromatography, and somewhat less in aqueous sodium hydroxide solutions [82]. Primary, secondary, and tertiary amines and ammonium hydroxide were found to dissolve silica only slightly [82]. In addition; it was noted [82] that the amount of silica dissolved in the eluant in parts per million was approximately the same as the percent of water in the eluant. This study [82] also showed that while the silica was being dissolved, the bonded phase remained intact. It has been shown [82, 84] that bonded-phase packings dissolve more slowly than silica, the rate of dissolution decreasing with the percent of silanol coverage.

Ionic components (i.e., KH_2PO_4, NaCl, etc.) increase the rate of dissolution of silica [84]. Citrate buffers should be avoided because they dissolve silica through chelating properties [85]. Pure water [83] was found to dissolve silica at the rate of 38 $\mu g/\mu l$ from a 4.6 mm × 25 cm column of 10-μm silica with a 1-ml/min flow rate at room temperature. Increasing the temperature increased the dissolution rate. In a static experiment [86] the concentration of silica in pure water at pH 5.5 after 27 days was found to be 118 ppm. In this experiment with the addition of 0.05% potassium phthalate at pH 4.2, the concentration of silica was found to be 49 ppm after 27 days.

The occurrence of column voids with accompanying loss of efficiency results from dissolution of the silica. This problem can be avoided by the use of a precolumn [83] of silica which is inserted between the pump and the injector. Since this column is prior to the injector and contributes nothing to the chromatography, it can be of any dimensions and contain a large-diameter silica adsorbent. The function of the precolumn is to saturate the mobile phase with silica so that it will not remove silica from the analytical column, thus maintaining its efficiency at a high level over its lifetime.

E. General Precautions

In the practice of historic open column adsorption chromatography, it was a rule that the column not be allowed to dry out. The adsorbent would crack and cause channeling if it were allowed to dry out. This same rule applies to the HPLC columns in use today. If a silica column is to be removed from the chromatograph for storage, it should contain a low-activity solvent such as hexane. The end fittings should then be tightly capped. At least one column comes from the manufacturer with a spring-loaded diaphragm device for storing the column with solvent under pressure. It would be a good idea to reuse one of these devices, especially if the column were to be stored for a long period of time.

If the solvent in use contains a dissolved salt or other dry chemical, do not allow these solvents to remain static on the column and in the rest of the chromatograph as well. Two potentially harmful situations can occur. First, the salt or dry chemical may crystallize either in the column or somewhere in the chromatograph's fine-bore tubing, causing a plug. Second, some solutions are favorable to the growth of microorganisms, which also clog the column. Normally, the latter would not be encountered in adsorption chromatography. The way to avoid these problems is to flush or purge the system at the end of the day with 10-20 column volumes of a second solvent, which could be the same solvent or, more correctly, mobile phase minus the dry chemical. This should give shorter reequilibration time on startup the next day.

In the slurry packing procedures employed in the commercial preparation of packed columns, pressures of the order of 10,000-20,000 psi are utilized [84]. In practice, most analytical LC columns are subjected to less than 3000 psi, so that there is little danger of damage to the column; even so, sudden changes in pressure should be avoided [85, 87]. Column manufacturers caution against mechanical shocks such as dropping, bending, or jarring, which can cause disruption of the particle alignment, resulting in poor column performance.

One additional caution concerns the reinstallation of a column once it has been removed from the chromatograph. Make sure that the column head fitting can be identified and that it is reconnected to the solvent delivery tube. Accidental reversal of the column could result in damaging an expensive detector due to particulate material trapped at the inlet frit of the column being backflushed into the detector.

F. Cleaning and Reactivation

The preceding discussion concerned prophylactic measures to extend the life of the column. As previously stated, the performance of any chromatographic column, no matter how carefully it has been used, will eventually deteriorate. When this happens there are

some steps that can be taken to regenerate the column to a point
somewhat less than its initial efficiency.

The procedure generally used is to elute the column with 10-20
column volumes of miscible solvents of increasing solvent strength,
such as chloroform to methylene chloride to methanol to water, and
then to reactivate the column in the reverse order. To obtain the
highest degree of activation the final eluant should be dry hexane.
An alternative procedure to reactivate the silica [88] is to pump a
solution of 2,2-dimethylpropane (DMP) in hexane, methylene chloride,
chloroform, and so on, with an acid catalyst such as acetic acid
at about 2% (v/v).

Instead of water as the strongest eluant, a 1-5% solution of
acetic acid in water [75] can be used. This is particularly applicable to regeneration of columns used in forensic drug analysis
since it is more effective in removing strongly retained basic components.

After regeneration and reequilibration of the column, the standard test mixture should then be chromatographed so that the regeneration process can be evaluated. If the column performance parameters
have not been improved, a new column is needed.

It is possible to remove from the head of the column a small
amount of packing material suspected of being damaged or defective.
The void must then be filled with some sort of packing, preferably
the same as the original adsorbent. Glass wool, micro glass beads,
or Teflon wool are less desirable, but can be used as a substitute
packing material. This would normally be a last resort procedure
and require a lot of patience as well as technique in order to avoid
the presence of the slightest dead volume after repacking.

VIII. APPLICATIONS

A search of the recent scientific literature shows very few forensic application papers utilizing liquid-solid adsorption chromatography. Nearly all of these are drug applications.

Prior to the introduction of ion-pairing reversed-phase chromatography, LSC was probably the chromatographic mode most widely applied in forensic drug analysis. The current widespread interest in this field is in a large part due to the many excellent applications of ion-pairing reversed-phase chromatography that have appeared in the recent literature. This interest in HPLC is indeed exciting to those of us who have been engaged in the practice of HPLC from its early days.

It is, however, unfortunate that LSC is largely being ignored and regarded by many as an outdated technique. As will be shown, excellent separations can be achieved using this technique and it is hoped that some renewed interest in LSC will occur.

Some of the application papers reviewed here are included because of their historical value, and in some cases represent the initial reports of the particular application found in the literature.

A. Depressants

D. H. Rodgers [89] investigated the chromatography of several phenothiazine drugs on a 13-μm pellicular adsorbent, Sil-X-1. Chloroform extracts of ammoniacal solutions of the phenothiazine drugs were chromatographed using a mobile phase of chlorobutane-isooctane (1 + 1) containing 1% diethylamine, at a flow rate of 1 ml/min. A UV detector at 254 nm was used. The phenothiazines were also chromatographed on a strong cation-exchange packing, ION-XOSC. Ion-exchange chromatography was also utilized in the separation of some rauwolfia alkaloids.

Sil-X-II, a 43-μm pellicular adsorbent, was used in the chromatography of some benzodiazepines. Sil-X-1 and Sil-X-II are partially deactivated silicas. The chromatography of a mixture of four benzodiazepines obtained with a mobile phase of 1% methanol in chloroform is shown in Fig. 18.

In addition, a quantitative procedure for chlordiazepoxide in Librax capsules was described. A weighted aliquot of capsule

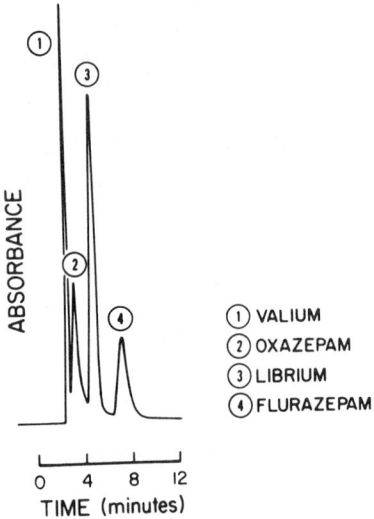

FIG. 18. Separation of mixtures of four benzodiazepine drugs by
adsorption chromatography. Column: 500 mm × 2.6 mm ID, Sil-X-II;
mobile phase: 1% methanol in chloroform; flow rate: 0.8 ml/min;
detector: UV at 254 nm. (From Ref. 89. Reproduced from the
Journal of Chromatographic Sciences, by permission of Preston Publications, Inc.)

contents equivalent to about 2 mg of chlordiazepoxide is transferred
to a 15-ml centrifuge tube. The tube is shaken for several minutes
with 3.0 ml of chloroform containing 4% diethylamine and 1.5 mg/ml
4'-hydroxyacetanalide as internal standard, centrifuged, and 5.0 μl
of supernatant solution injection injected. A standard-internal
standard solution is similarly prepared. A mobile phase of 1.5%
methanol in chloroform was used for this assay. A relative standard deviation of 0.86% was obtained for five sample charges.

An undesirable feature of this system, as shown in Fig. 18,
is that Valium elutes in the solvent front. Although the quantitative results are very good, a chromatogram of the sample-internal
standard would have improved this report.

The effects of water on solute retention were studied by Gonnet
and Rocca [67]. Benzodiazepines, barbiturates, phenothiazines,
pyrimidines, and 5H dibenzazepines were chromatographed on 5-μm

Lichrosorb Si 60 and 6-μm Partisil columns using dichloromethane-
isopropanol mixtures saturated with water. Ammonia was added to
the mobile phase used for the 5H dibenzazepines. TLC systems were
also studied.

The solute-silanol group interactions were found to decrease
in the water-saturated systems, resulting in very little tailing.

Capsule contents and tablets were ground and then extracted
with either dichloromethane or methanol, depending on the solubility.

As can be seen from Figs. 19 and 20, the systems show overall
good separation of the benzodiazepines and barbiturates that were
chromatographed. Note that two pairs of benzodiazepines, Nobrium
(medazepam)/Clobazam and Valium/Myolastan, coelute. It is unlikely
that such mixture would be encountered, but the possibility should
not be overlooked.

Although no quantitative data were presented, the systems appear to be suitable for quantitative analysis.

In a more recent study, Wittwer [90] described the separation
of 11 benzodiazepines on a 10-μm particle size μPorasil column.
The mobile phase was composed of 90 parts cyclohexane and 10 parts
of the mixture NH_4OH-methanol-w/w $CHCl_3$ (1 + 200 + 800). The
chloroform was water washed and filtered through a sintered glass
filter. Dual-wavelength UV detection at 254 and 280 nm was utilized
to increase the qualitative discrimination of the chromatography.
Figure 21 shows the chromatogram of a mixture of the 11 benzoidaxe-
pines. Table A1 in the appendix shows the retention, relative to
fluorazepam and the absorbance ratio, A_{254}/A_{280}, for each of the 11
benzoidazepines. As can be seen in Fig. 21, cyprazepam and diazepam
coelute, but as shown in Table A1 have widely different adsorption
ratios, so they can be differentiated.

The following extraction procedure can be used for all the
benzodiazepines, except for the dipotassium salt of clorazepate,
which is insoluble. An aliquot of ground tablet, capsule contents,
or powder, equivalent to about 5 mg of benzodiazepine, is accurately
weighted into a 100-ml volumetric flask. Chloroform is added to

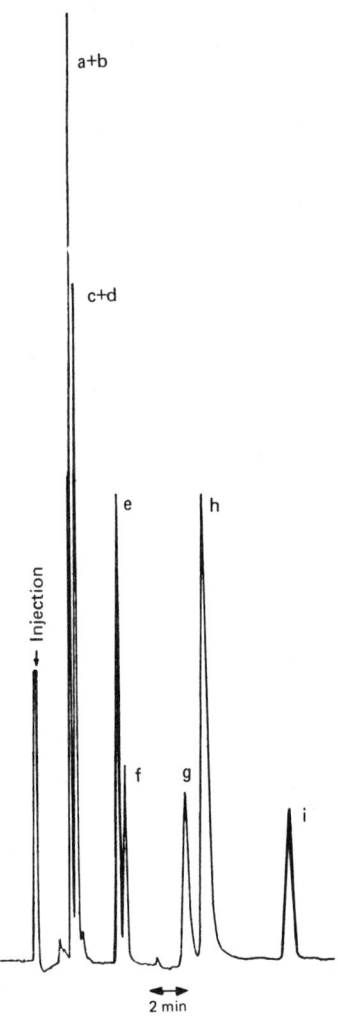

FIG. 19. Separation of benzodiazepines. Column: LiChrosorb Si 60,
5 μm, 20 cm × 4.6 mm ID; isocratic elution at ambient temperature;
mobile phase: dichloromethane-isopropanol (96:4) saturated with
water; pressure: 75 bar; chart speed: 0.5 cm/min; detector: UV
at 254 nm; sensitivity: 0.16; linear velocity: 0.26 cm/s.
a, medazepam; b, clobazepam; c, diazepam; d, tetrazepam; e, nitrazepam; f, clorazepate; g, lorazepam; h, oxazepam; i, chlordiazepoxide.
(From Ref. 67.)

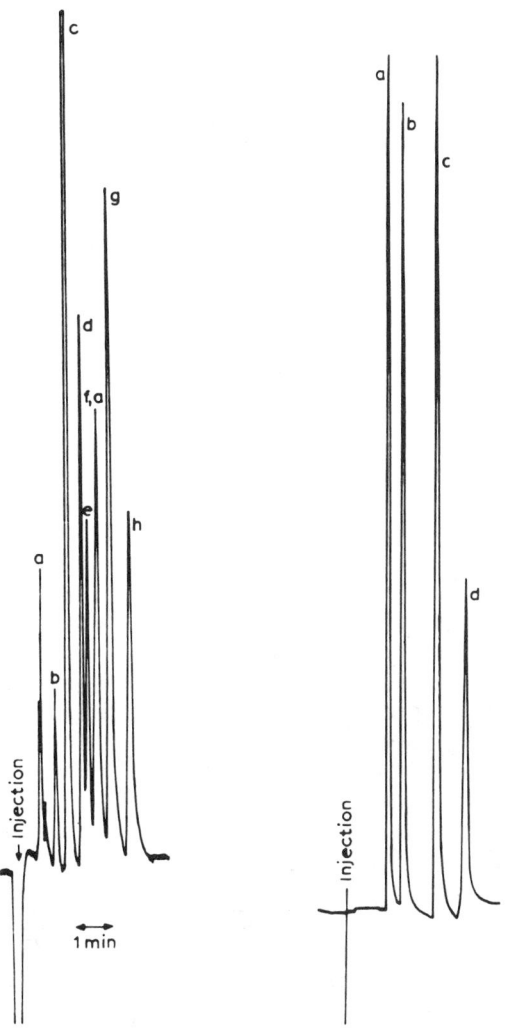

FIG. 20. Separation of barbiturates. Column: Partisil, 6 μm,
15 cm × 4.6 mm ID, isocratic elution at ambient temperature; mobile
phase: dichloromethane-isopropanol (97.5:2.5) saturated with water;
chart speed: 1 cm/s; detector: UV at 254 nm; sensitivity: 0.02;
linear velocity: 0.62 cm/s. a, Allobarbital; b, mephobarbital;
c, hexobarbital; d, secobarbital; e, tetrallobarbital; f + a,
amobarbital; g, phenobaroital; h, barbital. (From Ref. 67.)

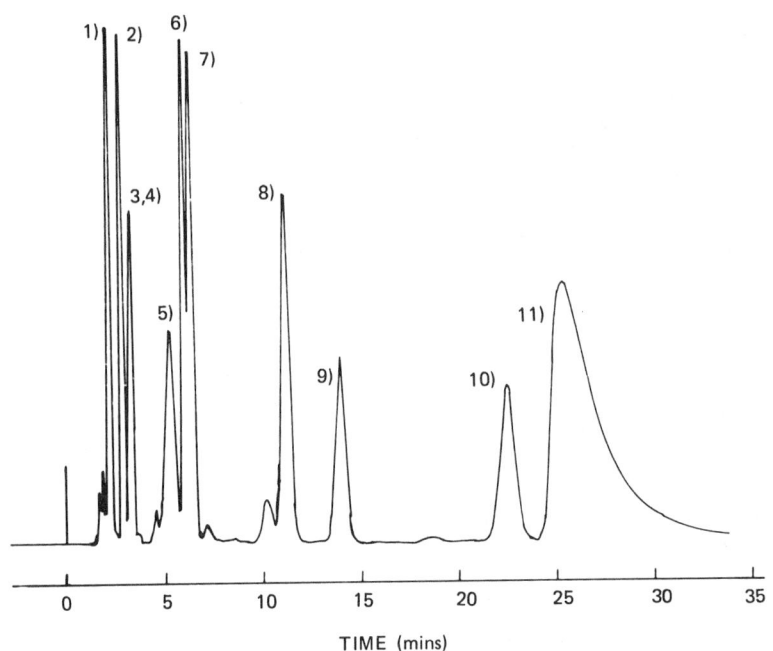

FIG. 21.   Chromatogram of a mixture of 11 benzodiazepines determined
at 254 nm wavelength.   1, medazepam; 2, prazepam; 3, cyprazepam;
4, diazepam; 5, flurazepam; 6, chlordiazepoxide; 7, N-desmethyldiazepam; 8, nitrazepam; 9, clonazepam; 10, demoxapam; 11, oxazepam.
(From Ref. 90.)

volume and an ultrasonic bath is used to aid solution. Replicate
5-µl injections of the clarified sample solution and of the appropriate benzodiazepine standard solution at a concentration of 0.05-
0.10 mg/ml are made. Duplicate analysis of a 5-mg diazepam tablet
showed 5.01 and 5.02 mg per tablet by area integration and 5.02 and
4.96 mg per tablet by peak height measurements.

   Clorazepate decarboxylates to N-desmethyldiazepam in acidic
solution, which can then be extracted into chloroform. An aliquot
of the powdered sample equivalent to about 5 mg of clorazepate is
accurately weighted and transferred to a 125-ml separatory funnel.
Then 25 ml of 0.1 N HCl or 0.1 N H_2SO_4 is added. After 10 min the
N-desmethyldiazepam is extracted with 3 to 30-ml portions of chloro-

form and filtered directly into a 100-ml volumetric flask. Chloroform is added to 100.0 ml. Five-microliter injections of the sample and standard prepared in the same manner are injected. Using this method, 95.5% recoveries of N-desmethyldiazepam from the dipotassium salt of clorazepate were obtained.

This report would have been improved with the actual analysis of a clorazepate sample.

B. Stimulants

In this early study of phenethylamines, Cashman et al. [91] investigated both ion exchange on DA-X4 resin and adsorption on Corasil II. Chloroform-methanol (4 + 1) was employed as the mobile phase in the adsorption system.

The phenethylamines as their salts were dissolved directly in the mobile phase, which was made "weakly" basic by the addition of triethylamine and NaOH. Table A2 shows the retention of nine phenethylamines, relative to methamphetamine. Figure 22 shows the chromatogram of a mixture of methamphetamine, methoxyphenamine, and ephedrine.

This adsorption system is primarily of historic interest and would not be particularly useful. As can be seen in Table A2, methamphetamine, at a flow rate of 0.4 ml/min, elutes in 9.5 min while amphetamine was not eluted. In this study an ion-exchange system was found to be superior to this adsorption system.

The addition of an amine, such as diethylamine, to the mobile phase would probably have improved the chromatography.

Without question, the most comprehensive study of forensic drug analysis using silica as the column packing is that of Jane [92]. Several specific applications from this work will be cited and it will also be discussed as a general screening tool. The first application is that of phenethylamines. The chromatography was done on a 6-μm Partisil column utilizing a mobile phase of methanol-2 N ammonia-1 N ammonium nitrate (27 + 2 + 1) at a flow rate of 1 ml/min with UV detection at 254 nm.

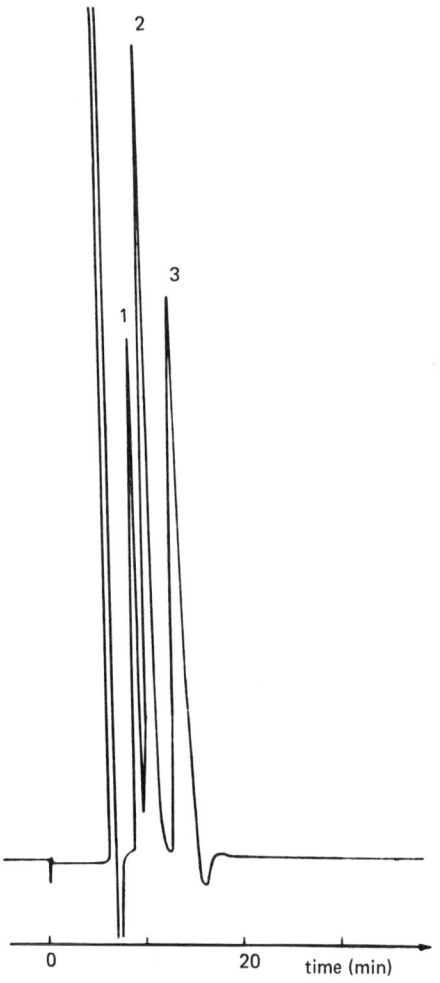

FIG. 22.   A 20-μl sample of a mixture of (1) methamphetamine,
(2) methoxyphenamine, and (3) ephedrine at concentrations of approximately 0.1 mg/ml each, dissolved in the developing solvent made
weakly basic by the addition of triethylamine and NaOH.   Column:
2 mm × 1500 mm stainless steel packed with Corasil II; solvent:
chloroform-methanol (4:1); temperature: ambient; flow rate: 0.8
ml/min at 400 psi; detector: UV at 254 nm at 4 × attenuation;
chart speed: 0.2 in/min.   (From Ref. 91.   Reproduced from the
Journal of Chromatographic Science, by permission of Preston Publications, Inc.)

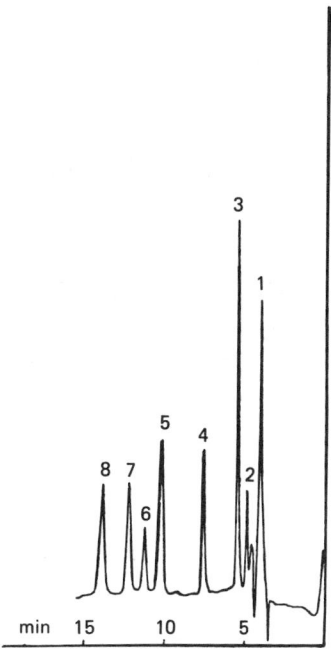

FIG. 23. Separation of 1 µl of a synthetic mixture of phenethylamines, containing: 1, benzphetamine; 2, phendimetrazine; 3, phenmetrazine; 4, dexamphetamine; 5, N-methylephedrine; 6, ephedrine;
7, methylamphetamine; 8, mephentermine. Solvent: .methanol-2 N
ammonia solution-1 N ammonium nitrate solution (27:2:1); flow rate:
1 ml/min; detector: UV at 254 nm. (From Ref. 92.)

Samples are dissolved in an amount of water or dilute hydrochloric acid to yield a concentration of about 10 mg/ml. Injections
of 1-5 µl were then made. Figure 23 is a chromatogram of a mixture of phenethylamines.

Table A3 shows the retention of 27 phenethylamines expressed
relative to methamphetamine. Although there are possible interferences, the most widely abused phenethylamines are well resolved.
Specific quantitative results were not presented, so that it is not
possible from the report to make a comparison with other techniques.

Cartoni and Natalizia [93] have applied adsorption chromatography to the analysis of pemoline in dosage form and also in urine.

The mobile phase used was n-hexane-isopropanol-concentrated ammonia (37 + 12 + 1) at a flow rate of 0.5 ml/min. A UV detector at 254 nm was employed.

The powdered product, 1-10 mg of pemoline, was dissolved in 100 ml of water, which was then filtered. A 10-ml aliquot plus 0.5 g of NaCl and NH_4OH, added dropwise to pH 8-9, was extracted with three portions of methyl acetate, dried over sodium sulfate, and "an appropriate amount of an internal standard was added for quantitative determination." The solution was concentrated to 100 μl and 1-2 μl was injected. Figure 24 shows the chromatogram of 0.1 μg of pemoline. The detector sensitivity is not stated.

The sensitivity of this procedure could be improved by the extraction of a larger portion of the aqueous sample solution, if not all. As it is written, only 10% of the aqueous sample solution is utilized. It would also be better to make the sample to 1.0 ml and inject 10 μl than the 100-μl volume and the 1- to 2-μl injection described. If an internal standard is to be used, it should certainly be identified and the standard/internal standard ratio should be stated.

It would appear from the data presented that only about 51% of the pemoline is extracted by methyl acetate from the aqueous phase described (1:1), and three extractions would yield only an 88% recovery rate. Two additional extractions would increase the recovery to 97%.

In a recent paper, Lewin et al. [94] describe an adsorption system for the analysis of cocaine, allococaine, pseudococaine, and allopseudococaine. A Partisil 10 PXS column is used with a mobile phase composed of isopropanol-heptane-diethylamine (25 + 75 + 0.1). An exponential flow program, increasing from 0.48 ml/min to 4.0 ml/min over 12 min was used so that the late eluting components would be eluted in a reasonable length of time. N,N-Dibenzylbenzamide was used as the internal standard.

Unknown cocaine samples were prepared for analysis by dissolving in isopropanol containing 1 mg/ml N,N-dibenzylbenzamide. Four

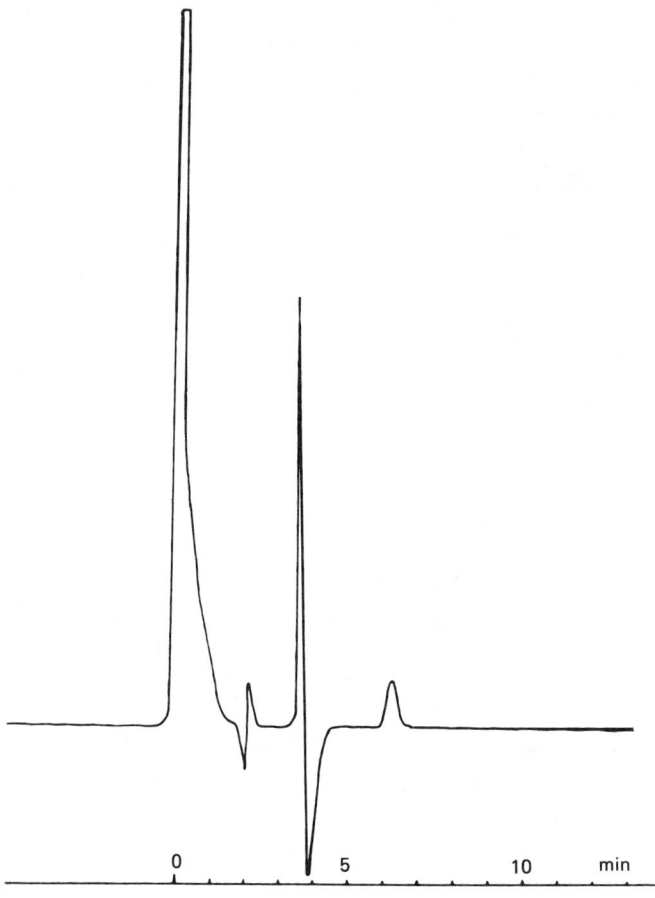

FIG. 24.   Chromatogram of 0.1 μg of pemoline from a pharmaceutical
product.   Column: 25 cm × 2.1 mm ID packed with Zorbax; solvent:
n-hexane-isopropanol-concentrated ammonia (37:12:1); pressure: 120
atm; flow rate: 0.5 ml/min; chart speed: 1.25 cm/min.   (From Ref.
93.)

solutions of each of the cocaine isomers containing from 1×10^{-3} to 9×10^{-3} mmol/mg internal standard were prepared and each of the 16 solutions was chromatographed three times. The area ratio of cocaine isomer to internal standard was then related to the concentration ratio, millimoles of cocaine isomer to milligrams of internal standard, using linear regression analysis. Figure 25 shows the separation of the four isomers plus the internal standard.

HPLC OF ISOMERIC COCAINES

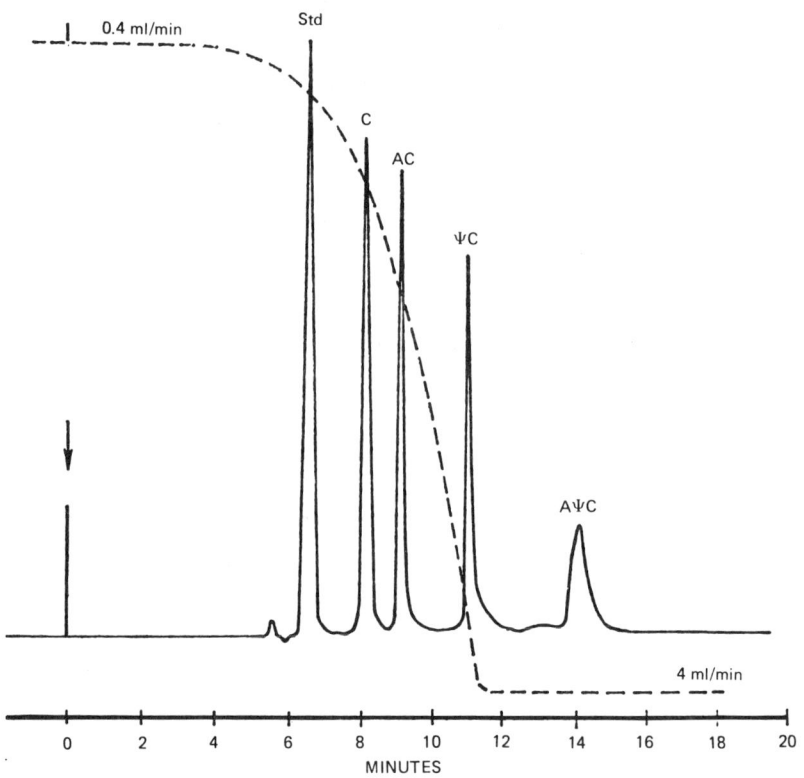

FIG. 25. Separation of isomeric cocaines by HPLC using Partisil 10 PXS. Solvent: 75% heptane, 25% isopropanol, with 0.1% diethylamine. The flow program is indicated by the dashed line. The solution was made up of 0.822 mg of cocaine HCl, 1.982 mg of pseudococaine HCl, 1.000 mg of allococaine, and 2.030 mg of allopseudococaine HCl in 1 ml of isopropanol containing 1.068 mg of N,N-dibenzylbenzamide as internal standard. (From Ref. 94.)

Cocaine and pseudococaine were found to be stable in isopropanol for at least 6 months, while allococaine and pseudoallococaine
were stable for only 1 week.

The authors concluded that gas-liquid chromatography (GLC) on
2% OV-17 or 3% SP-2250 columns was not a satisfactory procedure to
use for the analysis of mixtures of the four cocaine isomers (CI).
Cocaine and pseudococaine did not separate well and allococaine and
allopseudococaine were subject to thermal degradation yielding benzoic acid and products having m/e values of 182 by gas chromato-
graph-mass spectrometry (GC-MS) (CI-isobutane). Direct-probe MS
using CI-isobutane was also subject to misidentification since
pseudococaine and allopseudococaine showed insignificant m/e 304,
presumably due to enolization and then loss of benzoic acid. The
proposed HPLC procedure is not subject to these pitfalls and was
proposed as a positive identification of the cocaines. This was an
unfortunate conclusion because no chromatographic technique alone
will provide conclusive identification of a substance. It would
seem that LC traps of the components followed by GC-MS and direct-
probe MS would yield specific identification of the isomeric cocaines. Infrared (IR) spectrophotometry of the LC traps would also
lead to positive identification of the cocaines.

C. Opiates

The initial report in the literature describing heroin analysis is
that of Cashman and Thorton [95]. A 37- to 50-μm Porasil T column
was used with a mobile phase of chloroform-methanol (4 + 1). A
Beckman DB recording spectrophotometer utilizing a flow cell with
a volume of 300 μl served as the detector. Sample preparation consisted of dissolving the sample in chloroform-methanol (1 + 1).

Figure 26 is the chromatogram of heroin. 0^6-monoacetylmorphine,
and morphine. Each peak represents 200 μg of the component. Procaine was also chromatographed, but it could not be separated from
heroin with this system. A less polar chloroform-methanol system
(12.5% methanol) was investigated, but was discarded because the
time of 25 min was considered to be excessive.

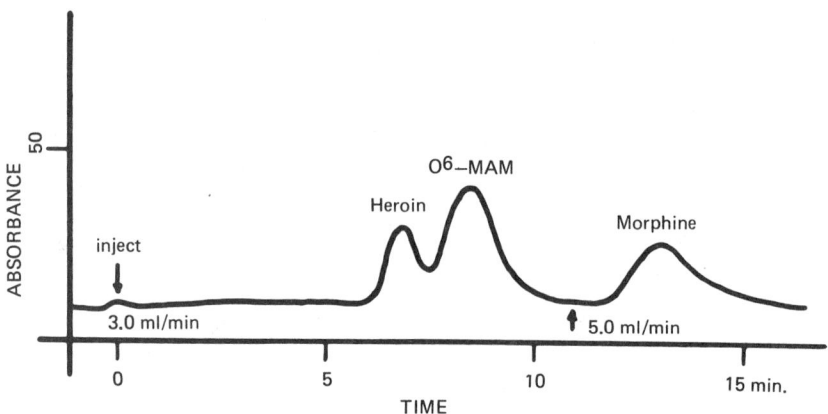

FIG. 26. Sample of heroin, O^6-monoacetylmorphine, and morphine in
1:1 chloroform-methanol. Column: 4.5 mm × 1 m stainless steel
Porasil T, 37-50 μm; solvent: chloroform-methanol (4:1); pressure:
300-500 psi; flow rate: 3.0-5.0 ml/min; chart speed: 1/2 in./min;
sample: 50 μl. (From Ref. 95.)

On alumina, procaine was separated from heroin, but not O^6-
monoacetylmorphine. In order to separate heroin, O^6-monoacetyl-
morphine, and procaine it was necessary to use the two columns
separately. Obviously, this is not a practical solution for this
chromatographic problem.

This report is interesting in that it points out one of the
problems often encountered with self-constructed liquid chromatographs (i.e., poor sensitivity). In this case it is largely due to
the very large volume (300 μl) flow cell used. It was stated that
on a commercial instrument a total sample concentration (heroin,
O^6-monoacetylmorphine, and morphine) of 100 ppm gave a full-scale
response. UV wavelength and detector sensitivity are not stated.
It was recognized that a three- to fivefold increase in sensitivity
could be obtained in an alkaline solution because of greater extinction coefficients.

This report also describes use of the UV spectrophotometer to
obtain a UV scan, 400-210 nm, of each component using the stop-flow
technique, thus providing an additional criterion for the identification of the component.

In the report of Jane [92] previously cited, reference was
made to the analysis of "Chinese heroin." As in the prior application from this work, a 6-μm Partisil column was used and the mobile
phase was methanol-2 N ammonia-1 N ammonium nitrate (27 + 2 + 1) at
a flow rate of 1 ml/min. The UV detector was at 278 nm. The sample
was dissolved in water or dilute hydrochloric acid at a concentration of 10 mg/ml. One to five microliters of the clear sample solution was injected. Figure 27 is a chromatogram of a Chinese heroin
sample. The chromatogram is very good, but the resolution of heroin

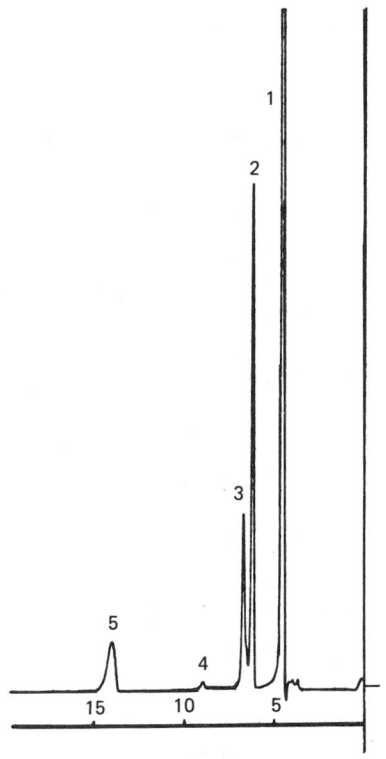

FIG. 27. Separation of a Chinese heroin sample, the peaks identified being: 1, caffeine; 2, heroin; 3, monoacetylmorphine; 4, morphine; 5, strychnine. Solvent: methanol-2 N ammonia solution-1
N ammonium nitrate solution (27:2:1); flow rate: 1 ml/min; detector: UV at 278 nm. (From Ref. 92.)

and O^6-monoacetylmorphine could be improved. The applicability of this system to other types of heroin, such as "Mexican brown heroin" or white heroin containing quinine or quinidine as adulterants, is questionable.

Tables A4 and A5 show the retention, expressed relative to morphine, for opium alkaloids and other compounds of forensic interest, respectively.

Huizer et al. [96] describe the analysis of heroin siezures in the Netherlands during 1975 and in September 1976. A GLC procedure was used in 1975 to quantitate heroin and caffeine only. The heroin samples encountered were described as "Hong Kong No. 3," which contained primarily heroin, O^6-monoacetylmorphine HCl, acetylcodeine HCl, caffeine, and occasionally strychnine.

The GLC procedure was abandoned in favor of the HPLC procedure because the other alkaloids were readily quantitated by this procedure.

A 5-μm LiChrosorb Si 60 column was used with a mobile phase composed of diethyl ether-isooctane-methanol-diethylamine (52.8 + 35 + 12 + 0.2) at a flow of 1.5 ml/min. About 40 mg of homogeneous sample was dissolved in 25.0 ml of chloroform which contained 160 mg of codeine per liter as the internal standard. Ten-microliter injections were made.

Figure 28 is a chromatogram of a heroin sample obtained with this system, and shows excellent separation. The precision of the quantitative determination of heroin is quite good with a relative standard deviation of 1.8% for seven determinations of a granular sample.

It is difficult to be too critical of a system giving such high-quality chromatograms, but there are some problem areas. The ethers as a class of solvents have many undesirable properties, including their susceptibility to peroxide formation, high degree of flammability, and high vapor pressure, which makes them difficult to pump in some LC pumps. It may be possible to substitute isopropyl ether for ethyl ether. Isopropyl ether has better pumping

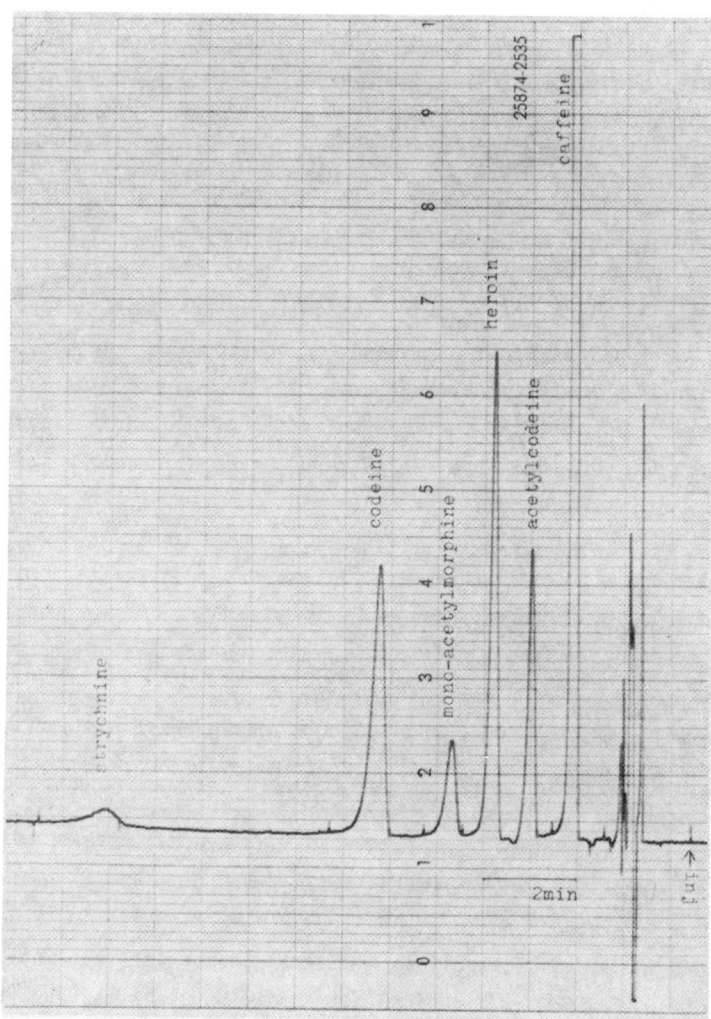

FIG. 28. High-pressure liquid chromatogram of a heroin sample.
(From Ref. 96.)

characteristics, but it still has the remaining undesirable properties of the ethers.

The advantages of an internal standard in a well-designed system with a properly functioning LC are debatable. If an internal
standard is to be used, it should be carefully chosen and should
not be a substance that could be present in the sample. Codeine is
such a substance and was a poor choice. All samples had to be
analyzed with and without the added codeine internal standard to
ensure the validity of the quantitation.

Although morphine is mentioned in the body of the report, no
retention data for it or other possible adulterants is presented.
This again makes it difficult to assess the applicability of this
system in the analysis of other heroin types.

It should also be noted that the potencies of the heroin samples
examined were very high, 20-50%, so that adjustments in sample size
and volume would be necessary in most cases.

In a recent report, Wittwer [97] describes the analysis of
"Brown" heroin samples of the "Mexican process" type. These samples
usually contain, in addition to heroin, acetylcodeine, 0^6-monoacetyl-
morphine, procaine, small quantities of noscapine and papaverine,
and lesser quantities of codeine and morphine.

A 10-μm μPorasil column was used. The mobile phase was composed of 750 ml of cyclohexane + 250 ml of the mixture of NH_4OH-
methanol-w/w $CHCl_3$ (1 + 200 + 800). The chloroform was water washed
and filtered through a sintered glass funnel before use. The flow
rate was 2.0 ml/min with UV detection at 254 nm.

Sample preparation consisted of dissolving a 1-g sample, accurately weighed and transferred into a 100-ml volumetric flask, in
10 ml of methanol with the aid of an ultrasonic bath for 1-2 min.
Then 50 ml of chloroform is added and sonicated an additional 1-2
min, cooled, and made to 100 ml with chloroform. The solution is
filtered through Whatman No. 1 filter paper before injection. Dilu-
tions are made if necessary so that a final concentration of 0.5-
0.6 mg/ml heroin HCl was obtained. Replicate 5-μl injections of the
sample solution and standard heroin HCl solution are made.

In this report the effects of loss of ammonia from the mobile
phase and the increase of water in the mobile phase were studied.
Solvents S-1, S-2, and S-3 were prepared using 28% ammonia, 14%
ammonia, and 7% ammonia, respectively. Heroin and 26 other substances found in heroin samples were chromatographed in each of
these mobile phases. Progressively increasing retention times with
essentially constant relative retentions were observed as is shown
in Table A6. When the retention times of heroin were monitored over
several days in the analysis of the samples, an inconsistent pattern
of daily increase in heroin retention was observed. It was conjec-
tured that this was due to a combination of pump malfunction caused
by air bubbles in the pump head, the fact that the solvent was not
degassed prior to use, and by the loss of ammonia from the mobile
phase. The system was considered usable in spite of this problem.

Since methanol and water are totally miscible, the effects of
any added water from this source was investigated by preparing
solvents S-4 and S-5 using methanol containing 1 and 2% added water.
As can be seen in Table A7, some rather striking selectivities
were observed. Heroin, acetylcodeine, and codeine showed decreased
retention with increasing water content, whereas antipyrine and
O^6-monoacetylmorphine showed increased retention. Thus codeine and
O^6-monoacetylmorphine, which are unresolved in solvent S-1, are
totally resolved in solvents S-4 and S-5. Caffeine and antipyrine
elute before heroin in S-1 and after heroin in S-5. Although S-1
was selected as the primary solvent, the sample composition may indicate S-4 or S-5 as more suitable.

Three "brown heroin" samples were analyzed by the proposed HPLC
procedure and also by GLC with octacosane as internal standard.
The chromatogram of one brown heroin sample is shown in Fig. 29.
Good agreement was obtained between the HPLC and the GLC results,
but poor precision was obtained for two of the three samples by
both procedures, suggesting a possible homogeneity problem.

The two major problems with this report are the observed increase in retention with time and the rather poor precision of the
quantitative results. Possibly a less volatile, non-UV-absorbing

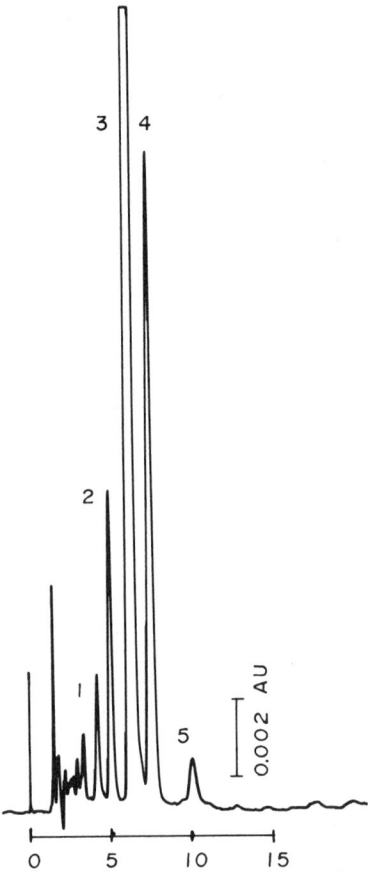

1. Acetylcodeine
2. Heroin
3. Procaine
4. Acetylprocaine
5. O^6-Monoacetylmorphine

FIG. 29.   Chromatogram of a brown heroin sample.   (From Ref. 97.)

amine can be found to replace the ammonia in this system.   The poor
precision may also be a result of variation in integrated area due
to the increased peak width associated with longer retention times.

The analysis of opium as proposed by Beasley, Smith, Ziegler,
and Charles [98] is interesting in that two modes of chromatography
are utilized: ion exchange in the sample preparation procedure and
adsorption chromatography in the quantitation.   The six major alkaloids in opium, narcotine, papaverine, thebaine, codeine, morphine,
and cryptopine are reported to be analyzed in approximately 3 h.

In the sample preparation 2 g of the opium is accurately weighed into a 250-ml beaker and 20 ml of dimethyl sulfoxide (DMSO) is added. The beaker is placed on the steam bath for 15 min and then an ultrasonic bath for 15 min. Thirty milliliters of water is added to the cooled solution and the pH is adjusted to 3-4 with either 10% HCl or 4% NaOH and then transferred to a 100-ml volumetric flask and made to 100.0 ml. A 10.0-ml aliquot of this solution, after filtering, is placed on an ion-exchange column containing 8 g of 50W-X2 ion-exchange resin. The sample is passed through the column at about 2.0 ml/min and then washed with 50 ml of water. The opiates are then eluted from the ion-exchange resin with 50 ml of ammoniacal methanol (68 ml of 58% NH_4OH in 700 ml of methanol plus water to make 1 liter), followed by 100 ml of 70% methanol in water. The eluates are evaporated using reduced pressure and dehydrated using anhydrous ethanol, and finally, 5.0 ml of chloroform-methanol (3 + 1) containing 2.0 mg/ml of brucine as internal standard is added. Five-microliter injections of the sample and a prepared standard mixture were injected onto a Corasil II column. Gradient elution elution is utilized using a homemade gradient-forming device. The program solvent was 100 ml of chloroform plus 400 ml of methanol plus 1 ml of diethylamine. The initial mobile phase consisted of 30 ml of program solvent in 3 liters of n-hexane. The program used is equivalent to a linear gradient, and including column reequilibration required approximately 40 min.

Figure 30 is a chromatogram of Indian opium plus internal standard.

Even though the gradient device is rather crude, the precision of the quantitative measurements (peak heights) of the alkaloids was acceptable, as was the precision of the relative retention volumes.

The chromatography would be improved with the use of a microparticle silica and a solvent delivery system utilizing electronically generated gradients.

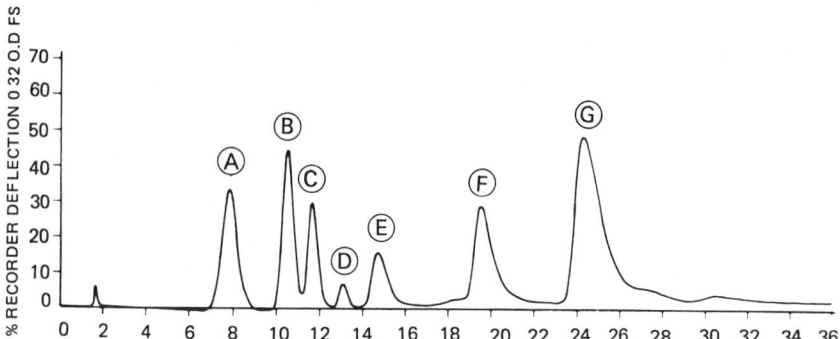

FIG. 30. Typical chromatogram of Indian opium: A, narcotine;
B, papaverine; C, thebaine; D, cryptopine; E, codeine; F, morphine;
G, brucine (internal standard). (From Ref. 98.)

The determination of morphine in opium is the third application cited from the work of Jane [92]. The chromatographic conditions are the same as cited previously, as is the sample preparation.

The chromatogram of an opium sample is shown in Fig. 31.
Codeine and morphine are only partially resolved and narcotine and
papaverine coelute (see also Tables A4 and A5). Although quite
rapid, the chromatography in this application could be better.

The quantitative recovery of morphine from opium can be difficult, and in this report there are no comparisons with other procedures or other recovery data presented to validate the very simple
sample preparation.

Vincent and Engelke [99] determined narcotine, papaverine,
thebaine, codeine and morphine in capsular tissue of *Papaver somniferum*. A 5-μm μ-Porasil column was used with a mobile phase composed of n-hexane-methylene chloride-ethanol-diethylamine (300 +
30 + 40 + 0.5) at a flow rate of 2.4 ml/min. A UV detector at 285
nm at 0.1 AUFS was employed.

The capsular tissues are ground with a ball mill and then 1 g
of the ground sample and 15 ml of 5% aqueous acetic acid are placed
in a glass-stoppered 50-ml centrifuge tube, thoroughly agitated,

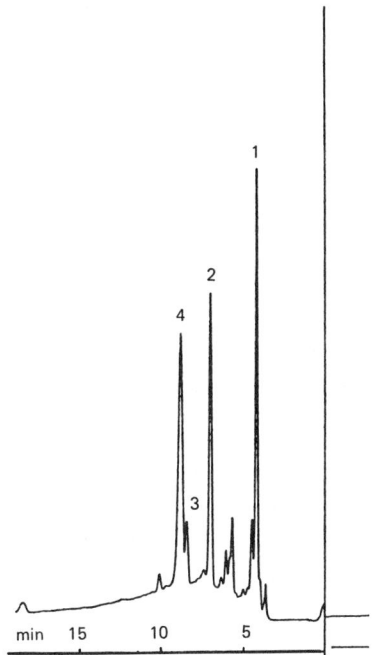

FIG. 31. Separation of an opium sample, the major components being: 1, narcotine; 2, thebaine; 3, codeine; and 4, morphine. Solvent: methanol-2 N ammonia solution-1 N ammonium nitrate solution (27:2:1); flow rate: 1 ml/min; detector: UV at 287 nm. (From Ref. 92.)

and then placed in an ultrasonic bath for 30 min. The pH is adjusted to 8.5 with NH_4OH and 20 ml of chloroform-2-propanol (3 + 1) is added and agitated and then placed in the ultrasonic bath for 10 min. The test tube is then centrifuged for 10 min at 4°C and the organic layer is removed using a Pasteur pipette. The extraction is repeated three times or until a negative test for opiates is obtained on the extract using Marquis, Meckes, and Ferreira reagents. The combined extracts are evaporated to dryness under vacuum. The residue is dissolved in 1 ml of ethanol, filtered with a Swinney filtering syringe. Ten-microliter injections of the prepared sample and standard solution were made using a valve and loop injector. Figure 32 is a chromatogram of P. somniferum capsular extract.

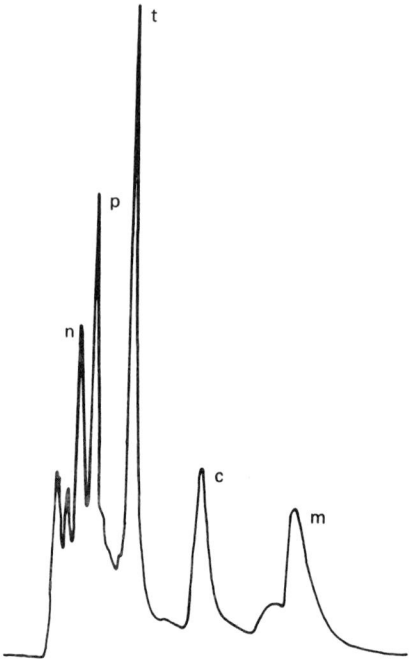

FIG. 32.   Chromatogram of *P. somniferum* capsular extract.   (From
Ref. 99.)

The quantitative results for morphine, codeine, and thebaine
agreed within 5-7% with prior United Nations (UN) analysis of the
samples.   Narcotine and papaverine were not quantitated by the UN
laboratory.

Recoveries of about 99% for codeine, morphine, and thebaine
added to capsular tissues were obtained by aqueous acetic acid extraction.   Good precision for each of the five alkaloids was observed.

The applicability of this procedure to gum or powdered opium
samples was not determined, but this would appear to be a viable
procedure for all types of opium samples.

The reason for centrifugation at 4°C is not stated in this
report and would apparently require specialized equipment.   In
addition, the solvent transfer of the extracted opiates could be

better handled if the extraction was done in a separatory funnel,
which could also be centrifuged if necessary.

D. Hallucinogens

In this early forensic application, Wittwer and Kluckhohn [100]
described the quantitative analysis of lysergic acid diethylamide
(LSD) tablets. Sil-X and Corasil II columns were used in this study
with mobile phases consisting of acetonitrile-isopropyl ether (40 +
60) and acetonitrile-isopropyl ether (25 + 75), respectively. UV
detection at 254 nm was used in each case.

The LSD tablets were ground in a mortar, and an accurately
weighed amount, of X number of ground tablets, was mixed with a
suitable amount of aqueous $NaHCO_3$ solution, mixed with Celite 545,
and packed on a chromatographic column. The column was eluted with
water-saturated diethyl ether, 100-200 ml. The ether was evaporated
to dryness, then the residue was evaporated to dryness and taken
up in a suitable volume of chloroform to give a concentration of
about 0.4-0.5 mg/ml LSD.

LSD was separated from 14 other ergot alkaloids and phencyclidine (PCP), 2,5-dimethoxy-4-methylamphetamine (STP), and strychnine
on both systems. The quantitative results were compared with those
obtained by conventional UV and fluorescence procedures. The results agreed very well with a UV procedure which used a two-column
separation that isolated iso-LSD and LSD.

Figures 33 and 34 are the chromatograms of one of the samples
on Corasil II and Sil-X, respectively. The chromatography, although
quite good in its time, could be improved through the use of a
microparticle silica. The same criticism noted earlier concerning
the use and problems with ethers applies to this mobile phase.

Wittwer [101] has demonstrated the separation of LSD and
lysergic acid methylpropylamide (LAMPA). A µPorasil column was
used with a mobile phase composed of cyclohexane and 0.4% NH_4OH
in tetrahydrofuran (1 + 1) with UV detection at 254 nm and a flow
rate of 2.0 ml/min. The sample preparation consisted of either a

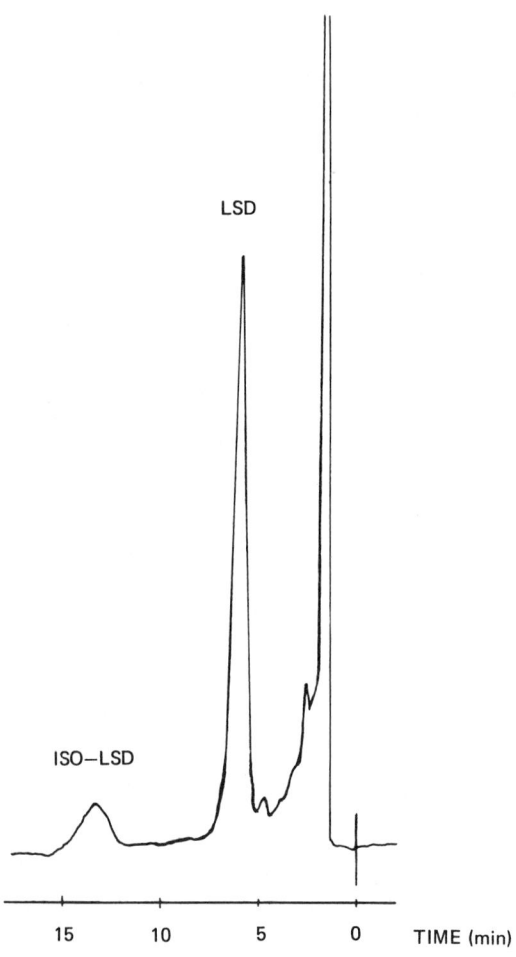

FIG. 33.   Chromatogram of sample 2, double-domed yellow-green tablets, on Carasil II.   (From Ref. 100.   Reprinted from the *Journal
of Chromatographic Science,* by permission of Preston Publications,
Inc.)

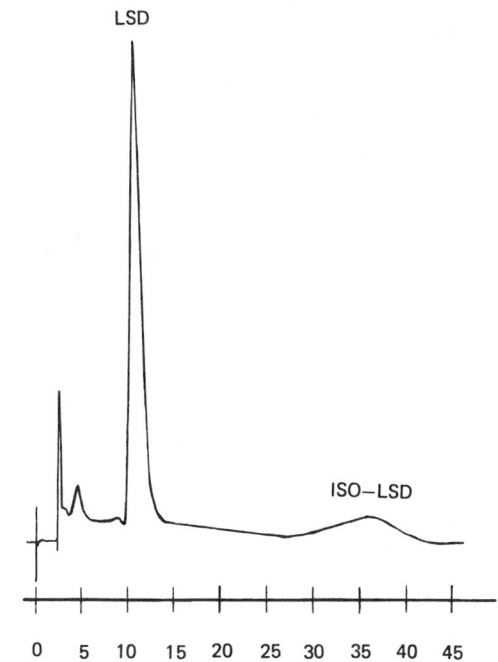

TIME (min)   0   5   10   15   20   25   30   35   40   45

FIG. 34. Chromatogram of sample 2, double-domed yellow-green tablets, on Sil-X. (From Ref. 100. Reproduced from the *Journal of
Chromatographic Science,* by permission of Preston Publications, Inc.)

$NaHCO_3$-Celite column extraction or an extraction from $NaHCO_3$ in a
separatory funnel with chloroform.

The chromatogram of LSD and LAMPA standards is shown in Fig.
35. It was also shown that 75-100 μg of the standards could be
injected, the eluates collected and evaporated to dryness, and
an infrared spectrum obtained using a 1.5-mm microdie and a beam
condenser. Thus the presence or absence of LSD and/or LAMPA can be
unequivocally shown.

The analysis of LSD is the final specific application taken
from the work of Jane [92]. Figure 36 shows the separation of
lysergic acid, lysergamide, LSD, and iso-LSD. The column is a 6-μm
Partisil and the mobile phase is methanol-0.2 N ammonium nitrate
(3 + 2) with UV detection at 320 nm. The sample preparation is the

CHROMATOGRAM OF LSD–LAMPA MIXTURE

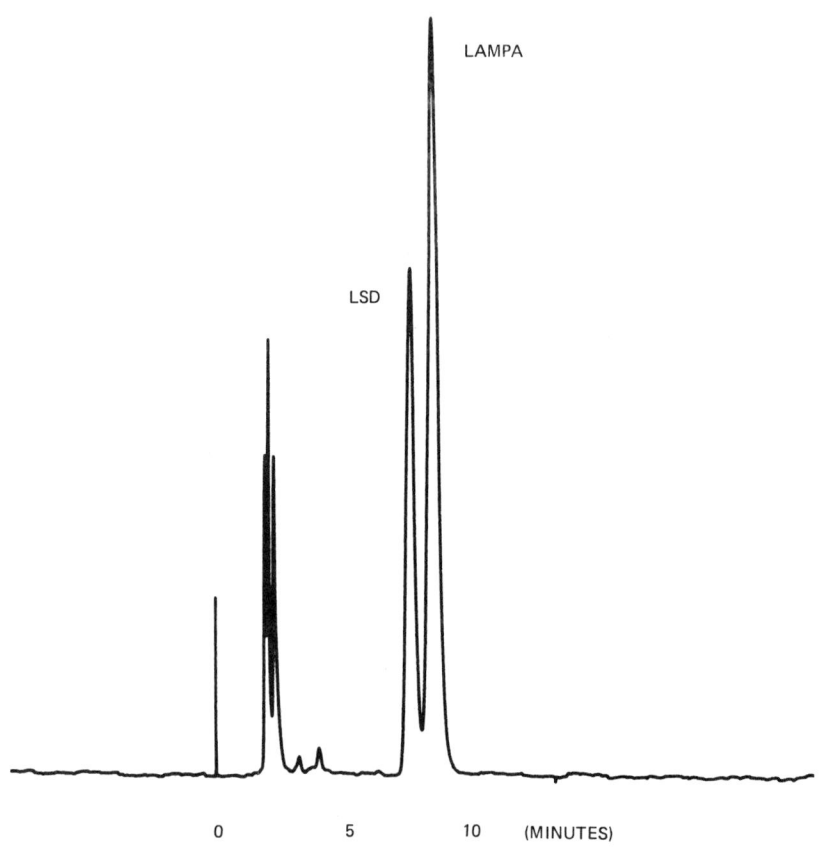

FIG. 35. Separation of LSD and LAMPA.

same as that described in prior applications from this work. Although no quantitative results are presented, the chromatogram is
excellent. Retention data for additional ergot alkaloids and possible adulterants would have improved this application.

E. General Screening

Chan et al. [102] described the use of both isocratic and gradient
systems as a screening technique for several street drugs. The

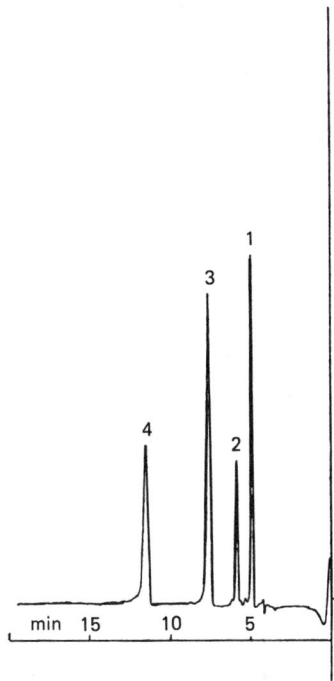

FIG. 36. Separation of 1 µl of a synthetic mixture of ergot alkaloids containing: 1, lysergic acid; 2, lysergamide; 3, LSD; 4,
iso-LSD. Solvent: methanol-0.2 N ammonium nitrate solution (3:2);
detector: UV at 320 nm; flow rate: 1 ml/min. (From Ref. 92.)

primary adsorbent was Corasil II, and isocratic solvent A was 0.22%
cyclohexylamine in cyclohexane (8 drops per 100 ml). Isocratic
solvent B was 0.22% cyclohexylamine in cyclohexane containing 1.5%
methanol. The flow rate was 1.45 ml/min.

The gradient solvent system was (1) 0.5% ethanol (95%), 0.25%
dioxane, and 0.13% (5 drops in 100 ml) cyclohexylamine in Skelly B;
and (2) 10% ethanol (95%), 20% dioxane, and 1.4% cyclohexylamine in
Skelly B. A linear gradient was formed at a flow rate of 1.45
ml/min.

A portion of the sample was powdered, extracted with 1 ml of
methanol, and the filtrate was then partitioned between 5 ml of
chloroform and 5 ml of N/10 HCl. The chloroform, containing acidic

and neutral drugs, was evaporated to dryness and taken up in 100 µl
of methanol for injection. The aqueous layer was then made basic
with NaOH pellets to pH 10-11 (pH paper) and extracted with 5 ml
of chloroform. The chloroform containing the basic drugs was evaporated to dryness and the residue taken up in 100 µl of methanol
for HPLC. The extracts were initially examined using the more polar
solvent B and then using the less polar A system if the unknown
substance eluted rapidly. Figure 37 is a chromatogram of early
eluting, low-polarity drugs obtained on system A. Table A8 shows
relative retention times for the drugs chromatographed isocratically

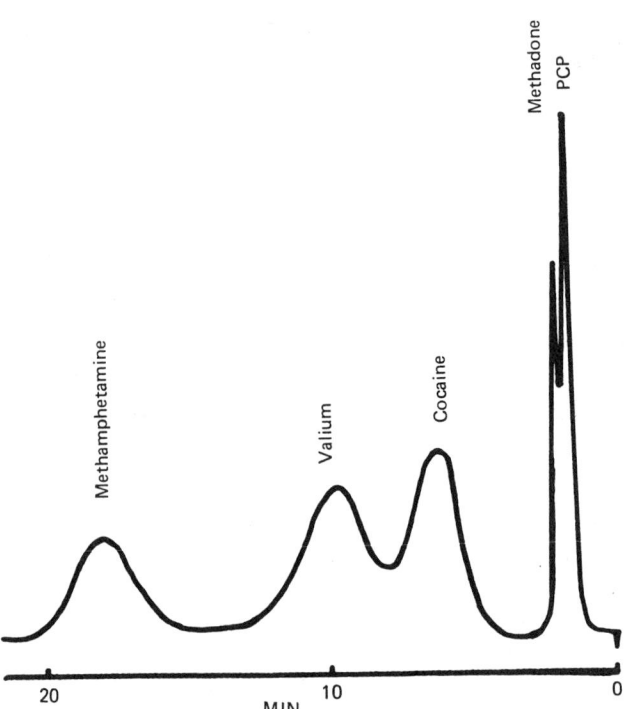

FIG. 37. Separation of low-polarity drugs. Column packing material:
Corasil II; column dimensions: 2.3 mm × 50 cm; solvent: 0.2%
cyclohexylamine in cyclohexane; flow rate: 1.45 ml/min; temperature: 23°C; sample size: about 5 µg each substance in total injection volume of 5 µl. (From Ref. 102. Reproduced from the *Journal
of Chromatographic Science,* by permission of Preston Publications, Inc

in systems A and B. Although retention data for the gradient system was presented, it will not be included here because there was
no statistical evaluation of the reproducibility of these data.

This report demonstrates the potential usefulness of LSC as a
screening tool, but there are several problem areas that will be
mentioned: (1) Some drugs are amphoteric and will be partially extracted by chloroform from N/10 HCl and will thus appear in both
acidic and basic fractions. (2) NaOH is not the reagent of choice
for rendering a solution containing unknown drugs basic prior to extraction. Heroin, for example, if present in the powder, would
certainly be at least partially hydrolyzed to O^6-monoacetylmorphine.
Cocaine is also subject to hydrolysis at such a high pH if allowed
to sit for an extended length of time. $NaHCO_3$ or Na_2CO_3 are much
better choices. (3) The procedure does not spell out the conditions
for evaporation of the alkaline chloroform extract. If heat is
used, the more volatile drugs, amphetamine, methamphetamine, and so
on, may be lost. It would be much better to add a drop or two of
HCl in methanol (10 ml in 100 ml), thus ensuring that HCl salts are
formed. (4) The use of methanol or ethanol as the solvent for introduction of the sample is not good LC technique with such nonpolar
mobile phases. Ideally, the solvent should be the mobile phase itself, but a solvent such as chloroform or even chloroform-methanol
(9 + 1) is a reasonable compromise. (5) As noted with the other
applications utilizing Corasil II, the chromatography could be
improved by using a microparticle silica.

Several specific applications have been previously cited from
the work of Jane [92]. It should also be pointed out that this work,
because of the variety of basic drugs chromatographed, is very useful in preliminary screening of unknown drug samples. The sample
preparation and mobile phases have already been described and the
retention data presented in Tables A3-A5.

Although no mechanism was proposed, solute retention was found
to be dependent on the methanol-water ratio, ammonia concentration,
and the concentration of ammonium nitrate. As the basicity of the

substance increases, a less aqueous, more alkaline solvent is required.

Sugden et al. [103] have noted the active silanol groups of
silica ionize above pH 4, so that silica behaves as a weak cation
exchanger. They proposed the following equilibria involving ion
exchange on the surface with the silanol groups;

$$SiOH \rightleftharpoons SiO^- + H^+ \tag{A}$$
$$SiO^- + C \rightleftharpoons SiOC \tag{B}$$
$$SiO^- + NR_3H^+ \rightleftharpoons SiONR_3H \tag{C}$$

Wheels [104] has proposed that there are two separation mechanisms, ion exchange and adsorption (dipole interactions, van der
Waals forces, and hydrogen bonding), that contribute to separation
in this system. The pH of Janes primary solvent (methanol-2 N
ammonium hydroxide-1 N ammonium nitrate, 27 + 2 + 1) is about 10.3
[106]. The pK_a values of most of the drugs listed in Tables A3-A5
lie below this value, so that at this relatively high pH (10.3),
they exist predominantly as the free bases and it would appear that
the ion-exchange mechanism would play a minor role in the separation mechanism. Sugden et al. [103], in a similar study using 70%
methanol in water at various pH levels, none exceeding pH 9, and
with different ionic modifiers, found little evidence of the ion-
exchange mechanism for cocaine, butacaine, and amylocaine on silica.
It should be noted that at this pH level these three substances
should exist primarily in the protonated form necessary for the ion-
exchange mechanism and since little evidence of this mechanism was
found, it would seem unlikely that it would be the predominant mechanism under less favorable conditions. It would thus appear that
the exact mechanism of retention in this system remains ill defined.

Achari and Theimer [105] also studied the retention of several
drugs on silica. A Partisil 10 column was used with a mobile phase
composed of methanol-methylene chloride (3 + 1) containing 1% (w/v)
NH_4OH at a flow rate of 9.7 ml/min. A UV detector at 254 nm was
used. The effects on solute retention of an increase of each com-

ponent of the mobile phase was investigated. As with the findings
of Jane [92], an increase in the volume percent of NH_4OH and methanol
resulted in decreased retention for strongly basic drugs, while an
increase in water resulted in longer retention. Higher concentrations of water resulted in longer retention for the less basic drugs
as well. Table A9 shows retention volumes for the 29 drugs investigated. The number of drugs chromatographed from a given family,
phenethylamines, for example, was insufficient, making a meaningful
assessment of this system difficult.

Baker et al. [106] investigated the usefulness of absorbance
ratios in conjunction with chromatographic retention volumes as an
aid in the identification of drug substances. A reversed-phase
system and two systems utilizing silica were used. In one of the
adsorption systems, the mobile phase of Jane [92] was used in conjunction with a μ-Porasil column. Fifty-two drugs were chromatographed and the absorbance at 254 and 280 nm was determined for each
drug. The absorbance ratios, 254 nm/280 nm, were then calculated,
Table A10 shows the retention, relative to morphine, the A_{254}/A_{280}
ratio, and the A_{254} for each drug substance.

In this system a drug was considered identified if its retention time differed by more than the sum of its standard deviation
and the standard deviation of the other drugs in the data set, or
using both parameters, if either the retention time or absorbance
ratio of the drug differed by more than the sum of the two standard
deviations of each of the remaining drugs. In a study of a small
group of drugs, the relative standard deviation for retention time
was determined to be 3.3% and the relative standard deviation of
the absorbance ratio was 1.9%. Applying these criteria to Table
A10 shows that only one of the 52 drugs, less than 2%, could be
identified by retention time alone, while 54 of the 52, 85%, could
be identified using both parameters. It should be pointed out that
a long-term study of retention times and absorbance ratios showed
relative standard deviations of 3.2% and 21%, respectively. It was
believed that the latter value could be improved with more frequent

calibration. The authors concluded that this technique was slightly
more useful than other commonly used paired techniques.

APPENDIX: TABLES A1 TO A10

TABLE A1. Retention Expressed Relative to Flurazepam

Benzodiazepine	Relative Retention Time	A_{254}/A_{280}
1. Medazepam	0.40	3.77
2. Prazepam	0.53	4.47
3. Cyprazepam	0.61	1.15
4. Diazepam	0.62	4.62
5. Flurazepam	1.00	4.97
6. Chlordiazepoxide	1.14	1.32
7. N-Desmethyldiazepam	1.19	4.58
8. Nitrazepam	2.07	1.38
9. Clonazepam	2.57	1.79
10. Demoxapam	4.18	2.98
11. Oxazepam	4.71	4.20

Note: Retention time of flurazepam, 308 s. Absorbance ratios
254 nm/280 nm.

Source: Ref. 90.

TABLE A2. Retention Times of Various Phenethylamines Relative to
Methamphetamine

	Relative Retention Time
Dextroamphetamine	Not eluted
Methamphetamine	1
Mescaline	Not eluted
Ephedrine	2.40
Methoxyphenamine	1.50
Mephentermine	2.50
2,5-Dimethoxy-4-methylamphetamine (DOM, STP)	1.25
p-Hydroxyamphetamine	1.50
3,4-Dihydroxyphenethylamine	8.25

Note: Retention time of methamphetamine: 9.5 min., Adsorbent:
Corasil II, mobile phase: $CHCl_3$:MeOH (4:1).

Source: Ref. 91. Reproduced from the *Journal of Chromatographic
Science*, by permission of Preston Publications, Inc.

TABLE A3. Retention of Amphetamine-Type Stimulants Relative to Methylamphetamine

Stimulant	R_f
Dexamphetamine	0.62
Methylamphetamine	1.00 (12.2 ml)
Ethylamphetamine	0.72
3-Chloropropylamphetamine	0.38
Benzphetamine	0.33
Phentermine	0.58
Mephentermine	1.14
Chlorphentermine	0.57
Hydroxyamphetamine	0.71
Ephedrine	0.92
Pseudoephedrine	0.98
Norephedrine	0.58
Norpseudoephedrine	0.76
N-Methylephedrine	0.84
Diethylpropion	0.40
Fencamfamin	0.52
Fenfluoramine	0.62
STP	0.67
Bromo-STP	0.67
Phendimetrazine	0.40
Phenmetrazine	0.45
Pipradol	0.53
Prolintane	0.70
Amitriptyline	0.47
Desipramine	0.98
Imipramine	1.06
Methylphenidate	0.39

Source: Ref. 92.

TABLE A4. Retention of the Opium Alkaloids Relative to Morphine

Alkaloid	R_f
Morphine	1.00 (8.9 ml)
Codeine	0.95
Thebaine	0.79
Papaverine	0.47
Narcotine	0.47
Narceine	0.92
Cotarnine	4.10
Dilaudid	1.50
Dicodid	1.32
Dionine	0.87
Paracodeine	1.44
Eucodal	0.60
Protopine	0.61
Laudanosine	0.68
Acedicon	0.74

Source: Ref. 92.

TABLE A5. Retention of Other Compounds of Forensic Interest Relative
to Morphine

Compound	R_f
Amethocaine	0.58
Antazoline	1.21
Atropine	2.35
Benzocaine	0.45
Benztropine	2.88
Bromodiphenhydramine	0.65
Butacaine	0.58
Caffeine	0.52
Chlordiazepoxide	0.48
Chlorpheniramine	1.02
Chlorpromazine	0.67
Cocaine	0.51
Diazepam	0.45
Dextropoxyphene	0.49
Diethazine	0.64
Dihydrohydroxymorphinone	0.68
Dihydromorphine	1.57
Diphenhydramine	0.64
Ethopropazine	0.61
Heroin	0.69
Lignocaine	0.46
Meclophenoxate	0.57
Methadone	0.74
Methapyrilene	0.59
Methaqualone	0.45
6-Methyldihydromorphine	1.26
6-Monoacetylisopropylmorphine	0.66
6-Monoacetylmorphine	0.75
Morphine	1.00 (8.9 ml)
Nalorphine	0.55
Nicotine	0.57
Nitrazepam	0.46
Paracetamol	0.46
Pethidine	0.62
Phenacetin	0.45
Phenbutrazate	0.45
Phencyclidene	0.66
Procaine	0.56
Quinidine	0.63
Quinine	0.65
Salicylamide	0.46
Strychnine	1.57
Theophyline	0.49

Source: Ref. 92.

TABLE A6. Effects of Ammonia Concentration on Solute Retention Relative to Heroin

Solute	Solvent Systems[a]		
	S-1	S-2	S-3
Methaqualone	0.38	0.35	0.33
Diazepam	0.40	0.37	0.35
Lidocaine	0.41	0.34	0.33
Noscapine	0.44	0.39	0.36
Cocaine	0.46	0.37	0.38
Papaverine	0.48	0.44	0.41
Aminopyrine	0.54	0.51	0.49
Benzocaine	0.60	0.54	0.52
Meperdine	0.63	0.62	0.62
Methapyrilene	0.64	0.64	0.66
Methadone	0.68	0.69	0.75
Caffeine	0.72	0.68	0.66
Barbital	0.75	0.65	0.61
Phenacetin	0.75	0.71	0.68
Phenobarbital	0.78	0.67	0.63
Tetracaine	0.79	0.76	0.77
Acetylcodeine	0.82	0.81	0.81
Antipyrine	0.90	0.86	0.82
Heroin	1.00	1.00	1.00
Procaine	1.32	1.28	1.34
Acetylprocaine	1.56	1.54	1.60
Codeine	1.98	1.90	1.94
O^6-Monoacetylmorphine	2.06	2.05	2.11
Quinidine	2.21	2.39	2.44
Quinine	2.48	2.66	2.67
Strychine	2.86	3.00	3.03
Morphine	5.72	5.35	5.56
Retention time of heroin(s)	323	353	377

Source: Ref. 97.
[a] Flow rate: 2.0 ml/min. For details of composition, see p. 135.

TABLE A7. Effects of Water on Solute Retention Relative to Heroin

Solute	Solvent Systems[a]		
	S-1	S-4	S-5
Aminopyrine	0.54	0.61	0.72
Caffeine	0.72	0.87	1.12
Acetylcodine	0.82	0.80	0.77
Antipyrine	0.90	1.05	1.31
Heroin	1.00	1.00	1.00
Procaine	1.32	1.46	1.86
Acetylprocaine	1.56	1.78	2.27
Codeine	1.98	2.00	2.11
O^6-Monoacetylmorphine	2.06	2.31	2.67
Quinidine	2.21	1.87	1.60
Quinine	2.48	2.09	1.73
Strychine	2.86	2.66	2.48
Morphine	5.72	6.68	8.38
Retention time of heroin(s)	323	293	245

Source: Ref. 97.
[a] Flow rate: 2.0 ml/min. For details of composition, see p. 135.

TABLE A8. Relative Retention Times of Representative Drugs of
Abuse on Corasil II

Drug	Solvent A	Solvent B
Phencyclidine (PCP)	0.20	0.24
Methadone	0.23	0.24
Cocaine	1.0[a]	0.28
Tetrahydrocannabinol (THC)	1.20	0.38
Valium	1.33	0.38
Methamphetamine	2.66	0.62
2,5-Dimethoxy-4-methylamphetamine (STP)	--[b]	0.81
3,4-Methylenedioxy-amphetamine (MDA)	--	1.0[c]
Butabarbital	--	1.48
Secobarbital	--	1.48
Amobarbital	--	1.48
Heroin	--	2.14
Dilantin	--	2.43
N,N-Dimethyltryptamine (DMT)	--	2.52
Phenobarbital	--	3.47
Lysergic acid diethylamide (LSD)	--	3.81
Mescaline	--	6.66

[a]Retention time = 5.91 min.

[b]-- indicates that the drug did not elute within 30 min of the
injection.

[c]Retention time = 4.14 min.

Source: Ref. 102. Reproduced from the Journal of Chromatographic
Science, by permission of Preston Publications, Inc.

TABLE A9. Retention Data of Various Drugs

Drug	Drug Source	k'	V_R(ml)
Theophylline	Henley	0.08	1.81
Sodium sulfacetamide	USP	0.16	1.95
6.Hydroxydopamine	Regis	0.54	2.59
Ethaverine	K & K	0.92	3.23
Papaverine	Merck	0.92	3.23
Tropicamide	Hoffman LaRoche	0.92	3.23
Caffeine	Eastman	0.92	3.23
Theobromine	Eastman	0.96	3.29
Scopolamine	Aldrich	1.00	3.36
Pyrilamine	Merck	1.33	3.91
Epinephrine	USP	1.50	4.20
Phenylpropanolamine	Chemo Puro	1.66	4.47
Quinidine	Falleck Chem. Co.	1.71	4.55
Quinine	Amend	1.71	4.55
Codeine	Merck	1.88	4.84
Brompheniramine	Gyma Lab	1.92	4.91
Chlorpheniramine	Lemke	1.92	4.91
Hydroquinidine	George Uhe	2.50	5.88
Amphetamine	Cooper, Tinsley	2.66	6.15
Phenylephrine	Gane's Chem. Works	2.66	6.15
Ephedrine	Merck	3.25	7.14
Strychnine	Aldrich	3.33	7.27
Dextromethorphan	Hoffman LaRoche	3.58	7.70
Antazoline	Ciba	4.00	8.40
Atropine	Henley & Co.	4.13	8.62
Homatropine	Aldrich	4.42	9.11
Naphazoline	Ciba	9.58	17.80
Xylometazoline	Ciba	13.58	24.50
Oxymetazoline	Schering	15.25	27.30

Note: Chromatographic conditions: 25 cm × 4.6 mm ID stainless steel Partisil 10 (dp 10 μm) column; mobile phase: MeOH:CH_2Cl_2, 3:1 (v/v), 1% (v/v) NH_4OH (29% NH_3) added to the mobile solvent, flow: 0.7 ml/min; detection: UV 254 nm.

Source: Ref. 105. Reproduced from the *Journal of Chromatographic Science*, by permission of Preston Publications, Inc.

TABLE A10. Retention at A_{254}/A_{280} and A_{254}

Drug	Relative Retention Time	A_{254}/A_{280}	A_{254}[a]
Noscapine	0.53[b]	0.61	0.035
Phenacetin	0.53	0.84	0.051
Naloxone	0.56	0.82	0.025
Papaverine	0.56	1.06	0.070
Benzphetamine	0.58	2.10	0.069
Piminodine	0.58	3.02	0.16
Cocaine	0.61	0.86	--
Phenazocine	0.61	0.24	0.0081
Procaine	0.61	0.44	0.048
Nylidrin	0.61	0.75	0.013
Levallorphan	0.64	0.12	0.0026
Methylphenidate	0.67	9.50	0.0062
Pentazocine	0.67	0.16	0.0049
Phendimetrazine	0.67	8.00	0.010
Ethinamate	0.70	1.00	0.0020
Phenmetrazine	0.72	31.0	0.040
Meperidine	0.75	30.7	0.0030
Quinine	0.75	0.62	0.0042
Promethazine	0.76	2.21	0.14
Diphenhydramine	0.77	90.0	0.059
Methapyrilene	0.77	1.89	0.13
Phenylpropanolamine	0.78	65.0	0.0021
Heroin	0.80	0.64	--
Methadone	0.83	1.57	0.072
Phencyclidine	0.83	22.0	--
Thioridazine	0.83	2.08	0.20
Amphetamine	0.86	60.0	0.0039
Oxymorphone	0.86	1.17	0.0044
Doxylamine	0.89	17.7	0.0017
Ethylmorphine	0.92	1.15	0.020
Hydroxyamphetamine	0.92	0.38	0.011
Propylhexedrine	0.92	4.0	0.0005
Oxycodone	0.92	1.13	0.0029
Codeine	1.00	0.88	--
Morphine	1.00[c]	1.09	0.016
Dimethyltriptamine	1.09	0.77	--
Methamphetamine	1.19	31.0	0.0034
Ephedrine	1.20	52.0	0.0034
Phenylephrine	1.22	0.50	0.0045
Hydrocodone	1.28	0.93	0.0085
Ethoheptazine	1.31	27.4	0.0022
Mescaline	1.31	2.93	--
Xylometazoline	1.33	8.67	0.0034
Mephenteramine	1.36	36.3	0.0038
Dihydrocodeine	1.36	0.53	0.0065
Oxymetazoline	1.36	0.31	0.012
Tetrahydrozoline	1.42	16.1	0.038
Hydromorphone	1.43	1.09	0.0091
Strychnine	1.54	3.22	0.0045
Dextromethorphan	1.56	0.14	0.0022
Naphazoline	1.61	0.49	0.062
Levorphanol	1.64	0.12	0.0013

[a]Absorbance of a 10-µl injection of a 1.0-mg/ml solution.

[b]The column void volume was slightly less than 0.53.

[c]Morphine was used as standard, retention time 3.5 min.

Source: Ref. 106.

REFERENCES

1. L. R. Snyder, *Principles of Adsorption Chromatography*. Marcel Dekker, New York, 1968.
2. L. R. Snyder and H. Poppe, *J. Chromatogr.* 184:363 (1980).
3. L. R. Snyder, *Anal. Chem.* 46:1384 (1974).
4. E. Soczewinski, *Anal. Chem.* 41:179 (1969).
5. E. Soczewinski and W. Golkiewicz, *Chromatographia* 5:431 (1972).
6. E. Soczewinski and W. Golkiewicz, *Chromatographia* 6:269 (1973).
7. E. Soczewinski, *J. Chromatogr.* 130:23 (1977).
8. R. P. W. Scott and P. Kucera, *Anal. Chem.* 45:749 (1973).
9. R. P. W. Scott and P. Kucera, *J. Chromatogr.* 112:425 (1975).
10. R. P. W. Scott, *J. Chromatogr.* 122:35 (1976).
11. R. P. W. Scott and P. Kucera, *J. Chromatogr.* 149:93 (1978).
12. R. P. W. Scott and P. Kucera, *J. Chromatogr.* 171:37 (1979).
13. R. P. W. Scott, *J. Chromatogr. Sci.* 18:297 (1980).
14. P. Jandera and J. Churacek, *J. Chromatogr.* 91:207 (1974).
15. P. Jandera and J. Churacek, *J. Chromatogr.* 91:223 (1974).
16. P. Jandera and J. Churacek, *J. Chromatogr.* 93:17 (1974).
17. P. Jandera and J. Churacek, *J. Chromatogr.* 104:9 (1975).
18. P. Jandera and J. Churacek, *J. Chromatogr.* 104:23 (1975).
19. P. Jandera, M. Janderova, J. Churacek, *J. Chromatogr.* 115:9 (1975).
20. E. H. Slatts, J. C. Kraak, W. J. T Brugman, and H. Poppe, *J. Chromatogr.* 149:255 (1978).
21. M. Jaroniec, J. K. Rozylo, and B. Oscik-Mendyk, *J. Chromatogr.* 179:237 (1979).
22. J. P. Thomas, A. Brun, and J. P. Bounine, *J. Chromatogr.* 139:21 (1977).
23. J. P. Thomas, A. Brun, and J. P. Bounine, *J. Chromatogr.* 172:107 (1979).
24. S. Hara, *J. Chromatogr.* 137:41 (1977).
25. S. Hara, Y. Fujii, M. Hirasawa, and S. Miyamoto, *J. Chromatogr.* 149:143 (1978).
26. S. Hara, N. Yamauchi, C. Nakae, and S. Sakai, *Anal. Chem.* 52:33 (1980).
27. S. Hara, A. Ohsawa, and A. Dobashi, *J. Liq. Chromatogr.* 4:409 (1981).
28. L. R. Snyder and J. J. Kirkland, *Introduction to Modern Liquid Chromatography*, 2nd ed. Wiley-Interscience, New York, 1979.
29. J. M. Bather and R. A. C. Gray, *J. Chromatogr.* 122:159 (1976).
30. R. P. W. Scott and S. Traiman, *J. Chromatogr.* 196:193 (1980).
31. R. P. W. Scott and P. Kucera, *J. Chromatogr. Sci.* 13:337 (1975).
32. L. R. Snyder and J. J. Kirkland, *Introduction to Modern Liquid Chromatography*. Wiley-Interscience, New York, 1974 (first edition).
33. R. E. Majors, *J. Chromatogr. Sci.* 15:334 (1977).
34. G. R. Laird, J. Jurand, and J. H. Knox, *Proc. Soc. Anal. Chem.* 12:311 (1974).
35. R. Endele, I. Halasz, and K. Unger, *J. Chromatogr.* 99:377 (1974).

36. L. R. Snyder, *J. Chromatogr. Sci.* 7:352 (1969).
37. H. N. M. Stewart, R. Amos, and S. G. Perry, *J. Chromatogr.* 38:209 (1968).
38. J. J. Kirkland, *J. Chromatogr. Sci.* 10:129 (1972).
39. H. C. Bechell and J. J. DeStefano, *J. Chromatogr. Sci.* 10:481 (1972).
40. J. J. Kirkland, *J. Chromatogr. Sci.* 9:206 (1971).
41. R. E. Majors, *Anal. Chem.* 44:1722 (1972).
42. J. J. Kirkland, *J. Chromatogr. Sci.* 10:593 (1972).
43. R. M. Cassiday, D. S. Legay, and R. W. Frei, *Anal. Chem.* 46:340 (1974).
44. H. R. Linder, H. P. Keller, and R. W. Frei, *J. Chromatogr. Sci.* 14:234 (1976).
45. G. B. Cox, C. R. Liscombe, M. J. Slucutt, K. Sugden, and J. A. Upfield, *J. Chromatogr.* 117:269 (1976).
46. T. J. N. Webber and E. H. McKerrell, *J. Chromatogr.* 122:243 (1976).
47. S. H. Chang, K. M. Gooding, and F. E. Regnier, *J. Chromatogr.* 125:103 (1976).
48. J. J. Kirkland and P. E. Antle, *J. Chromatogr. Sci.* 15:137 (1977).
49. R. McIlwrick, *Spectra-Phys. Chromatogr. Rev.* 3(1):5 (1977).
50. C. J. Little, A. P. Dale, D. A. Ord, and T. R. Marten, *Anal. Chem.* 49:1311 (1977).
51. L. R. Snyder and J. J. Kirkland, *Introduction to Modern Liquid Chromatography,* 2nd ed. Wiley-Interscience, New York, 1974, pp. 207-217.
52. W. Trappe, *Biochem. Z.* 305:150 (1940).
53. K. N. Trueblood and E. W. Malmberg, *Anal. Chem.* 21:1055 (1949).
54. K. N. Trueblood and E. W. Malmberg, *J. Am. Chem. Soc.* 72:4112 (1950).
55. E. D. Smith and A. L. LeRosen, *Anal. Chem.* 23:732 (1951).
56. L. M. Kay and K. N. Trueblood, *Anal. Chem.* 26:1566 (1954).
57. H. J. Cahnmann, *Anal. Chem.* 29:1307 (1957).
58. L. R. Snyder, *Anal. Chem.* 39:698 (1967).
59. D. L. Saunders, *J. Chromatogr.* 125:163 (1976).
60. L. R. Snyder, *J. Chromatogr. Sci.* 7:595 (1969).
61. H. Englehardt, *J. Chromatogr. Sci.* 15:380 (1977).
62. H. Englehardt and W. Boeme, *J. Chromatogr.* 133:67 (1977).
63. K. Chmel, *J. Chromatogr.* 97:131 (1974).
64. J. J. Kirkland, *J. Chromatogr.* 83:149 (1973).
66. C. Gonnet and J. L. Roca, *J. Chromatogr.* 109:297 (1975).
66. J. H. Knox and A. Pryde, *J. Chromatogr.* 112:171 (1975).
67. G. Gonnet and J. L. Rocca, *J. Chromatogr.* 120:419 (1976).
68. Z. El Rassi, C. Gonnet, and J. L. Rocca, *J. Chromatogr.* 125:179 (1976).
69. J. A. Riddick and W. B. Bunger, *Techniques of Chemistry,* vol. 2: *Organic Solvents,* 3rd ed. Wiley-Interscience, New York, 1970.
70. M. L. Moskovitz, *Am. Lab.,* p. 142 (Dec. 1980).
71. D. L. Saunders, *J. Chromatogr. Sci.* 15:372 (1977).

72. L. R. Snyder, *J. Chromatogr.* 92:223 (1974).
73. L. R. Snyder, *J. Chromatogr. Sci.* 16:223 (1978).
74. D. L. Saunders, *Anal. Chem.* 46:470 (1974).
75. J. H. Knox, *J. Chromatogr. Sci.* 15:352 (1977).
76. P. A. Bristow and J. H. Knox, *Chromatographia* 10:279 (1977).
77. L. R. Snyder and J. J. Kirkland, *Introduction to Modern Liquid Chromatography*, 2nd ed. Wiley-Interscience, New York, 1974, pp. 219-225.
78. A. T. James and A. J. P. Martin, *Analyst* 77:915 (1952).
79. R. V. Vivilecchia, B. G. Lightbody, N. Z. Thimot, and H. M. Quinn, *J. Chromatogr. Sci.* 15:424 (1977).
80. J. Kwok, L. R. Snyder, and J. C. Sternberg, *Anal. Chem.* 40:118 (1968).
81. C. Horvath, W. Melander, and I. Molnar, *Anal. Chem.* 49:142 (1977).
82. A. Wehrli, J. C. Hildebrand, H. P. Keller, and R. Stamplfi, *J. Chromatogr.* 149:199 (1978).
83. J. G. Atwood, G. J. Schmidt, and W. Slavin, *J. Chromatogr.* 109:171 (1979).
84. F. M. Rabel, *Am. Lab.*, p. 81 (Jan. 1980).
85. *µBondapak and µPorasil Liquid Chromatography Columns; Care and Use Manual*, Manual No. CU84588 Rev. D., Waters Associates, Millford, Mass., Jan. 1978.
86. P. E. Barker, B. W. Hatt, and S. R. Holding, *J. Chromatogr.* 206:27 (1981).
87. F. M. Rabel, *J. Chromatogr. Sci.* 18:394 (1980).
88. R. A. Bredeweg, L. D. Rothman, and C. D. Pfeiffer, *Anal. Chem.* 51:2061 (1979).
89. D. H. Rodgers, *J. Chromatogr. Sci.* 12:742 (1974).
90. J. D. Wittwer, *J. Liq. Chromatogr.* 3:1713 (1980).
91. P. J. Cashman, J. I. Thornton, and D. L. Shelman, *J. Chromatogr. Sci.* 11:7 (1973).
92. I. Jane, *J. Chromatogr.* 111:227 (1975).
93. G. P. Cartoni and F. Natalizia, *J. Chromatogr.* 123:474 (1976).
94. A. H. Lewin, S. R. Parker, and F. I. Carroll, *J. Chromatogr.* 193:371 (1980).
95. P. J. Cashman and J. I. Thornton, *J. Forensic Sci. Soc.* 12:417 (1972).
96. H. Huizer, H. Logtenberg, and A. J. Steenstra, *Bull. Narc.* 29:65 (1974).
97. J. D. Wittwer, *Forensic Sci. Int.* 18:215 (1981).
98. T. H. Beasley, D. W. Smith, H. W. Ziegler, and R. L. Charles, *J. Assoc. Off. Anal. Chem.* 57:85 (1974).
99. P. G. Vincent and B. F. Engelke, *J. Assoc. Off. Anal. Chem.* 62:310 (1979).
100. J. D. Wittwer and J. H. Kluckhohn, *J. Chromatogr. Sci.* 11:1 (1973).
101. J. D. Wittwer, private communication, 1974.
102. M. L. Chan, C. Whetsell, and J. D. McChesney, *J. Chromatogr. Sci.* 12:512 (1974).

103. K. Sugden, G. B. Cox, and C. R. Loscombe, *J. Chromatogr.*
     149:377 (1978).
104. B. B. Wheels, *J. Chromatogr.* 187:65 (1980).
105. R. G. Achari and E. E. Theimer, *J. Chromatogr. Sci.* 15:320
     (1977).
106. J. K. Baker, R. E. Skelton, and C. Y. Ma, *J. Chromatogr.*
     168:417 (1979).

4 Use of Bonded-Phase Columns in Drug Analysis

Ira S. Lurie*

Northeast Regional Laboratory
Drug Enforcement Administration
New York, New York

I. INTRODUCTION

Since approximately 80% of all high-performance liquid chromatographic separations are carried out today using bonded-phase columns
in the reversed-phase mode [1], it would stand to reason that this
type of chromatography would be of major importance in forensic
drug analysis. For the most part in reversed-phase chromatography,
columns are employed with a nonpolar moiety bonded to a silica substrate with an aqueous-organic mobile phase. The organic modified
is usually methanol, acetonitrile, or tetrahydrofuran. Why does
this type of chromatography presently enjoy such widespread use?

    1. Many compounds, such as biologically active substances,
have limited solubility in the nonpolar mobile phases that are employed in normal-phase chromatography.

    2. Ionic or highly polar compounds have high heats of adsorp-
tion on straight silica or alumina columns and therefore can elute
as tailing peaks. Most of the active adsorption sites can be removed
by bonding silica with a nonpolar moiety such as monochlorodimethyl

*Present affiliation: Special Testing and Research Laboratory,
Drug Enforcement Administration, McLean, Virginia.

octadecylsilane. In addition, in reversed-phase solvents, secondary
equilibrium such as that resulting from ion-pairing phenomena can
alleviate the problems associated with chromatography of ionized
compounds on highly polar silica or alumina columns.

3. Column deactivation from polar modifiers is a problem in
liquid-solid chromatography which frequently can lead to irreduci-
bility in chromatographic systems. The latter phenomenon is not encountered in reversed-phase columns.

4. Long reequilibration times during gradient elution are
common in adsorption chromatography. The use of reversed-phase
chromatography with bonded-phase columns is advantageous because of
the short reequilibration times required.

5. Ionic compounds can be chromatographed via ion-exchange
chromatography. This mode of chromatography is tedious because precise control of variables such as pH and ionic strength is required
for reproducible chromatography. Also, the lifetimes of these
columns tend to be relatively short compared with bonded-phase
columns. The latter column can last over two years, whereas it is
not uncommon for an ion-exchange column to be unusable after only 2
months of use.

6. Liquid-liquid partition chromatography in the reversed-
phase mode is disadvantageous because the nonpolar stationary phase
has a finite solubility in the mobile phase which results in gradual
deterioration of the stationary phase. This problem can be circum-
vented by the use of a precolumn containing mobile phase saturated
with stationary phase and/or saturating the mobile phase with stationary phase in the solvent reservoir. The use of a precolumn requires careful temperature control. Because of the problems encountered in liquid-liquid chromatography, this technique has, for
the most part, been replaced by bonded-phase columns.

In view of the considerations discussed above, it is not surprising that reversed-phase chromatography with bonded-phase columns
enjoys widespread popularity in forensic drug analysis since many of
the drugs analyzed are highly polar. In this chapter we discuss

drugs encountered in street samples, clandestine laboratories, and
compliance exhibits. The analysis of drugs of forensic interest in
body fluids is discussed in Chap. 6. Since the compounds of interest that are described in this chapter are usually present at moderate levels, that is, at the parts per million level, and possess an
ultraviolet (UV) chromophore, a fixed UV detector at 254 nm or 280
nm is generally employed. In certain instances for greater selectivity or sensitivity, electrochemical detection or a variable-
wavelength UV detector at wavelengths other than 254 nm or 280 nm
are utilized.

This chapter discusses primarily the use of bonded-phase columns
in the reversed-phase mode for analytical separations. The use of
bonded-phase columns in the normal-phase and ion-exchange modes is
also discussed. Diluent analysis is reviewed as well as the identification of dyes present in forensic samples. In addition, there
will be a discussion of procedures for the isolation and identification of drugs.

Before the role of reversed-phase chromatography in forensic
drug analysis is examined, let us look at various general theoretical and practical considerations that are related to this mode of
chromatography. Since the normal-phase mode closely parallels adsorption chromatography, which is outlined in detail in Chap. 3, a
general discussion of this mode of separation is not presented.

II. THEORY AND PRACTICE OF REVERSED-PHASE CHROMATOGRAPHY

Reversed-phase chromatography refers to the use of a polar eluant
with a nonpolar stationary phase, in contrast to normal-phase
chromatography, where a polar stationary phase is employed with a
nonpolar mobile phase. Since reversed-phase chromatography is by
far the most popular of the two techniques, it was suggested by
C. Horvath at a recent talk to the New York Chromatography Society
that normal-phase chromatography should be referred to as reversed-
reversed-phase. Since reversed-phase chromatography requires a nonpolar stationary phase, bonded phases with nonpolar moieties are

employed. The most widely used bonded phase is made from bonding an
organochlorosilane group to a silica substrate. The organic portion
of this substrate is typically a C_{18}, C_8, or alkylphenyl moiety.
Mono-, di-, and trichlorosilanes are employed, leading to monomeric
or bristle-type coverage [2], as depicted in Fig. 1. The reaction
for this process is described as follows:

$$
\begin{array}{ccc}
& \text{R}_1 & \text{R}_1 \\
& | & | \\
\text{Si-OH} + \text{Cl-Si-R} & \rightleftharpoons & \text{Si-O-Si-R} + \text{HCl} \\
& | & | \\
& \text{R}_2 & \text{R}_2
\end{array}
$$

If di- or trichlorosilanes are utilized and ether is not totally
excluded, polymeric phases can be formed which are disadvantageous
because of poor mass transfer [3]. Because of better mass transfer
the use of monomeric bonded phases results in higher column efficiencies. The bonded phases described are generally stable in the pH
range 1-7. At a pH of less than 1, the silica carbon linkage is
affected, while at a pH of greater than 7, dissolution of the silica
backbone could occur. Higher pHs could be tolerated depending on
the source of hydroxyl ions and the amount of methanol in the mobile
phase [4]. Bonded phases consisting of polar moieties such as
cyano, amino, and sulfonic acid groups could be employed using reactions similar to those described above. Cyano and amino bonded

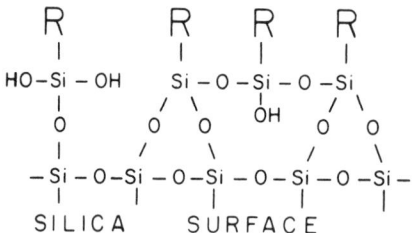

FIG. 1. Illustration of the molecular fur consisting of the hydrocarbonaceous functions, R, covalently bound via siloxane bridges to
the silica surface. (From Ref. 1. Reproduced from the *Journal of
Chromatographic Science*, by permission of Preston Publications, Inc.)

groups would normally be utilized in normal-phase chromatography, while sulfonic acid groups could be used in ion-exchange chromatography. Other ways of bonding organic moieties to silica substrates consist of the reaction of the silanol group to an alcohol to from an ester linkage. This bonded phase is generally not employed because it is easily hydrolyzed. Finally, the silanol group can react with an amine to form a Si-N-C linkage. These phases have a lower stability range (pH 3-8) than the Si-O-Si-C linkages described earlier [5]. The most popular columns today employ particle sizes of 10 µm or less, for reasons described in Chap. 1.

The solvents generally employed in reversed-phase chromatography consist of water and an organic modifier for neutral compounds. For basic and acidic substances an aqueous buffer system plus an organic modifier is employed. If secondary equilibrium is desired, as in reversed-phase ion-pairing chromatography, counterions of varying degrees of polarity are added. The organic modifiers typically consist of acetonitrile, methanol, and tetrahydrofuran. Other solvents having potential use in reversed-phase chromatography are acetone, dioxane, ethanol, isopropanol, and n-propanol. A summary of relevant properties in reversed-phase chromatography is presented in Table 1. From this table it is readily apparent that acetone would be unsuitable for use with a UV detector because of its high UV cutoff. The major disadvantage of dioxane would be slow mass transfer in the mobile phase because of its high viscocity, leading to reduced column efficiencies. Isopropanol and n-propanol would suffer from a similar disadvantage because of their high viscocities. Tetrahydrofuran can form peroxides, which can jeopardize its UV transparency. Although acetonitrile is the best solvent in terms of kinetic considerations because of its lowest viscocity, certain solubility problems can arise with ionic compounds such as ion-pairing reagents. Items e, f, and g from Table 1 are of theoretical significance in reversed-phase chromatography and will be discussed in Sec. IV.E. Item h, which has been discussed in Chap. 3, is a measure of solvent strength in this chromatographic mode.

TABLE 1. Relevant Properties in Reversed-Phase Chromatography

	MW	B.P. (°C)	n^a	UV^b (nm)	ρ^c (g/cm^{-1})	η^d (cP)	ε^e	μ^f (debye)	γ^g (dyn/cm^{-1})	E^h
Acetone[i]	58.1	56	1.357	330	0.791	0.322	20.7	2.72	23	0.56
Acetonitrile	41.0	82	1.342	190	0.787	0.358	38.8	3.37	29	0.65
Dioxane	88.1	101	1.420	215	1.034	1.26	2.21	0.45	33	0.56
Ethanol	46.1	78	1.359	205	0.789	1.19	24.5	1.68	22	0.88
Methanol	32.0	65	1.326	205	0.792	0.584	32.7	1.66	22	0.95
iso-Propanol	60.1	82	1.375	205	0.785	2.39	19.9	1.68	21	0.82
n-Propanol	60.1	97	1.383	205	0.804	2.20	20.3	1.65	23	0.82
Tetrahydrofuran	72.1	66	1.404	210	0.889	0.51	7.58	1.70	27.6	0.82
Water	18.0	100	1.333	170	0.998	1.00	78.5	1.84	73	0.45

[a]Refractive index at 25°C

[b]UV cutoff; the wavelength at which the optical density of a 1-cm-thick neat sample is unity as measured against air.

[c]Density at 20°C.

[d]Viscosity at 20°C.

[e]Dielectric constant.

[f]Dipole moment.

[g]Surface tension.

[h]Elutropic value on alumina according to L. R. Snyder, Principals of Adsorption Chromatography, New York: Marcel Dekker, Inc., 1968, pp. 194-195.

[i]Not suitable for use with UV detector.

Source: Ref. 1. Reproduced from the *Journal of Chromatographic Science,* by permission of Preston Publications, Inc.

The mechanism of reversed-phase chromatography is very much in dispute. Basically, this process can be thought of as a solubility phenomenon. Locke believes that the selectivity of two similar solutes is a function of the difference in solubility of two solutes in the mobile phase and is independent of the stationary phase [6]. Since all reversed-phase solvents contain water, theoretical considerations are based on water solution chemistry. Hermann, for instance, suggests that the solubility of hydrocarbons in water results from the formation of a cavity of water molecules around the hydrocarbon [7]. This results in a loss of entropy for the hydrocarbon molecules which would favor the removal of these species from the aqueous environment. Such a phenomenon could be the driving force for the association of these species with the hydrocarbon stationary phase. As Horvath points out when describing a similar hydrophobic effect, retention in reversed-phase chromatography is not due mainly to the attraction of the stationary phase for the nonpolar solute, but the tendency for these compounds to dissociate from water [8]. Horvath describes this retention in terms of solvent properties such as surface tension and dielectric constant and solute properties as surface area, dipole moment, and charge. This theory is based on the "solvophobic effect" presented by Sinanoglu and Abdulrur, which is a general treatment of the hydrophobic effect which can be extended to other solvents [9]. According to Horvath's theory [8], the solute binds to the nonpolar ligands of the stationary phase to form a complex, and the free energy of this binding process is given by the difference between the "solvophic effects" arising from bringing the complex into solution and placing the individual components into the solvent. Therefore, the solute retention could be described as follows:

$$k' = \frac{KV_s}{V_m} \tag{1}$$

where K is the partition coefficient for the binding process and V_s and V_m are the volumes of the stationary phase and mobile phase, respectively.

$$\ln K = \frac{-\Delta G}{RT} \tag{2}$$

where ΔG is the free-energy change for the binding process. These
solvophic effects involve the formation of a cavity for both the
solute-ligand complex and solute itself in solution, which depends
on solute and ligand properties such as surface area and solute
properties such as surface tension and dielectric constant. In
addition, solute-solvent interactions, which depend primarily on
the dipole moment and charge of solute and dipole moment and dielectric constant of solvent, play a role in retention. Horvath et al.
[8] derived an overall expression for retention in reversed-phase
chromatography:

$$\ln k' = \frac{V_s}{V_m} + \frac{1}{RT}[\Delta A(N\gamma + a)$$
$$+ NA_s\gamma(\lambda^e - 1) + W - \frac{\Delta Z}{\varepsilon}] + \frac{\ln RT}{P_0 V} \tag{3}$$

where ΔA is the contact area between the solute and the complex,
whose magnitude can be approximated by the nonpolar surface area
of the solute molecule; γ is the surface tension of the mobile
phase; λ^e is a factor that adjusts the macroscopic surface tension
to the molecular dimensions; Z represents the charge values for the
solute, ligand, and complex, respectively; ε refers to the dielectric constant of the mobile phase; and R, T, P_0, and V are the usual
gas law constants. The overall process described in Eq. (3) is illustrated in Fig. 2. Two widely known phenomena in reversed-phase
chromatography can be explained in terms of Eq. (3).

1. The greater the lipophilic character of a compound, the longer
   the retention in a reversed-phase chromatographic system. This
   effect is manifested in Eq. (3) by the increase of the capacity
   factor with the contact area.

2. The smaller the lipophilic character of the mobile phase, the
   longer the retention. Equation (3) predicts that retention will
   increase with the surface tension of the mobile phase. The
   larger the amount of water in the mobile phase relative to the
   organic or lipophilic component, the greater the surface tension.

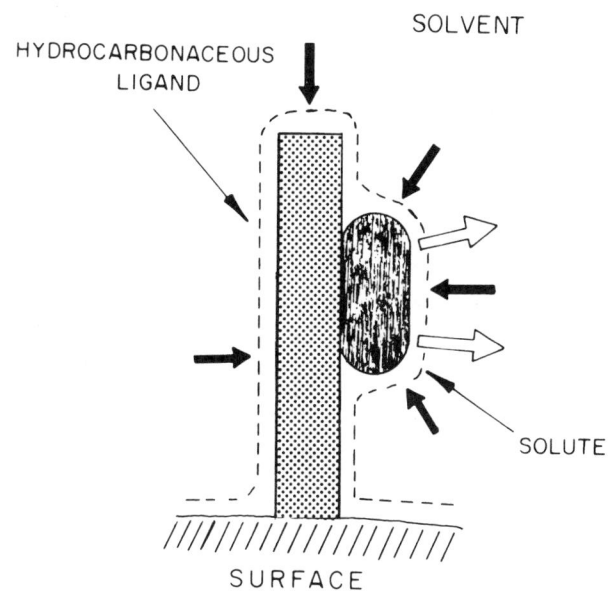

SOLVENT

HYDROCARBONACEOUS LIGAND

SOLUTE

SURFACE

FIG. 2. Association between the solute and nonpolar ligand at the surface of the stationary phase in reversed-phase chromatography. The binding is essentially due to solvent effects. As suggested by the solid arrows, the association of the two species is facilitated by the decrease in the molecular surface area exposed to the solvent upon complex formation. Attractive interactions with the solvent, which are symbolized by open arrows, have a countervailing effect. The magnitude of the nonpolar interaction between the solute and ligand, which determines solute retention, is given by the difference between the two effects. Polar functions in the solute enhance interaction with polar solvents, therefore reduce binging. On the other hand, the binding is stronger when the contact area in the complex is greater and/or the surface tension of the solvent is higher. (From Ref. 1. Reproduced from the *Journal of Chromatographic Science*, by permission of Preston Publications, Inc.)

Karger et al. also believed that retention in reversed-phase chromatography is due to a hydrophobic effect and explored the role of hydrophobic selectivity in reversed-phase chromatography [10]. The role of such factors as organic modifier, temperature, and solute structure on selectivity were studied. Retention was estimated based on a topological index. This index, known as molecular connectivity, is proportional to the cavity surface area the molecule

resides in an aqueous environment. This index is a measure of the
water solubility of nonelectrolytes. The molecular connectivity is
defined as follows:

$$\chi = \sum_{k=1}^{k} \frac{1}{(\delta_i \delta_j)_k^{1/2}} \tag{4}$$

where χ = 1, 2, 3, 4, which relates to the number of atoms attached
to atoms i and j, respectively, and k equals the number of bonds
in the group or molecule. In calculating the molecular connectivity,
only the skeleton is important and hydrogen atoms are neglected.
For saturated cyclic structures a decrease of 0.5 in the index is
employed to account for ring closure. Corrections are also made for
aromaticity and unsaturation. The molecular connectivity, which relates to the degree of branching, can be shown to be proportional
to log k'. This is expected since the hydrophobic effect would predict that log k' is proportional to the solubility of the compound
in water. For a given mobile-phase solute, selectivity can be related to structure by the molecular connectivity index. The change
in log α_{CH_2} (methylene group increments) versus volume percent of
organic component in water-methanol, water-acetonitrile, and water-
acetone mixtures were investigated. In all three mixtures it was
found that the log α_{CH_2} decreased with increased volume percent
organic component. These relationships are shown in Fig. 3. The
linearity of the water-methanol curve over most of the volume percent of methanol can be attributed to the fact that methanol can act
as both a proton donor or acceptor and thus act as a dilution medium
for water until very high concentrations of methanol are present.
On the other hand, acetone and acetonitrile are only proton acceptors and thus cause changes in the hydrogen-bonding network of water
more readily. It is apparent from Fig. 3 that values of log α_{CH_2} for
normalized time conditions will tend to be equal because of opposing
trends. For example, in comparing a water-methanol mixture with a
water-acetonitrile mixture, for equal retention a higher water

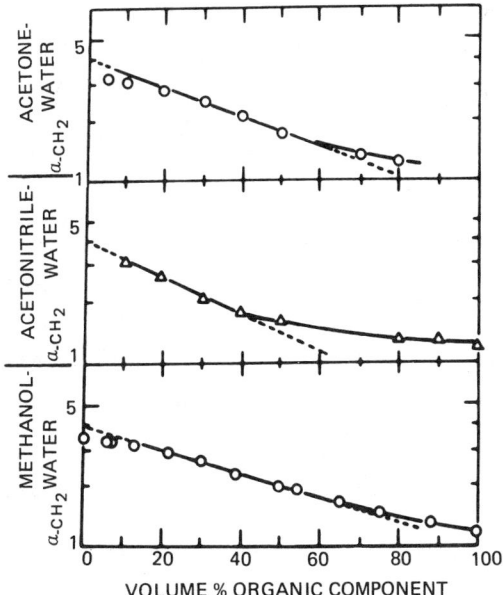

VOLUME % ORGANIC COMPONENT

FIG. 3. Semilogarithmic plots of a_{CH_2} (obtained from log k' versus carbon number plots for the n-alcohols) versus volume percent organic component for water-acetone, water-acetonitrile, and water-methanol mobile phases, on a chemically bonded octadecylsilane stationary phase at ambient temperature. (From Ref. 10.)

content is required in the water-acetonitrile mixture. Since acetonitrile is more lipophilic than methanol, it is a stronger solvent in reversed-phase chromatography. The greater selectivity at the higher water content compensates for the lower selectivity at a given composition for the acetonitrile-water mixture.

The retention of alcohols decreases by approximately a factor of 2 with a 30°C increase in temperature. Hydrophobic selectivity decreased slightly with this increase in temperature. Under normalized time conditions, that is, when temperature and mobile phase composition are simultaneously varied to give similar k' values, similar selectivities were obtained. The smaller selectivity from the use of higher temperature is compensated by the employment of a more aqueous mobile phase, which gives a higher selectivity.

A semiempirical approach which allows the prediction of k'
values for a given chromatographic system in reversed-phase chromatography from octanol-water partition coefficients is illustrated by
the equation [11]

$$\log k'x = c_1 \log k_{Oct}x + c_2 \qquad (5)$$

where $k_{Oct}x$ is the octanol-water partition coefficient and c_1 and
c_2 are the slope and y intercept of the straight line that defines
the relationship. This correlation is based on the premise that
the partition of a compound between an organic-aqueous mobile-phase
column and a nonpolar bonded-phase column is approximated by octanol-
water partition coefficients. These coefficients are the measurement of the distribution of a compound between an octanol phase and
a given mobile phase as measured by ultraviolet spectroscopy. The
relationship illustrated by Eq. (5) is limited since a separate
equation must be derived for a given column-mobile phase combination. Baker and Ma were able to derive retention indexes that are
independent of chromatographic conditions for reversed-phase chromatography [12]. Furthermore, these retention indexes could be directly
related to octanol-water partition coefficients [13]. These indexes
and their utility to forensic analysis are discussed in Sec. IV.C.

As mentioned in Sec. II, for acidic and basic solutes buffer
systems are employed. The typical buffers consist of acetate and
phosphate salts at a pH value that favors the formation of the un-
ionized form of a solute. Most acidic drugs of forensic interest
would exist as the free acid at a pH value of 1-3.5 since their pK_a
values tend to be greater than 4. The ionized form of a drug would
have a marked decrease in affinity for the reversed-phase packing,
which would lead to inadequate retention and resolution [14]. In
addition, tailing could arise from interaction of the ionized acid
with the unbonded silanol sites [15]. If both the un-ionized and
ionized species of a drug are present, which occurs if the pH of the
mobile phase is close to the pK_a value of the drug, a tailing peak
could result. This phenomenon occurs because of the difference in

retention between the free acid and the ionized species. For these reasons the chromatography of the un-ionized species is desirable in reversed-phase chromatography. A similar argument could be presented for the chromatography of basic drugs. The process where the mobile phase is buffered to favor the formation of the unprotonated species is referred to as ion suppression. In order to chromatograph basic drugs by ion suppression it would require in most instances a pH greater than 7, which is generally unsatisfactory for bonded phases. For bases that are completely ionized at the mobile phase, pH mixed mechanisms such as ion pairing and ion exchange which lead to tailing peaks could occur. Ion pairing adsorption occurs between the positively charged ionized drug and the anionic species in the buffer, while ion exchange occurs between the ionized drug and weakly acidic unbonded silanol sites. When the pH of the mobile phase is near the pK_a of the drug of interest, an additional mechanism would involve the retention of the free base.

Ion-pairing chromatography, which is a type of reversed-phase chromatography employing secondary equilibrium, is an excellent alternative for the chromatography of basic solutes. In ion-pairing chromatography a counterion is added to the mobile phase which is available to form an ion pair with the oppositely charged ionized form of a solute. This ion pair could be retained on a bonded-phase column by one of three proposed mechanisms. The first two mechanisms can be depicted by the following diagram [16]:

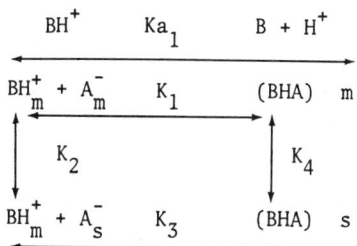

where B represents a basic drug and A^- depicts a negative counterion. The subscript s depicts stationary phase and m refers to mobile phase. Horvath has shown the ion-pairing mechanism to proceed by

equations K_{a1}, K_1, and K_4, which represents ion-pairing formation
occurring in the mobile phase followed by adsorption of the ion
pair onto the stationary phase [17]. Equations K_{a1}, K_2, and K_3
illustrate ion pairing occurring by an ion-exchange mechanism proposed by Kissinger [18]. Recently, Bidlingmeyer presented an ion-
interaction mechanism which is not based on either of the two
mechanisms previously proposed [19]. In this mechanism a primary
ion layer and an oppositely charged secondary layer are formed on
the nonpolar surface. In the case of positively charged solutes,
the primary ions would be the negatively charged counterion described
in the two previously proposed mechanisms, while the secondary layer
would consist of positively charged species originating from the
ion-pairing reagent. For example, if heptanesulfonic acid sodium
salt was the ion-pairing reagent, the primary layer would consist of
heptanesulfonate ion and the secondary layer would be sodium ions.
Positively charged basic solute ions are first attracted to the
secondary layer by electrostatic forces. Then because of the attraction of the nonpolar bonded phase for the lipophilic portion of
the solute molecule, a reagent counterion would be replaced on the
primary layer by the positively charged sample molecule. In order
to restore electrostatic equilibrium another negatively charged
counterion will be adsorbed onto the stationary phase. The net result is that a pair of ions (not necessarily an ion pair) has been
adsorbed onto the bonded-phase column. Tomlinson et al. have published an excellent review on the techniques of ion pairing [20].
Regardless of the true mechanism, this technique offers great ver-
satility because of the ability to control retention by adjusting
variables such as type and concentration of counterion, pH, and
organic-aqueous ratio. Reversed-phase ion-pairing chromatography
is described in detail in Sec. IV.E when discussing the role of
this methodology in forensic drug analysis.

     Now that we have examined general principles of reversed-phase
chromatography employing bonded-phase columns, let us discuss the
great utility of this mode of chromatography in forensic drug

analysis. In addition, a few examples of normal-phase chromatography employing bonded-phase columns will be cited. The literature related to forensic drug analysis can be divided basically
into two categories. First, there are reports discussing the analysis of a particular class of drugs (e.g., cannabis and opium alkaloids). The second type of paper presents generalized schemes for
the analysis of drugs of forensic interest. We begin with applications of reversed-phase chromatography to the analysis of specific
types of drugs.

III. DRUG APPLICATIONS

A. Cannabis

Smith and coworkers have performed extensive work on the analysis
of cannabis samples employing an octadecyl column with reversed-
phase mobile phases [21-23]. Color tests, microscopy, and thin-
layer chromatography are usually employed for the identification of
cannabis samples. If comparative analysis is required in order to
identify common origins or to elucidate distribution chains, some
type of chromatography, such as thin layer, gas, or liquid, could
be employed. The method of choice would depend on the discriminating power of a particular technique. Wheals and Smith examined the
relative merits of using thin layer chromatography (TLC), gas-
liquid chromatography (GLC) without silylation, or high-performance
liquid chromatography (HPLC) for the comparative analysis of cannabis samples [21]. These authors found that HPLC offers the greatest
discriminating power and therefore has the greatest utility for comparative cannabis analysis. Thirty-four cannabis samples of known
geographical origin were examined. These exhibits consisted of 12
cannabis resins and 22 herbal samples. As seen in Table 2, the
greatest number of cannabis samples could be distinguished by HPLC.
The four samples that could not be distinguished by HPLC were of
common geographical origin with other samples examined. None of
these four samples could be distinguished by GLC. In addition, GLC

176   *Lurie*

TABLE 2.   A Comparison of Different Methods of Cannabis
Discrimination

Group	Visual	TLC	GLC	HPLC
Resin	5	9	7	10
Herbal	6	3	18	20
Total	11	11[a]	25	30

[a]Two resins and two herbal samples comprised a single category.
Note: Twelve cannabis resin samples and 22 herbal samples were
examined, making a possible total of 34 distinct groups attainable
if complete discrimination could be obtained.
Source: Ref. 21.

failed to distinguish in three instances samples of widely different origins. Different portions of the same exhibit gave identical
HPLC chromatograms. Thus HPLC has the potential for indicating the
origin of cannabis exhibits. A 25-cm-long, 4.9-mm column with
Partisil 5 packing material bonded to octadecyltrichlorosilane was
employed with the mobile phase consisting of methanol-0.02 N sulfuric acid (80:20) with peak detection at 254 nm. Complex chromatograms of good peak shape with an analysis time of 10-15 min were
generated. Starting mobile phase was used to dissolve the cannabis
samples prior to injections of 0.5-2 ul. When methanol was used
there was a loss of resolution after the analysis of 15-20 samples
probably due to partial precipitation of the sample at the head of
the column.

The major components of the chromatograms used in comparative
analysis were identified by Smith in a subsequent communication [22].
Employing preparative liquid chromatography in conjunction with
chromatographic and spectrographic techniques, Smith using identical chromatographic conditions described in his previous paper
identified nine major components. As Fig. 4 illustrates these compounds consist of Δ9- and Δ8-tetrahydrocannabinol, Δ9-tetrahydro-
cannabinolic acid, cannabidiol, cannabidiolic acid, cannabinol,
cannabinolic acid, cannabicromene, and cannabichromenic acid. The
identity of these compounds were confirmed by gas chromatography-
mass spectrometry (GC-MS). Sulfuric acid was employed in the mobile

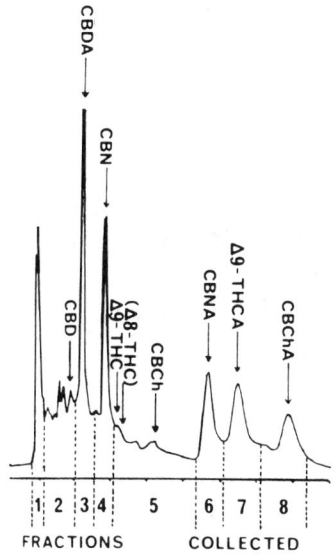

FIG. 4. HPLC of cannabis resin (100 mg extracted with 1 ml of
eluting solvent), illustrating fractions collected and identity of
peaks. 1 μl injected onto column. Detector wavelength: 254 nm;
adsorbance: 0.1. Scale graduations represent 0.5-min intervals.
(From Ref. 22.)

phase to improve the chromatography of the acidic components by an
ion-suppression mechanism. The great utility of HPLC in both comparative and quantitative analysis of cannabis samples arises from
the ability to analyze neutral and acidic compounds simultaneously
without the use of silylation procedures. Silylation is required
for the GC analysis of acidic compounds to prevent decarboxylation.
In addition, the separation of cannabidiol and cannabichromene requires silylation by gas chromatography. Smith also illustrated
that the quantitative analysis of cannabinol, cannabidiol, and Δ8-
and Δ9-tetrahydrocannabinol should be conducted at low-UV wavelengths
(210-220 nm) owing to their greater extinction coefficients at these
wavelengths. The use of a fixed 254-nm wavelength detector being
relatively insensitive for these compounds is not recommended for
quantitative analysis. However, since a maximum number of com-

ponents in a cannabis sample could be detected at 250-260 nm, a
fixed 254-nm detector is recommended for comparative analysis.

In a third report Smith and Vaughan presented data on the
quantitative analysis of the components of cannabis using the previously developed HPLC system [23]. In addition, the location and
quantitation of cannabigerol and cannabigerolic acid employing the
same mobile phase was utilized. Excellent precision was obtained for
the analysis of cannabis substituent content for both the same and
similar resins. In this paper Smith and Vaughan described an interesting method for quantitating cannabigerolic acid in the presence of cannabinol which is unresolved from the former component.
This methodology employing dual-wavelength detection allows the
quantitation of any component A in the presence of component B if
these components are exactly coincident. The concentration of compound A is given by the formula

$$C_A = \frac{(F_{1B}/F_{2B})(R_1 - R_2)}{F_{1B}(1/F_{1A} - 1/F_{2A})} \tag{6}$$

where R_1 and R_2 are calculated from peak height measurements as
follows:

$$R_1 = \frac{H_1}{S_1}$$

$$R_2 = \frac{H_2}{S_2} \tag{7}$$

where H_1 and H_2 are the peak heights of the mixture of A and B at
wavelengths 1 and 2 and S_1 and S_2 are the internal standard peak
heights at wavelengths 1 and 2. F_{1A} and F_{2A} are the calibration
factors for components at wavelengths 1 and 2 found from the calibration curves. F_{1B}/F_{2B} is evaluated from the following relationship:

$$\frac{F_{1B}}{F_{2B}} = \frac{H_{2B}/S_2}{H_{1B}/S_1} \tag{8}$$

For the quantitative procedures above di-n-octyl phthalate was used
as an internal standard. A chloroform-methanol solution (1:9) was
used as an extraction and injection solvent. This solvent was employed instead of starting mobile phase because of several limitations of the latter solvent for quantitative analysis. Problems associated with the use of starting mobile phase included: (1) limited
solubility of the internal standard, (2) incomplete extractions of
the cannabis samples, (3) adsorption of the internal standard by
particulate matter in the sample. According to Smith and Vaughan,
no long-term stability problems have been encountered in the analysis of a wide range of cannabis samples.

Knaus et al. also investigated the reversed-phase HPLC analysis of the neutral cannabinoids: Δ^8-tetrahydrocannabinol, cannabinol, and cannabidiol [24]. A µBondapak C_{18} column with a mobile
phase consisting of 75% methanol and 25% water was employed with
fixed UV detection at 254 nm. This mobile phase was similar to the
one utilized by Wheals and Smith except that no acid was present
since only neutral compounds were assayed. In addition to the advantages of HPLC analysis of cannabinoids that were already described by Smith [22] such as no decarboxylation of the acidic
cannabinoids and ease of collecting separated components for further
analysis, other benefits were discussed. The chromatography was
carried out at room temperature in the absence of air, which pre-
vents isomerization of Δ^9-tetrahydrocannabinol to the thermo-
dynamically more stable Δ^8-tetrahydrocannabinol. In addition, the
oxidation of Δ^9-tetrahydrocannabinol to cannabinol is prevented.
Knaus et al. also described the HPLC of various derivatives such
as t-butyldimethylsilyl ethers and trimethylsilyacetates which
could offer the following advantages:

1. The derivatization of the carboxyl and/or phenolic moieties in
    cannabinoids would provide an additional retention time for
    these components which would make their chromatographic identi-
    fication more selective.

2. These derivatives having longer retention times are better
    separated from natural components present in body fluids.

3. Good parent ions via mass spectrometry are obtained with these
   derivatives because of their excellent thermal stability. Therefore, their use in a direct HPLC-MS interface system would be
   desirable. In addition, these derivatives could be collected
   from the liquid chromatography and then run by mass spectrometry.

B. Tryptamines

Balandrin et al. described methodology for the analysis of hallu-
ciongenic tryptamine derivatives in a leaf extract of *Acacia podal-
rieaefolia* A. Cunn. [25]. Although GLC has been employed for the
analysis of tryptamines, difficulties have been encountered. If
these compounds are chromatographed without being derivatized,
primary amines are found to tail badly. The HPLC methodology described employing reversed-phase chromatography requires no derivatization. A μBondapak C_{18} column was utilized with a mobile phase
consisting of 1,4-dioxane-0.1 M ammonium carbonate (4:5). A typical
chromatogram is illustrated in Fig. 5. The tailing peaks are possibly due to mixed mechanisms occurring, such as partition, ion
exchange, and ion pairing. About 2 g of dried ethanolic extract of

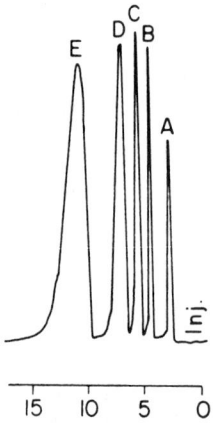

FIG. 5. Liquid chromatogram of standard tryptamines. For operat-
ing conditions, see the text. A, solvent impurity; B, tryptamine;
C, bufotenine; D, N-methyltryptamine; E, N,N-dimethyltryptamine.
(From Ref. 25.)

plant material was thoroughly moistened with 28% ammonium hydroxide
and dried on a steam bath. After cooling, the sample was placed in
a 100-ml boiling flask with a condenser to which 20 ml of chloroform
was added. This mixture was refluxed and cooled to room temperature
prior to being filtered. After the residue was returned to the
flask, an additional 15 ml of chloroform was added and the mixture
refluxed for 30 min. After filtering and cooling, the filtrates
were combined and partitioning occurred with 10 ml of a 1% (w/v)
HCl solution. The chloroform layer was discarded. The acidic
aqueous layer was adjusted to pH 8.5 with 28% ammonium hydroxide
prior to extracting the free base into 10 ml of chloroform three
times. Under reduced pressure the chloroform was removed and the
residue was taken up in 6.0 ml of 0.1 M ammonium carbonate (1:1).
Triplicate 20-μl injections were employed on the liquid chromatograph. The use of methanol-0.1 M ammonium carbonate (1:1) as an
injection solvent resulted in better resolution than employing
starting mobile phase. The wavelength of choice for UV detection
was 280 nm because of the greater selectivity obtained for the
analysis of the tryptamine compounds in the presence of other plant
constituents. As often occurs, the use of higher wavelengths results in decreased detection sensitivity. Tryptamine was found to
consist of 0.06% (w/w) of the plant extract and the recovery of
known amounts of the hydrochloride salt of this compound was 95.3%.

C. Ergot Alkaloids

Because gas chromatographic analysis of ergot alkaloids is unsatisfactory because of their low vapor pressure and thermal instability, silylation procedures are required. HPLC is ideal for the
assay of these compounds since no derivatization is required and
chromatography can be carried out at ambient temperatures. Jane
and Wheals studied the GC and HPLC analysis of lysergic acid diethylamide (LSD), the ergot alkaloid of greatest forensic importance
[27]. A major drawback of the GC analysis of LSD by silylation
techniques was the presence of peaks that often partially or com-

pletely obscured the LSD peak. A similar problem was encountered
in the reversed-phase HPLC analysis of LSD when an ultraviolet detector was employed. The interferring peaks were eliminated by the
use of a fluorimetric detector which proved to be highly selective
for the detection of LSD. The sensitivities of the two detectors
were found to be approximately equal, with a lower limit of 2-10
ng of LSD. A Corasil C_{18} column was employed with a mobile phase
consisting of 6:4 methanol-0.1% ammonium carbonate. A methanolic
extract of the same was injected onto the liquid chromatograph. A
typical chromatogram is illustrated in Fig. 6. At this time a few
points are worth mentioning. In comparison to chromatograms generated on presently employed highly efficient reversed-phase columns,
the separation illustrated appears crude. In the early 1970s at
the time of this paper, the pellicular columns were the most efficient columns readily available. These columns generally gave plate

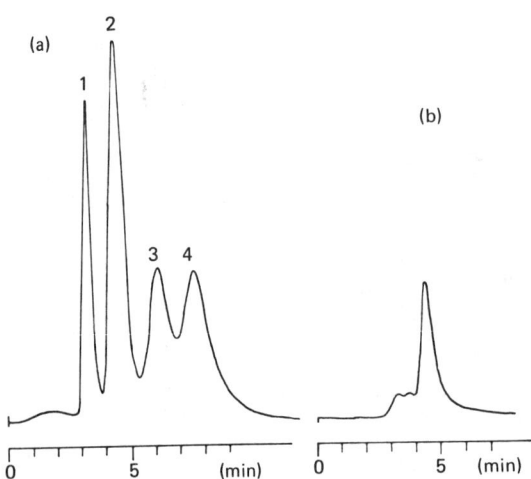

FIG. 6. LC separation of LSD on a 1.2-m column containing Corasil
C_{18} with 6:4 mathanol-0.1% ammonium carbonate solution as solvent.
(a) A synthetic mixture: 1, ergometrine maleate; 2, LSD; 3, iso-
LSD; 4, ergotamine tartrate. (b) A tablet extract. (From Ref. 27.)
Ref. 27.)

counts of less than 500. Using the presently employed microparticulate columns, efficiencies of greater than 2000 are commonplace.
As will be shown in Sec. IV.E, employing ion-pairing techniques, LSD
could be assayed without peak interferences using an ultraviolet
detector. As Jane and Wheals point out, one of the great advantages
of HPLC for LSD analysis would be the ease of collecting the chromatographic peaks for further analysis by spectroscopic techniques.
They routinely collected the fraction of the eluent ascribed to
LSD, diluted it with water, and determined its fluorescence characteristics via a spectrofluorimeter. Although the fluorescence spectrum they obtained was similar to other ergot alkaloids, they believed the combination of retention time and fluorescence spectrum
was sufficient to identify the presence of LSD. It is this author's
belief that a better analysis would be the combination of high-
performance liquid chromatographic and mass spectrometric analysis.
Fluorimetric procedures have been employed for the quantitation of
LSD in illicit formulations. Since many samples contain significant amounts of iso-LSD, which would be expected to exhibit similar
fluorescence to that of LSD, fluorimetric quantitation would lead
to erroneous results.

   For the analysis of the ergot alkaloids related to ergotamine
tartrate, both high- and low-pressure gradient elution techniques
have been employed [28, 29]. A μBondapak C_{18} column was utilized
for the separations employing high-pressure gradients [28], while a
Nucleosil C_{18} column was used for low-pressure gradients [29]. In
both systems the gradients were generated from mixtures of 0.01 M
ammonium carbonate, water, and acetonitrile. Faster analysis was
possible using low-pressure gradients. A comparison of the use of
high-pressure versus low-pressure step gradients is shown in Table
3. Bethke et al. employed 320 nm for analysis, which represents
the absorbance maximum for the 9,10-double bond of lysergic acid
derivatives [29]. Although these compounds have reduced adsorbance
at 320 nm versus 254 nm which was used by Frei et al. [29], the use
of the higher wavelength resulted in a more selective analysis.

TABLE 3.   Advantages and Disadvantages of the Step Gradient System
Compared with Conventional Two-Pump Gradient Systems

Step Gradient System	Two-Pump Gradient System
Advantages	Disadvantages
Mixing of any number of solvent components possible	Only mixing of two solvent components possible
Exact and reproducible composition of the mobile phase in all parts of the gradient profile	Exact control of the composition of the mobile phase at the start of the gradient (e.g., 1% of solvent A) is critical
Free from interference of small flow inhomogeneities (e.g., caused by air bubbles)	High standard of flow stability is necessary, especially for the first part of the gradient shapes with small concentrations of solvent A
Simple connection to automatic injection systems with reconditioning gradient if necessary	
Low cost	Connection to automatic injection systems in many systems not possible without additional interface
Disadvantages	High cost
Only step-shape gradients	Advantages
Flow limitations with slow switching valves	Linear, convex, or concave gradient shapes possible
	No flow limitations up to maximal pump specification

Source: Ref. 29.

Szepesy et al. [30] separated various ergot alkaloids by reversed-phase chromatography utilizing LiChrosorb RP-2 (5-μm particle size), LiChrosorb RP-8 (10-μm particle size), and silica RP-18 (12- and 40-μm particle sizes) columns with a mobile phase consisting of acetonitrile-0.01 M ammonium carbonate solution (2:3).   The number designations for these columns refer to the number of carbons on the bonded alkyl chain.   The k' and α values for these various separations are shown in Table 4.   The data in this table illustrates that by increasing the chain length of the bonded alkyl group, the capacity factor and to some extent the selectivity factor increases.   For reasons described earlier 320 nm was used for UV detection.

Sondack described a HPLC procedure for the analysis of ergonovine maleate utilizing a μBondapak C_{18} column and a mobile phase

TABLE 4. Capacity Ratios (k') and Separation Factors (α) for Ergot Alkaloids on Reversed-Phase Packings

Substance	LiChrosorb RP-2		LiChrosorb RP-8		Silica RP-18	
	k'	α	k'	α	k'	α
Lysergic acid	0.16		0.20		0.10	
Ergometrine maleate	1.54	1.47	1.11	2.11	0.81	2.27
Ergometrinine	2.27	2.35	2.34	3.21	1.84	3.95
Ergotamine tartarate	5.34	1.17	7.51	1.46	7.27	1.63
Ergocornine	6.24	1.25	11.00	1.37	11.86	1.57
Ergocryptine	7.8		15.14		18.65	
Ergocristine	8.45	1.08	16.66	1.10	20.50	1.10
Ergotaminine	9.87	1.53	—	1.75	29.6	1.86
Ergocorninine	12.91	1.12	29.17	1.47	38.2	1.58
Ergocryptinine	14.50		43.06		60.6	
Dihydroergotamine tartarate	—		—		7.94	
Dihydroergocormine methanesulfonate	7.15	1.27	9.43	1.44	9.7	1.50
Dihydroergocryptine methanesulfonate	9.11	1.08	13.60	1.10	14.6	1.20
Dihydroergocristine methanesulfonate	9.82		14.91		17.60	

Note: Eluant: acetonitrile-0.01 M ammonium carbonate solution (2:3).

Source: Ref. 30.

consisting of 20% (v/v) acetonitrile and 1% (v/v) acetic acid in
water [31]. An acidic mobile phase was employed because the author
felt the high pH of eluting solvents which contain ammonium carbonate buffers is usually not considered optimal for good column
life. Samples were dissolved in water prior to injection into the
liquid chromatograph. Since ergonivine was fully ionized at the pH
Sondack employed in his mobile phase, inadequate retention and or
excessive peak tailing could be expected.

Wurst et al. studied the chromatography of the ergot alkaloids
of the ergopeptine type using an amino bonded column in the norma-
phase mode [32]. Retention data for these compounds are presented
in Table 5.

TABLE 5. Relative Retentions of Clavine Alkaloids and Simple
Derivatives of Lysergic Acid

Alkaloid	Diethyl ether-ethanol		Diethyl ether-isopropanol		Chloroform-isopropanol	
	(84:16)	(80:20)	(70:30)	(60:40)	(90:10)	(80:20)
Paspaclavine	0.11	0.11	0.11	0.14	0.12	0.12
Isosetoclavine	0.25	0.26	0.27	0.26	0.70	0.70
Lysergene	0.38	0.39	0.46	0.45	0.58	0.63
Setoclavine	0.55	0.61	0.67	0.67	0.98	1.05
Erginine	0.83	0.80	0.69	0.64	0.58	0.55
Lysergine	0.88	0.87	1.20	1.14	1.24	1.30
Agroclavine	1.00	1.00	1.00	1.00	1.00	1.00
Pyroclavine	1.45	1.49	1.80	1.63	1.65	1.71
Festuclavine	2.08	1.98	2.37	2.01	1.46	1.47
Penniclavine	2.80	2.65	2.97	2.72	5.05	3.85
Ergine	2.90	2.51	2.29	1.85	4.00	2.95
Paliclavine	3.12	2.25	1.66	1.26	3.85	2.99
Elymoclavine	3.22	2.90	3.17	2.72	4.48	3.05
Lysergol	3.50	2.99	3.48	2.83	5.80	4.30
Chanoclavine	8.00	6.03	8.56	6.33	9.83	6.03
Retention volume of agroclavine (ml)	4.67	3.00	4.90	3.50	2.67	1.90

Note: Column: MicroPak NH$_2$; particle size, 10 μm; 25 cm × 2.0 mm
ID; detector: Variscan LC UV; wavelength: 225 (240) nm; flow rate:
1 ml/min; pressure: 3.0-8.5 MPa.

Source: Ref. 52.

D. Tranquilizers

Salmon and Wood presented a HPLC procedure for the determination of
nortriptyline in sugar-coated tablets [33]. The authors found that
the analysis of these tablets by GC were subject to severe interferences due to the presence of the sugar coating. For the HPLC
assay a phenyl Corasil column was employed with a mobile phase consisting of methanol-acetonitrile-0.25% (m/v) aqueous ammonium carbonate solution (40 + 40 + 20) v/v. A flow rate of 1.0 ml/min with
a variable-wavelength UV detector at 254 nm was utilized. The
mobile-phase pH of approximately 8 was probably too basic for prolonged column life. Since the pK_a of nortriptyline is 9.73, this
compound would be fully ionized at the mobile-phase pH. Therefore,
the tailing observed for the nortriptyline peak could be due to
mixed mechanisms such as ion pairing between nortriptyline cation
and carbonate; and ion exchange and adsorption of nortriptyline cation with unbonded silanol sites. A better approach for the analysis of this compound could be via ion pairing at an acidic pH where
ion exchange would not be expected to occur. Alternatively, porous
polymers which are commercially available from Hamilton could be
utilized at a basic pH where nortriptyline would exist as the free
base. In the tablet assays triflupromazine was used as an internal
standard with the tablet dissolved in an acidic aqueous solution.

Hulshoff and Perrin, in a study of the determination of partition coefficients of 1,4-benzodiazepines by high-performance liquid
chromatography and thin-layer chromatography, presented retention
data for these compounds on five different HPLC columns where liquid
stationary phases were employed [34]. Elution of the drugs occurred
with buffer solutions of ammonia and ammonium sulfate in methanol-
water mixtures (pH 9, ionic strength 0.1 M) at flow rates of 0.9-
1.0 ml/min. The results obtained are presented in Table 6. As
pointed out in Sec. I, the use of liquid stationary phases has certain drawbacks. Certain columns were silylated in order to prevent
adsorption on the active silica sites.

TABLE 6. Slopes (b'') and Intercepts (Log k'_w) of the Graphs of Log k'_w Against Methanol Concentration in the Mobile Phase from HPLC Experiments with Five Different Column Packing Materials

Compound	Porasil C (3% oleyl alcohol)[a]					Porasil C (5% oleyl alcohol)[a]					Porasil C (5% oleyl alcohol)[b]				
	n	Log k'_w	s	b''	s	n	Log k'_w	s	b''	s	n	Log k'_w	s	b''	s
Medazepam	9	3.73	0.12	-0.0548	0.0022						7	3.33	0.080	-0.0530	0.0014
Prazepam	15	2.81	0.026	-0.0439	0.0005						7	3.40	0.12	-0.0500	0.0021
Flurazepam	14	3.03	0.055	-0.0487	0.0011	11	3.11	0.039	-0.0494	0.0009	12	2.74	0.029	-0.0420	0.0006
Diazepam	18	2.08	0.009	-0.0360	0.0002	22	2.17	0.013	-0.0367	0.0003	16	2.36	0.014	-0.0380	0.0003
Chlordiazepoxide	25	1.82	0.006	-0.0336	0.0001	24	1.91	0.017	-0.0343	0.0004	19	2.36	0.014	-0.0407	0.0003
Lorazepam	19	1.72	0.023	-0.0344	0.0006	21	1.82	0.023	-0.0368	0.0005	19	2.24	0.016	-0.0378	0.0003
Oxazepam	20	1.62	0.007	-0.0328	0.0002	25	1.73	0.019	-0.0349	0.0004	20	2.05	0.014	-0.0375	0.0003
Clonazepam	21	1.47	0.010	-0.0324	0.0003						17	2.30	0.019	-0.0381	0.0004
Temazepam	22	1.67	0.007	-0.0326	0.0002						19	1.95	0.015	-0.0361	0.0003
Flunitrazepam	20	1.41	0.010	-0.0308	0.0002						18	1.86	0.011	-0.0341	0.0002
Nitrazepam	17	1.26	0.008	-0.0288	0.0002										
Bromazepam	14	1.04	0.008	-0.0273	0.0002						13	1.55	0.054	-0.0281	0.0011

Compound	C_{18}/Corasil (untreated)[a]					C_{18}/Corasil (silylated)[b]				
	n	Log k'_w	s	b''	s	n	Log k'_w	s	b''	s
Medazepam	7	4.16	0.034	-0.0535	0.0006	11	4.58	0.069	-0.0663	0.0012
Prazepam	12	3.92	0.026	-0.0578	0.0005	15	4.20	0.045	-0.0647	0.0008
Flurazepam	9	4.56	0.068	-0.0633	0.0012	16	4.46	0.036	-0.0694	0.0007
Diazepam	12	3.14	0.035	-0.0603	0.0007	23	3.24	0.015	-0.0549	0.0003
Chlordiazepoxide	16	2.95	0.023	-0.0480	0.0005	23	2.89	0.019	-0.0515	0.0004
Lorazepam	18	2.82	0.014	-0.0505	0.0003	23	2.88	0.015	-0.0553	0.0003
Oxazepam	14	2.53	0.023	-0.0450	0.0003	23	2.69	0.013	-0.0514	0.0003
Clonazepam	16	2.54	0.017	-0.0501	0.0004	20	2.50	0.018	-0.0522	0.0005
Temazepam						21	2.86	0.017	-0.0528	0.0004
Flunitrazepam	19	2.72	0.017	-0.0521	0.0004	21	2.58	0.019	-0.0531	0.0005
Nitrazepam	19	2.35	0.011	-0.0464	0.0003	21	2.26	0.018	-0.0475	0.0004
Bromazepam	16	2.35	0.031	-0.0473	0.0008	17	2.20	0.017	-0.0504	0.0005

[a] Silylated with dichlorodimethylsilane

[b] Silylated with hexamethyldisilane HMDS + trimethylsilyl chloride.

Note: n, number of determinations; s, standard deviation.

Source: Ref. 34.

Zagar et al. presented a method for the determination of chlordiazepoxide and two related compounds, 7-chloro-1,3-dihydro-5-phenyl-2H-1,4-benzodiazepin-2-one-4-oxide and 2-amino-5-chlorobenzophenone in bulk powders and capsules [35]. The compounds are separated on a Partisil 10 ODS column with a mobile phase consisting of acetonitrile-0.0035 M ammonium acetatate (65 + 35). A fixed UV detector at 254 nm was employed.

The reversed-phase ion-pairing chromatography of the antidepressive amines desipramine, imipramine, N-methylimipramine, nortriptyline, amitriptyline, N-methylamitriptyline, dixyrazine, perphenazine, and fluphenazine were discussed by Wahlund and Sokolowski [36]. A LiChrosorb RP-8 column coated with 1-pentanol was used as a support. Aqueous mobile phases containing dihydrogen phosphate and bromide as counterions were employed. N,N,N-trimethylnonylammonium ion or N,N-dimethyloctylamine was added to the mobile phases employed to reduce tailing, increase separation efficiency, and reduce k' values. Although it would be expected that the mechanism dominating retention would be the partition of the ion pair formed between the cationic form of the amine and the negative counterion partitioning into the 1-pentanol stationary phase, Wahlund and Sokolowski found that this was not the case. Instead, indications have been obtained that the ion-pairing mechanism possibly involves the hydrophobic surface of the support. The addition of long-chain amines to the mobile phase supposedly reduces dissociation of the ion pair in the stationary phase, which causes band tailing and reduces separation efficiency. Mobile phases containing bromide counterions gave larger k' values than the use of dihydrogen phosphate counterions, due to the higher extraction constants into the stationary phase of the bromide ion pairs.

Sokolowski and Wahlund studied peak tailing of tricyclic antidepressant amines on a number of alkyl-bonded silica columns with reversed-phase ion-pairing chromatographic systems [37]. Mobile phases that were utilized consisted of mixtures of methanol

and buffers containing phosphate or bromide as counterions. Unlike
their previous study, the stationary phase used was the column
packing itself. Similar to their earlier paper, the addition of
alkylammonium ions added to the eluant eliminated peak tailing and
decreased the retention of amines. In Table 7 are presented retention data and asymmetry values for several tricyclic amines,
amobarbital, and diphenylacetic acid on various stationary phases
with an eluant consisting of 1:1 methanol-phosphate buffer (pH 3).
Table 8 depicts the k' values and asymmetry factors for these compounds using the same eluants and stationary phases except that
0.005 M N,N-dimethyloctylamine (DMOA) was added to the mobile phase.
Several trends are apparent from these tables. Acidic compounds
such as amobarbital and diphenylacetic acid had satisfactory asymmetry values (less than 2) on all stationary phases employed with
or without DMOA present in the eluant. For these compounds the
addition of DMOA had a negligible effect on their asymmetry values.
In general, basic compounds except in the case of the µBondapak C_{18}
column, had poor asymmetry values in an eluant when DMOA was absent.
Their is a dramatic decrease in asymmetry values for these compounds
when DMOA is added to the eluant except in the case of the µBondapak C_{18} column. In addition, the retention of the basic compounds
was significantly reduced when DMOA was added to the mobile phase.
A retention model for ion-pair adsorption was used for evaluation
of the results. It was found that the stationary phase contained
adsorption sites with different ability to retain ammonium compounds
and uncharged compounds.

Noggle and Clark examined the HPLC retention of 10 benzodiazepines using reversed-phase chromatography [38]. A µBondapak C_{18}
column was employed with a mobile phase consisting of pH 8 phosphate buffer-methanol (100 + 75) for the separation of chlorazepate,
nitrazepam, clonazepam, oxazepam, lorazepam, chlordiazepoxide,
n-desmethyldiazepam, and diazepam. Because of the high lipophillic
character of prazepam and flurazepam, a stronger mobile phase consisting of water (pH 7)-methanol (75 + 150) was used. A chromato-

TABLE 7. Retention and Asymmetry on Different Solid Phases

Solutes	C°×10^4	Nucleosil C$_8$, 5 μm		LiChrosorb RP-8, 5 μm		μBondapak C$_{18}$, 10 μm		Nucleosil C$_{18}$, 5 μm		ODS-Hypersil, 5 μm		LiChrosorb RP-18, 5 μm		Spherisorb ODS, 5 μm[a]	
		k'	ASF	k'	ASF	k'	ASF	k'	ASF	k'	ASF	k'	ASF	k'	ASF
Desipramine	3.9-4.1	3.79	2.8	8.24	3.6	3.81	1.5	12.2[b]	3.1	12.5	5.93	17.3	4.4	39.9	3.3
Imipramine	3.6-3.7	3.60	2.2	8.00	5.6	3.43	1.5	11.4	3.6	13.6	5.9	20.8	3.6	79.6	2.7
Trimipramine	3.3-3.9	4.54	1.9	10.3	4.4	4.51	1.4	15.1	3.3	17.3	3.7	29.5		>150	
N-Methylimipramine	3.4	9.16	2.9	6.69	6.1	3.05	1.6	9.56	6.0	12.5	5.0	18.4	5.4		
Propranolol	16-17	1.61	3.0	2.75[c]	2.9	1.54	1.5	3.98	3.3	3.23	4.7	3.54	4.2	9.64	3.5
Amobarbital	39	3.06	1.4	4.21	1.3	2.70	1.3	6.94	1.2	4.23	2.0	4.72	1.7	3.29	1.2
Diphenylacetic acid	20-23	5.74	1.4	8.38[d]	1.3	4.82	1.4	15.7	1.4	8.72	1.9	10.3	1.6	8.11	1.2

[a]Column 150 × 4.5 mm ID

[b]C° = 8.4 × 10^{-4}.

[c]C° = 7.1 × 10^{-4}.

[d]C° = 9.4 × 10^{-4}.

Note: Eluant: 1:1 methanol-phosphate buffer (pH 3). C°, sample concentration.

Source: Ref. 37.

TABLE 8. Retention and Symmetry after Addition of DMOA to the Eluant

Solute	$C^o \times 10^4$	LiChrosorb RP-8, 5 μm		μBondapak C_{18}, 10 μm		ODS-Hypersil, 5 μm		LiChrosorb, RP-18, 5 μm		Spherisorb ODS, 5 μm [a]	
		k'	ASF	k'	ASF	k'	ASF	k'	ASF	k'	ASF
Desipramine	3.9	2.04	1.3	1.46	1.4	2.53	1.8	2.54	1.7	2.68	1.3
Imipramine	3.6	1.76	1.2	1.26	1.5	2.17	1.7	2.14	1.5	3.43	1.1
Trimipramine	3.5-3.9	2.28	1.1	1.57	1.3	2.82	1.6	2.78	1.6	4.15	1.1
N-Methylimipramine	3.4	1.00	1.3	0.90	1.5	1.29	2.0	1.23	1.8	3.70	1.5
Propranolol	16-17	0.69	1.3	0.49	1.5	0.79	2.8	0.77	2.0	0.93	1.4
Amobarbital	39	3.30	1.2	2.15	1.5	3.41	1.9	3.56	1.6	2.36	1.6
Diphenylacetic acid	9.4-20	9.91	1.1	5.02	1.4	11.7	1.9	12.1	1.7	10.9	1.5

[a] Column 150 × 4.5 mm. ID.

Note: Eluant 0.050 M DMOA in 1:1 methanol-phosphate buffer (pH 3).

Source: Ref. 37.

gram representing standard benzodiazepines detected at 254 nm is
shown in Fig. 7. A pH of 8 was required to prevent the decarboxylation of clorazepate to N-desmethyldiazepam, which occurs at a lower
pH. As previously mentioned, this high pH would not be generally
recommended for prolonged column use.

An interesting study on the liquid chromatographic identification of clorazepate in pharmaceutical products was presented in a
subsequent paper by Noggle and Clark [39]. In this paper was a detailed account of the chromatographic behavior of clorazepate under

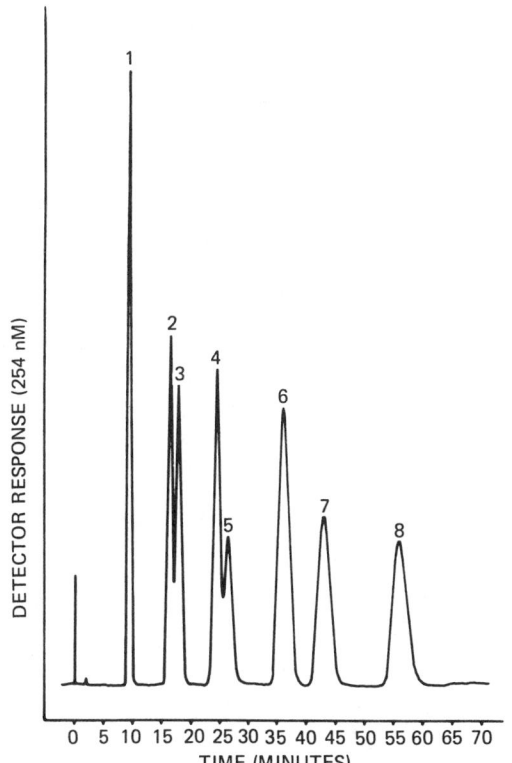

FIG. 7. HPLC separation of benzodiazepines using pH 8.04 phosphate
buffer-methanol (100 + 75) mobile phase and 1.5 ml/min flow rate.
Peaks: 1, clorazepate; 2, nitrazepam; 3, clonazepam; 4, oxazepam;
5, lorazepam; 6, chlordiazepoxide; 7, N-desmethyldiazepam;
8, diazepam. (From Ref. 38.)

both acidic and basic conditions. The chromatographic conditions
employed in their previous study was found to be optimum for this
compound.

Emery and Kowtko published a HPLC procedure for the determination of diazepam in tablets [40]. The manufacturing intermediate
7-chloro-1,3-dihydro-5-phenyl-2H-1,4-benzodiazepin-2-one and the
decomposition products 3-amino-6-chloro-1-methyl-4-phenylcarbostyril
and 2-methylamino-5-chlorobenzophenone are all well resolved from
each other and diazepam in approximately 30 min. A μBondapak C_{18}
column was employed with a mobile phase consisting of 65% methanol,
35% water, with a fixed UV detector at 254 nm. It is of interest
to note that the author's preliminary work was done with a mobile
phase of water, methanol with 0.1% ammonium carbonate. This approach was abandoned because of possible column deterioration from
ammonium carbonate and it was found that the addition of this buffer was superfluous. This was a very logical approach since diazepam, with a pK_a value of approximately 3.5 [41] would exist as the
free base in a mobile phase consisting of methanol and water. For
the tablet assay benzene was used as an internal standard and the
extracting solvent was methanol.

The methodology for the analysis of benzodiazepines such as
diazepam and chlorodiazepoxide that was previously examined all involved ultraviolet detectors. Lund et al. evaluated the use of
amperometric detectors for the analysis of these compounds [42].
A 5-μm Spherisorb ODS column was employed with a mobile phase comprised of methanol-water (60:40) containing 0.05 M ammonium acetate.
A chromatogram for the separation of diazepam, chlorodiazepoxide,
and nitrazepam is presented in Fig. 8. Amperometric detection can
offer certain advantages over UV detection in terms of selectivity,
sensitivity, and cost. Lund et al. were involved in the design of
a new amperometric detector that overcame the shortcomings of some
commercial detectors. In this study when using a glassy carbon
electrode, no increase in sensitivity was observed for amperometric
versus ultraviolet detection. In fact, for diazepam and chlordiaze-

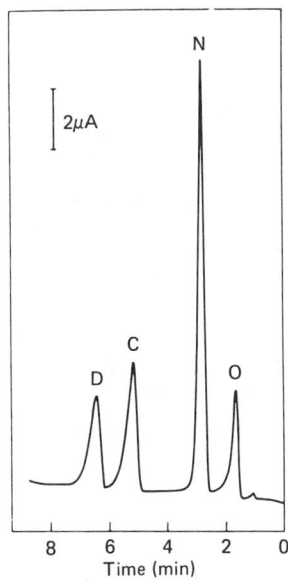

FIG. 8.   Separation of 20 μg (0.07 μmol) of nitrazepam (N), 30 μg
(0.1 μmol) of chlordiazepoxide (C), 28 μg (0.1 μmol) of diazepam
(D), and trace amounts of oxygen (O) using amperometric detection.
ODS column: mobile-phase methanol-water (60:40) containing 0.05 M
ammonium acetate; flow rate: 1 ml/min; reduction potential: 1:25
V versus silver. (From Ref. 42.)

poxide UV detection offered greater sensitivity. Amperometric
detection in series with UV detection offers great potential in
terms of selective information on the electroactive components of
complicated mixtures. Ammonium acetate was added to the mobile
phase to serve as a supporting electrolyte for the electrochemical
detector.

    Hanekamp et al. obtained a detection limit of 1 ng for nitraze-
pam employing a pulse dropping mercury electrode with an electrochemical scrubber [43]. A LiChrosorb RP-8 column was utilized with
a mobile phase consisting of 50:50 methanol-water (v/v), 0.1 M
KNO_3, 10^{-3} M HNO_3. The applied potential was -0.600 V.

    Burke and Sokoloff presented a HPLC procedure for the simul-
taneous determination of amitripyline hydrochloride and chlordiaze-

poxide in tablets in which both components were separated from their
impurities and decomposition products [44]. Fixed ultraviolet detection at 254 nm was employed with a μBondapak C_{18} column and a
mobile phase consisting of 0.01 M sodium lauryl sulfate in tetra-
hydrofuran-methanol-pH 2.5 Britton Robinson buffer (120:30:150).
A flow rate of 1 ml/min was utilized. The use of the ion-pairing
reagent sodium lauryl sufate appears to be a good choice since
amitriptyline is a strong base with a pK_a of approximately 9.4
[42]. Sulfanilamide was used as an internal standard. Samples
were dissolved in water-tetrahydrofuran-methanol (150:120:30) and
adjusted to pH 2.5 prior to chromatography. Spectrometric determinations of the Bromcresol Green complex of amitriptyline hydrochloride and of the diazotization and coupling product of the chlordiazepoxide were found to be tedious and time consuming. The HPLC
assay agreed well with spectrometric determinations.

HPLC methodology employing an amino-bonded microparticulate
column in the normal-phase mode for the determination of chlorpromazine hydrochloride and its two oxidation products was described by
Takahashi [45]. Spectrofluorometry has been employed for the analysis of phenothiazine drugs. This technique would have limitations
when assaying the oxidation products of chlorpromazine because of
their similar spectrofluorimetric characteristics. The mobile phase
utilized consisted of acetonitrile-benzene-water (16:14:1) with the
aqueous portion containing 0.01% of sodium bisulfite and d-ara-
boascorbic acid as antioxidants. Tablets, concentrates, syrups,
and suppositories were assayed using acetonitrile as solvent.
Takahashi preferred the bonded-phase column over straight silica
because of its stability toward hydrolysis and because it is less
sensitive to water adsorption. Ultraviolet and fluorescence were
both employed, with the latter being overall more selective and
sensitive for the assay of chlorpromazine and its oxidation products. Ultraviolet adsorption was carried out at 345 nm while
fluorometry consisted of excitation at 280 nm and emission at 385
nm for the oxidation products and an emission wavelength of 450 nm
for chlorpromazine.

E. Opium Alkaloids

Das Gupta presented a method for the determination of morphine in paregoric USP by reversed-phase chromatography [46]. The mobile phase consisted of 0.1 M KH_2PO_4 buffer solution in 7% (v/v) methanol in water employed with a µBondapak C_{18} column. The flow rate was 1.8 ml/min with peak detection at 254 nm with a fixed UV detector. For the determination of morphine in paregoric, 10 ml of sample was transferred to a 150-ml beaker to which was added 2.0 ml of 0.1 N sulfuric acid and 20 ml of water. The resultant solution was boiled on a hot plate until 8 ml of solution remained. This solution was cooled and diluted to volume (25 ml) with water from which 40 µl was injected onto the liquid chromatograph. For the preparation of a standard mixture similar to paregoric, 53.3 mg of morphine sulfate was dissolved in water containing 8 ml of 0.1 N sulfuric acid and brought to a volume of 250 ml. Good precision and agreement with the labeled claim was obtained. The HPLC method depicted is far superior to the tedious and time-consuming USP method.

Wu and Wittick described two similar isocratic systems for the analysis of morphine, codeine, and thebaine in gum opium [47]. Two µBondapak C_{18} columns in series were employed for each mobile phase. For the analysis of morphine a mobile phase consisting of 5% acetonitrile in 0.1 M NaH_2PO_4 was employed, while for codeine and thebaine the mobile phase was 25% acetonitrile in 0.1 M NaH_2PO_4. The pH of these mobile phases were approximately 5. The tailing that was observed for these components of opium is possibly due to mixed mechanisms occurring, such as ion pairing, adsorption, and ion exchange. The use of paired-ion chromatography at a lower pH where ion exchange with the unbonded silanol sites would not occur could result in better peak shape for these compounds. For the assay of opium 2 g of gum opium was soaked overnight in 20 ml of water. By stirring and ultrasonic agitation, the resulting slurry was completely dispersed. After centifugation the resulting supernatant was transferred to a 50-ml volumetric flask. Using 10-ml and 15-ml portions of water the solid residue was washed and after centifugation the

resulting supernatants were added to the volumetric flask. The 50-
ml volumetric flask was diluted to volume with water. For the
determination of morphine a 20-ml aliquot plus 0.2 g of calcium hydroxide was extracted with 10 ml of methylene chloride using high-
speed centifugation. The extract was cooled to room temperature
and a 15-ml aliquot of the aqueous phase containing morphine was
pipetted into a 25-ml volumetric flask containing 10 drops of glacial acetic acid and diluted to volume with water. The solution
was then filtered through a 0.45-μm Millipore filter prior to
chromatographic analysis. Morphine sulfate pentahydrate equivalent
to 120 mg of morphine base was dissolved in 25 ml of water. A 20-
ml aliquot containing 0.2 g of calcium hydroxide was extracted with
10 ml of methylene chloride. After high-speed centifugation a 15-
ml aliquot of the aqueous phase was pipetted into a 25-ml volumetric
flask containing 10 drops of glacial acetic acid and diluted to
volume with water. Approximately 10 ml of this solution was filtered through a Millipore filter prior to chromatographic analysis.
For the determination of codeine and thebaine a 25-ml aliquot of the
aqueous opium extract from the original volumetric flask was extracted using 0.2 g of calcium hydroxide and 5 g of sodium chloride
with 10 ml of methylene chloride. After centifugation a 4-ml aliquot was removed from the methylene chloride phase and diluted to
25 ml with presaturated methylene chloride. After backwashing with
20 ml of a presaturated calcium hydroxide solution the methylene
chloride solution was filtered through a Millipore filter. Fifteen
milliliters of the filtrate was evaporated to dryness and redis-
solved in 10 ml of the mobile phase with the aid of a few drops of
85% phosphoric acid. The resulting solution was filtered through
a Millipore filter and chromatographically analyzed. Codeine and
thebaine standards equivalent to 2 mg of codeine base and 12 mg of
thebaine base were weighed into a 25-ml volumetric flask. Ten
milliliters of mobile phase was added followed by 10 drops of 85%
phosphoric acid and the resulting solution was diluted to 25 ml with
mobile phase. The column was regenerated by washing the column with

approximately 150 ml of mobile phase containing 0.1 M NaH_2PO_4 +
0.05 M $NaClO_4$ in 25% CH_3CN/H_2O, pH 3.0 (adjusted with H_3PO_4) after
approximately 50 morphine determinations. The mobile phase that was
employed for the determination of codeine and thebaine could also be
used for the assay of narcotine and papaverine. A variable ultraviolet detector set at 286 nm was used for the determination of the
opium alkaloids. Good reproducibility was obtained for the opium
determinations over several days.

Soni and Dugar presented chromatographic conditions for the
reversed-phase ion-pair determination of illicit heroin samples
using a μBondapak C_{18} column [48]. Lurie has shown, using similar
systems and the identical chromatographic conditions, that heroin
will not separate from acetylcodeine (49-51). In this same article
by Soni and Dugar chromatographic data were presented for the reversed-phase ion-pair separation of legitimate preparations containing aspirin (a), phenacetin (p), caffeine (c), and codeine; apc
and oxycodone; apc, ephedrine, and codeine; and acetominophen and
codeine. For the first three preparations a μBondapak C_{18} column
was employed with a mobile phase consisting of 47% methanol with
0.01 M tetrabutylammonium phosphate, pH 7.5. Two points are worth
noting about the use of the described mobile phase. First column
stability may be a problem because the pH of the mobile phase is
greater than 7. In addition, the tailing observed for codeine,
ephedrine, and oxycodone could be attributed to these drugs being
ionized at the mobile phase pH. The use of tetrabutylammonium phosphate is reasonable for the analysis of aspirin via an ion-pair mechanism since aspirin is fully ionized at the mobile phase pH of 7.5.
However, for the simultaneous analysis of codeine in the presence of
aspirin, an acidic mobile phase at pH 2.3 where codeine could be
analyzed by ion pairing and aspirin by ion suppression would be
desirable. It has been shown for optimum stability of aspirin a
mobile phase pH of approximately 2.3 should be employed [51].

Das Gupta described the simultaneous quantitation of acetominophen, aspirin, caffeine, codeine phosphate, phenacetin, and saly-

cilamide using reversed-phase chromatography with an acidic mobile
phase [52]. A μBondapak C_{18} column was utilized with a mobile phase
containing 0.01 M KH_2PO_4 in water with 19% methanol, pH 2.3 (final
pH adjusted with 85% phosphoric acid). Under these chromatographic
conditions aspirin was assayed by an ion suppression mechanism while
the separation of codeine possibly involved ion pairing between
ionized codeine and phosphate ion. For commercial preparations containing the compounds listed above, excellent agreement was obtained
between the assayed amounts and the manufacturer's claim. Since
prior derivatization is required for the analysis of aspirin by gas
chromatography, the latter technique is more time consuming.

    Muhammad and Bodnar described a reversed-phase ion-pairing
chromatographic method for the determination of guaifensin, phenylpropanolamine hydrochloride, sodium bensoate, and codeine phosphate
in cough syrups [53]. A μBondapak alkylphenyl column was employed
with a mobile phase consisting of methanol, water (36:62) each with
0.004 M 1-heptanesulfonic acid salt and 1% glacial acetic acid.
The mode of detection was fixed UV at 254 nm and the internal standard was p-hydroxyphenoxy acetic acid. For sample preparation a
5-ml aliquot of syrup was pipetted into a 50-ml volumetric flask to
which had been added 5 ml of an internal standard solution. The
contents were diluted to volume with mobile phase. Excellent precision and accuracy was obtained for the methodology employed.
HPLC was particularly advantageous since no prior extractions were
required as is the case in GC analysis.

    A HPLC method for the analysis of illicit heroin exhibits was
described by Love and Pannell [54]. A μBondapak C_{18} column was
utilized and the mobile phase at a pH of 7 consisted of 65% acetonitrile, and a 35% aqueous buffer containing 0.75 g of ammonium acetate per 100 ml. An excellent separation was obtained between
heroin and acetylcodeine and acetylmorphine, two by-products in the
heroin synthesis present in most heroin exhibits. A chromatogram
of mixed opiates and caffeine is presented in Fig. 9. Retention
data for only three of approximately 40 possible adulterants

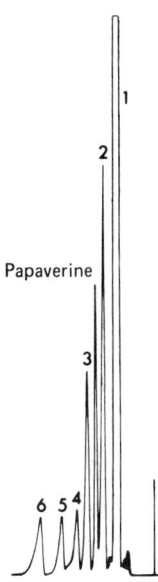

FIG. 9.   Chromatogram of mixed opiates, caffeine, and papaverine internal standard: 1, caffeine; 2, morphine; 3, monoacetylmorphine; 4, codeine, 5, heroin; 6, acetylcodeine. (From Ref. 54.)

present in heroin samples were presented.   In a evaluation of the method of Love and Pannell, Lurie et al. [51] found that heroin does not separate from quinine an adulterant present in the majority of the heroin exhibits analyzed in their laboratory.   Since the adulterants present can vary in different regions, the coelution of heroin with quinine may not present a problem.   Lurie et al. further found that acidic adulterants were poorly retained and exhibited significant tailing.   The system was found to lack specificity for the identification of many of the adulterants.   In addition, tailing was observed for heroin and basic drugs in general due to mixed mechanisms taking place such as partition, adsorption, ion exchange, and ion pairing.   It would appear that the problems mentioned above could be overcome by employing an ion-pairing chromatographic system at an acidic pH for the analysis of heroin and its adulterants. Later in this chapter in the section on generalized analysis schemes

by reversed-phase ion-pairing chromatography, such an approach is
presented [51]. In the methodology by Love and Pannell, samples
were assayed directly in methanol using a fixed UV detector with a
280-nm converter.

Poochikian and Cradock presented a procedure for the determination of heroin and its hydrolysis products monoacetylmorphine and
morphine [55]. A mobile phase consisting of acetonitrile-0.015 M
potassium dihydrogen phosphate, pH 3.5, was utilized in conjunction
with a μBondapak C_{18} column. Since the method was developed for
pharmacology studies rather than for forensic analysis, no retention data were given for acetylcodeine or the adulterants present
in heroin samples.

Retention data for various opium alkaloids, including heroin,
morphine, codeine, thebaine, dihydrocodeine, and related compounds,
were presented by Olieman et al. [56], who employed reversed-phase
ion-pairing chromatography. The liquid chromatographic conditions
consisted of a μBondapak C_{18} column and mobile phases consisting of
methanol or acetonitrile, water, acetic acid, and heptanesulfonic
acid. In an independent study a similar chromatographic system was
used for opium alkaloids as well as other drugs of forensic interest, as described in Sec. IV.E [49].

A high-performance liquid chromatographic method for the analysis of codeine in syrups using reversed-phase ion-pairing chromatography was developed by Kubiak and Munson [57]. Using a μBondapak
C_{18} column and a mobile phase consisting of 0.005 M dioctyl sodium
sulfosuccinate and 0.01 M ammonium nitrate in acetonitrile-water
(375:625), an excellent separation was obtained between the excipients in the syrup and codeine. Kubiak and Munson found that the
use of nitrate ions from ammonium or sodium salts improved peak
shapes and reduced tailing. A fixed UV detector at 254 nm was
utilized. For the quantitative determination of codeine in syrups,
both internal and external standard methods were employed. In the
internal standard method 5.0 ml of syrup was transferred into a
125-ml Erlenmeyer flask containing 40 ml of the internal standard

solution (diphenylamine, 0.05 mg/ml in the mobile phase) prior to injecting 25-µl aliquots of the resulting solution. Syrup, 5.0 ml, was added to 40 ml of the mobile phase for the external standard method.

A method for the separation and determination of opium alkaloids in opium by high-performance liquid chromatography was described by Nobuhara et al. [58]. The column employed consisted of a Nucleosil 10-CN with a mobile phase consisting of 1% ammonium acetate (pH 5.8)-acetonitrile-dioxane (80:10:10), flow rate 1.5 ml/min. Detection was carried out with a fixed UV detector at 254 nm. The separation for the opium alkaloids morphine, codeine, thebaine, narcotine, and papaverine is similar to the one reported by Lurie [50], who used employed reversed-phase ion-pairing chromatography with heptanesulfonate as counterions. It is interesting to note that the use of a Nucleosil 10-C_{18} column with a similar eluant that was used with the Nucleosil CN column resulted in asymmetric peaks for the various compounds. Majors points out that "better peak symmetry is often observed on shorter chain phases, possibly due to a better wetting of the surface by the mobile phase and resulting in better solute mass transfer" [15]. The Nucleosil 10-CN column has a much shorter alkyl chain length than the Nucleosil 10-C_{18} column. Although an excellent separation was obtained for the opium alkaloid standards on the system presented by Nobuhara et al., the morphine and codeine peaks were subject to interferences from coextractives present in opium. No such interferences were observed in the methodology employed by Wu and Wittick, who utilized extensive sample cleanup [47]. In the present paper the sample preparation consisted of mechanically shaking a 2-g portion of gum opium with 20 ml of 2.5% acetic acid for 20 min. After centifuging the mixture, the supernatant was separated and filtered. The extraction procedure was then repeated three times and the extracts combined and diluted to 100 ml with 2.5% acetic acid. A 5-µl aliquot of the resulting solution was diluted to 20 ml with methanol prior to injecting 6 µl of sample onto the liquid chromatograph.

F. Phenethylamines

Phenethylamines when present in their usual salt form exhibit ap-
preciable tailing on the stationary phases normally employed for
gas chromatographic analysis. Basic precolumns are required, which
require frequent repacking in order to obtain satisfactory chromatographic performance. Good chromatographic performance is obtained
for phenethylamine salts using bonded-phase columns.

Noggle presented an interesting method for the analysis of
phenethylamines and related compounds employing phenylisothiocynate
derivatives [59]. Use of these reagents increases ultraviolet detectability and improves chromatographic behavior for the compounds
examined. Primary and secondary amines can form phenylisothiocynate
derivatives, as illustrated in Fig. 10, which depicts a reaction
between propylhexidrine, a non-UV adsorber, and phenylisothiocynate.
Although this technique is very useful for the analysis of phenethylamines such as amphetamine, ephedrine, phenylpropanolamine, and
methamphetamine, I disagree with the author's premise that the low
extinction coefficient of phenethylamines is one of the major problems associated with HPLC analysis. These compounds can be readily
assayed using one of the higher-sensitivity settings on the ultraviolet detector. For a compound such as propylhexdrine that lacks
a UV chromophore, the use of derivatization in lieu of the less-

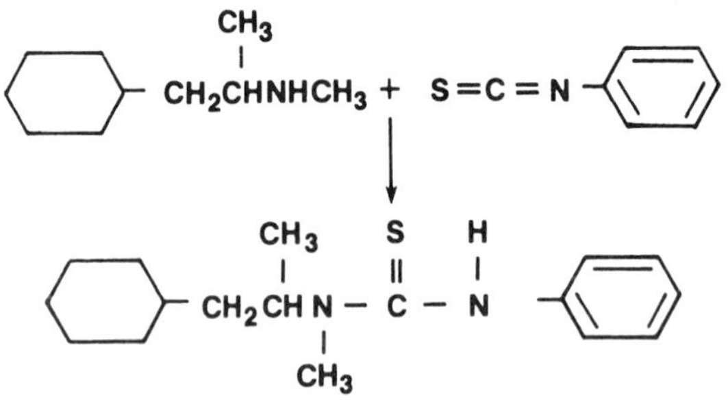

FIG. 10. Reaction of propylhexedrine and phenylisothiocyanate.
(From Ref. 59.)

sensitive refractive index detector would be desirable. The phenylisothiocynate derivatives were prepared by extraction of the amine
from dilute base into methylene chloride and adding approximately
10 μl of phenylisothiocynate. Prior to chromatography the methylene
chloride is evaporated to dryness and the resulting residue dissolved
in methanol. A μBondapak C_{18} column was employed with a mobile
phase consisting of methanol-water-acetic acid (45 + 54 + 1) for
the analysis of various primary and secondary amines, which is illustrated in Fig. 11. This system is excellent for the analysis of
"look-alike capsules and tablets," which are preparations containing combinations of caffeine, ephedrine, pseudoephedrine, and phenylpropanolamine which are sold on the street as amphetamines. A
mobile-phase consisting of methanol-water-acetic acid (60 + 39 + 1)
was recommended for phentermine, propylhexdrine, chlorphentermine,
and clortermine. Although the technique presented appears viable
as a qualitative tool, it would be of interest to study the feasibility of the use of phenylisothiocynate derivatives for quantitative
analysis.

G. Methadone

Beasley and Ziegler presented a dilute-and-shoot method for the
determination of methadone in cherry syrup [60]. The use of reversed-phase chromatography was particularly advantageous in this
instance since it was amenable to a dilute-and-shoot analysis of
the aqueous syrup sample. The analysis by other techniques such
as gas chromatography is cumbersome since prior extractions are required. Seventy milliliters of water was transferred to a 100-ml
volumetric flask prior to delivering from a 30-ml syringe the sample
syrup formulation to the mark. Fifty-microliter aliquots of the
resulting solution were injected into a chromatographic system containing a C_{18} bonded LiChrosorb column. A linear gradient that
consisted of 100% mobile phase A to 80% mobile phase B in mobile
phase A in 20 min was utilized. Mobile phase A was a 2-liter solution which consisted of 2.5 ml of concentrated ammonium hydroxide,

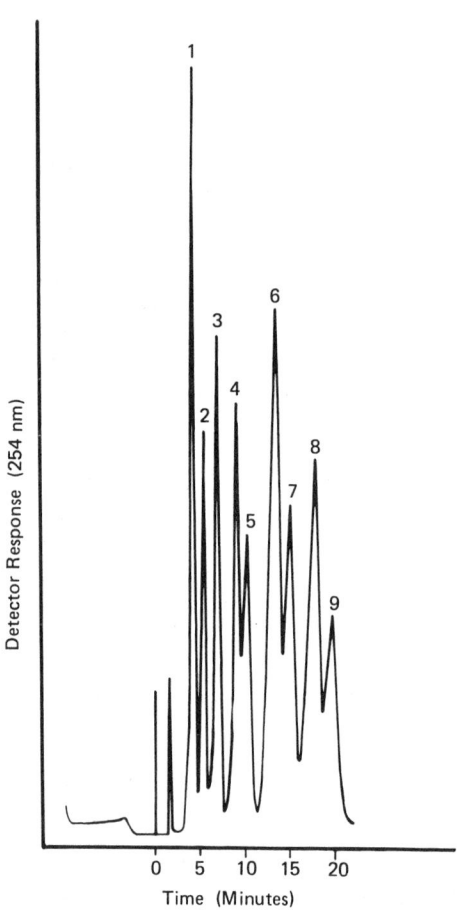

FIG. 11. Liquid chromatographic separation of phenylisothiocyanate
derivatives of some primary and secondary amines at 2.0 ml/min
methanol-water-acetic acid (45 + 54 + 1). Peaks: 1, isopropylamine;
2, piperidine; 3, ephedrine; 4, phenylpropanolamine; 5, pseudoephedrine; 6, phenethylamine; 7, phenmetrazine; 8, methamphetamine;
9, amphetamine. (From Ref. 59.)

100 ml of formic acid, and 1897.5 ml of water. Mobile phase B consisted of acetonitrile. A fixed UV detector with simultaneous
monitoring at 254 nm and 280 nm was employed. Excellent precision
and accuracy were obtained for the quantitative determination of
methadone. In the eluant the authors employ, methadone exhibits
significant tailing, probably due to this compound being ionized.

Hsieh et al. presented an ion-pairing method for the determination of methadone in time-release tablets [61]. The peak shape
as expected was better for methadone than that observed in the system employed by Beasley and Ziegler. A μBondapak C_{18} column was
utilized with a mobile phase consisting of methanol-water (75:25)
with 0.005 M pentanesulfonic acid. A variable UV detector operated
at 230 nm allowed for greater sensitivity than would be observed at
254 or 280 nm. For sample analysis, tablets were pulverized in a
mortar and dispensed in a volumetric flask which contained mobile
phase spiked with 0.2 μg/ml anthrancene which was used as an internal standard. Better resolution could have been obtained between
the peaks due to tablet excipients and the methadone peak by using
a more lipophilic counterion. As shown later in this chapter, increasing the carbon chain of the ion-pairing reagent selectively
increases retention of a basic compound such as methadone.

H. Diethylpropion

Walters and Walters developed methodology for the determination of
diethylpropion in tablets in the presence of the decomposition impurity 1-phenyl-1,2-propanedione [62]. An ODS SIL-X-1 column was
employed with a mobile phase consisting of 70% acetonitrile, 10%
methanol, and 20% aqueous 0.1% ammonium carbonate at a flow rate
of 1.0 ml/min. A variable UV detector at 255 nm was utilized. For
the determination of tablets a portion of ground tablet material
equivalent to 25 mg of diethylpropion hydrochloride was added to
a 50-ml volumetric flask and 25 ml of methanol was added. After
placing the flask on a sonic bath for 10 min the solution was diluted to volume with methanol. The solution was filtered and 2.0

ml of a procaine hydrochloride internal standard solution (2 mg/ml
in methanol) was added. Five-microliter injections of both standard
and sample were utilized. Excellent precision and good agreement
with an ultraviolet spectrographic procedure were obtained.

I. Propoxyphene

A method was developed by Souter for the quantitative analysis of
the α- and β-carbinol diastereoisomers of propoxyphene hydrochloride
[63]. The mobile phase consisted of a 30% diethylamine solution
(0.1% diethylamine in n-butylchloride) in n-hexane delivered at a
flow rate of 1.0 ml/min. A Micropak-NH_2 column was employed with a
20-μl injection loop for the determination of an n-hexane extract
of a sample at a UV detection of 254 nm.

J. Meprobamate

Lurie reported on a procedure for the quantitative determination of
meprobamate in tablets [64]. A μBondapak C_{18} column was utilized
with a mobile phase consisting of 40% methanol, 59% water, and 1%
acetic acid adjusted to pH 3.5 with 1 N sodium hydroxide. Meprobamate breaks down at the usual temperatures used in the injection
port in gas chromatographic analysis. Since this compound lacks a
UV chromophor, the usual method of analysis is nuclear magnetic
resonance (NMR) or infrared (IR), which is time-consuming for
multiple analysis. Excellent precision and accuracy were obtained
using a refractive index detector with 20-μl injections of a
methanolic extract of tablet material.

K. Psilocybin and Psilocin

A method has been published for the analysis of psilocybin and
psilocin in mushroom extracts employing reversed-phase high-performance liquid chromatography [65]. A μBondapak C_{18} column was
employed with a mobile phase consisting of 40:60:15 (v/v/w)
methanol-buffer-cetrimonium bromide solution. The aqueous portion
consisted of a phosphate buffer at pH 7 to 7.5 consisting of

approximately 0.25% disodium phosphate (Na_2HPO_4 w/v) and 0.15% mono-
basic sodium phosphate ($NaH_2PO_4 \cdot 2H_2O$ w/v). A fixed UV detector at
280 nm was utilized. The positively charged cetrimonium ion, which
is a long-chain aliphatic amine, is available to ion pair with the
psilocin anionic species which can be formed because of the presence
of a phenolic moiety on this molecule. The separation that was obtained for psilocin in a mushroom extract was less than optimum because this molecule was subject to interferences from mushroom
coextractives. One possible approach to improve the separation
would be to increase the concentration of the cetrimonium bromide
reagent which could selectively increase the retention of the psilocin peak to a region of the chromatogram free of interferences. For
sample preparation about 0.3 g of mushroom material was dried in an
oven for 16 h at 40° C and then with sand ground to a powder and
roller mixed with methanol for 24 h. After filtering the mixture,
the residue was washed with methanol. The resulting eluant was reduced in volume and made up to a total volume of 10 ml of methanol.
After the extract was filtered through a 0.45-μm filter, 10- to
20-μl aliquots were injected onto the liquid chromatograph. For
the analysis of the mushroom material preserved with sugar, approximately 3 g was roller mixed with methanol. The remaining extraction procedure was identical to the one employed for nonpreserved
mushroom material.

L. Cocaine

Sugden et al. studied the effects of the variation of pH, methanol
concentration, ionic strength, and types of ions present in the
mobile phase on the retention of amylocaine, benzocaine, butacaine,
and cocaine using reversed-phase chromatography [66]. An ODS silica
column derived from LiChrosorb Si 100 silica was employed. In general, mobile phases consisting of 70% methanol-water with various
salts added were employed. In the absence of any salt present in
the mobile phase, the local anesthetics described above (except
for benzocaine) are ionized at the mobile-phase pH and thus interact

with residual silanol groups by mechanisms such as adsorption and
possibly ion exchange. On adding salt to the mobile phase, the retention drastically decreases for amylocaine, butacaine, and cocaine
since the residual silanol sites are now shielded. In addition, it
was found that salt added to the mobile phase results in ion-pairing
effects. At high pH values the retention of amylocaine, butacaine,
and cocaine increases due to the formation of the un-ionized species
which has a high affinity for the nonpolar ODS stationary phase.
As the hydrogen ion concentration is increased, retention decreases
since ion-pair formation becomes more favorable and thus less interaction of the ionized drug species with the residual silanol sites.
Larger reduction in retention with decrease of pH was found with
nitrate ions versus formate ions since the former anion can form an
ion pair more readily. As expected, increasing the concentration
of methanol in the eluant decreased retention for the local anesthetics studied.

   Allen et al. presented HPLC data for the separation of the
cocaine diastereoisomers cocaine, pseudococaine, allococaine, and
pseudoallocaine [67]. A Partisil 10 PAC column was employed with a
mobile phase consisting of acetonitrile.

M. Barbiturates

Although gas chromatography has been successfully employed for the
analysis of barbiturates drawbacks exist. Certain barbiturates,
such as phenobarbital, can exhibit excessive tailing. In addition,
sodium salts of barbiturates cannot be assayed directly but require
prior extractions or acidification of the sample. HPLC analysis
of barbiturates suffers from none of the drawbacks noted above.

   Clark and Gelsomino presented chromatographic data for barbiturates on a μBondapak C_{18} column and a mobile phase consisting of
methanol, water (50:50) with the water 0.05 M in ammonium phosphate
[68]. The barbiturates included amobarbital, aprobarbital, barbital,
butabarbital, hexobarbital, mephobarbital, pentobarbital, phenobarbital, and secobarbital. A chromatogram of a standard mixture of

these compounds is presented in Fig. 12. At the mobile-phase pH,
sodium salts of barbiturates would be converted to the free acid,
which exhibits good chromatographic properties.

Clark and Chan presented an interesting method for increasing
the detectability of barbiturates which utilized postcolumn ionization [69]. To obtain good chromatographic performance for the

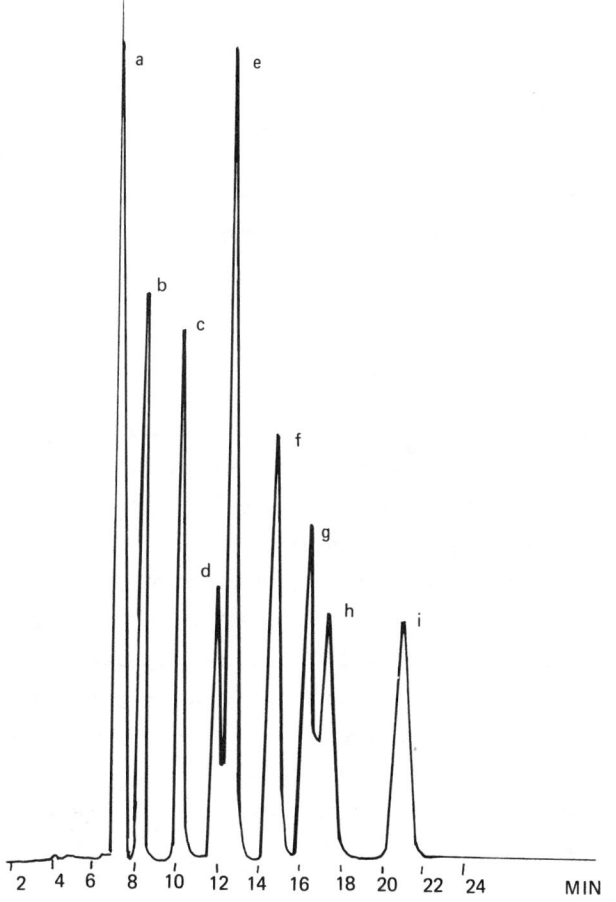

FIG. 12. Chromatogram of a standard mixture of barbiturates:
a, barbital; b, phenobarbital; c, aprobarbital; d, butabarbital;
e, mephobarbital; f, hexobarbital; g, amobarbital, h, pentobarbital; i, secobarbital.

analysis of barbiturates, it is necessary to analyze these compounds
as the free acid. However, at the commonly employed UV wavelength
of 254 nm, these compounds in the free acid form are weak UV adsorbers. Barbiturates that exist in the ionized form at pH 10 are
reasonably strong UV adsorbers. Thus by pumping in a pH 10 borate
buffer at a flow rate of 0.2 ml/min into the eluant stream after
the column by means of a union tee, a 20-fold increase in peak area
over the same concentration of un-ionized barbiturate was realized.
A μBondapak C_{18} column with a mobile phase consisting of 65% water
(pH 7) and 35% methanol was employed.

IV. GENERALIZED ANALYSIS SCHEMES

Now that we have examined applications of bonded-phase chromatography
to specific types of drugs, let us switch gears and look at generalized schemes for the chromatographic separation of a wide
variety of drugs of forensic interest.

A. Evaluation of μBondapak C_{18} Column

In an interesting evaluation of the utility of the μBondapak C_{18}
column for the analysis of drugs, Twitchett and Moffat found that
most basic drugs exhibited poor column efficiency on this stationary phase [70]. Acidic and neutral drugs displayed satisfactory
column performance. Thirty compounds were chromatographed using a
μBondapak C_{18} column with eluant solutions (pH 3.0, 5.0, 7.0, and
9.0) made from 0.025 M NaH_2PO_4 and/or 0.025 M Na_2HPO_4 solutions and
varying amounts of methanol (0-60%). These solutions were adjusted
to their final pH by the addition of sodium hydroxide or phosphoric
acid solutions. For the most part, detection was carried out with
a variable UV detector at 220 nm. Partition was shown to depend
on three physicochemical properties:

1. The relative proportion of the drug present in the un-ionized
    form, which is a function of the pH of the eluant and the pK_a
    value of the drug

2. The relative lipid solubility of the un-ionized species of the compound in both the stationary and mobile phases

3. The organic content of the eluant

For example, the retention of a strong acid such as salycilic acid (pK_a = 3) was significantly decreased when the pH rose above 3, while the retention of a weak acid [e.g., phenobarbital (pK_a = 7.4)] was constant until the pH exceeded the pK_a of the drug on which the retention decreased. The ionized form of the acid had less affinity for the relatively nonpolar stationary phase. The lipid solubility can be quantified in terms of the partition coefficient between water and an organic solvent such as n-octanol. The retention volumes of acidic drugs when they were present in the un-ionized state showed good correlation with the n-ocatanol partition coeff- icients. Increasing the organic content of the mobile phase de- creased the retention of the acidic compounds. Good column effici- encies and satisfactory peak shapes were obtained for the acidic compounds over a wide pH range.

Neutral drugs showed identical characteristics as to lipid solubility and effect of methanol content of eluant as acidic drugs.

In contrast to the acidic drugs, the retention volumes of the basic drugs studied, such as methamphetamine (pK_a = 10.1), in- creased as the pH increased, as illustrated in Fig. 13. Although not discussed in this chapter, based solely on pK_a considerations the rapid rise in retention when the pH was increased from 5 to 7 would be unexpected since this drug would be predominantly ionized at pH 7. However, this effect could be explained by an ion-exchange and adsorption mechanism between the cationic form of the basic drug and residual unbonded silanol sites on the bonded-phase column. This effect was consistent with the isoelectric point of silica being ap- proximately 5.5. When present in the un-ionized form, the retention of basic drugs, as is the case of acidic and neutral drugs, was cor- related with the n-octanol partition coefficient. In addition, in- creasing the amount of methanol in the mobile phase in general de- creased retention for basic drugs. In marked contrast to acidic and

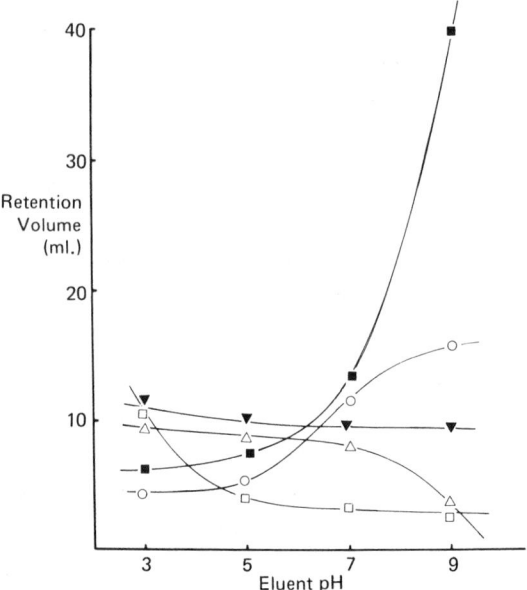

FIG. 13. Effect of variation of the eluant pH upon elution volumes
(methanol concentration 40%). □-□, Salicylic acid; Δ-Δ, phenobarbitone; ▼-▼, phenacetin; O-O, nicotine; ■-■, methylamphetamine.
(From Ref. 70.)

neutral drugs, basic compounds with high pK_a values (values of approximately 10) such as amphetamine and methamphetamine exhibited
poor efficiencies and peak shapes in the mobile phases studied, as
illustrated in Table 9. This poor column performance probably occurred because basic drugs such as amphetamine and methamphetamine
were ionized at the mobile pHs studied. Mixed mechanisms for the
retention of the cationic form of the drug could have occurred involving ion pairing with phosphate ions in the eluant, ion exchange
with the ionized unbonded silanol sites and dipole or hydrogen-
bonding interactions with silanol groups. In addition, a partition
mechanism could have involved any free base present. It is not
apparent why acidic drugs that were fully ionized had relatively
high efficiencies.

In light of the study by Twitchett and Moffet, it would appear
that the use of bonded-phase columns such as Bondapak C_{18}/Corasil
and Bondapak phenyl/Corasil with an eluant consisting of water/

TABLE 9. Column Efficiency for Some Drugs on a Microparticulate
Octadecysilane Column

Drug	pH 9; 60% Methanol N	pH 9; 60% Methanol k'	pH 3; 60% Methanol N	pH 3; 60% Methanol k'
Acetylsalicylic acid	1900	0.08	1300	1.3
Phenobarbitone	1300	0.25	1200	0.8
Nitrazepam	1550	1.50	1450	1.6
Sulphanilamide	2150	0.3	1250	0.3
Amphetamine	450	3.75	500	0.7
Methylamphetamine	350	6.50	300	0.75
Nicotine	1400	1.7	250	0.75
Amitriptyline	--	--	60	8.5
Tubocuraine	60	15.1	70	1.0

Note: Column efficiency, $N = 5.54 \, (R/W_{1/2})^2$, where R = retention
time (min) and $W_{1/2}$ = peak width at half height (min). Column capacity factor, $k' = [(V_r - V_0)/V_0]$, where V_r = elution volume of
solute and V_0 = column void volume.

Source: Ref. 70.

acetonitrile/0.1% by weight ammonium carbonate (pH approximately 8)
for the analysis of strongly basic drugs as suggested by Trinler et
al. [71, 72] would not be recommended. The chromatograms depicted
in these reports displayed poor peak shapes and efficiencies for
strongly basic drugs such as amphetamine, methamphetamine, ephedrine,
phencyclidine (PCP), cocaine, and methadone. Low efficiencies for
these compounds were due in part to the use of the pellicular column
packings.

B. Use of Dual-Wavelength Detection

Baket et al. presented a reversed-phase system consisting of a
µBondapak C_{18} column and a mobile phase containing 0.025 M NaH_2PO_4
in methanol-water (2:3) adjusted to pH 7.0 using 5% aqueous sodium
hydroxide for the analysis of 78 drugs of forensic interest [73].
For reasons discussed earlier, this system is not optimum for basic
drugs in general. In spite of the limitations of the methodology
above, this paper presented a very important concept for forensic
analysis: the use of dual-wavelength UV detection. A linear relationship between absorbance and concentration is predicted by

Beer's law. If a chromatographic peak is measured at more than one
wavelength, its concentration would not change, and thus the ratio
of the absorbances at any two wavelengths would be proportional to
the ratio of the extinction coefficients [74]. Since these coefficients are intrinsic characteristics of a compound, an absorbance
ratio coupled with a retention time could provide high specificity
in identifying a compound. Peak height or peak area, which are both
proportional to absorbance values, could be used to generate absorbance ratios. In the chromatographic system described by Baker et
al., peak height was utilized to determine 254 nm/280 nm absorbance
ratios in conjunction with relative retention times in order to
identify various drugs. In their study the specificity of the described technique was compared with other chromatographic methods of
analysis. Using relative retention times in a HPLC system, 9% of
all compounds could be uniquely identified versus 41% utilizing GC.
The use of absorbance ratioing coupled with relative retention time
significantly increased the percentage of compounds uniquely identified to 95%. As Table 10 indicates, the latter technique compares
favorably to other multiple chromatographic methods, such as GC
retention time plus two GC detectors in series, HPLC retention time
plus GC retention time, and retention times in two HPLC systems.

C. Use of Retention Indexes

As described in Sec. II, Baker et al. presented retention indexes
for the chromatographic analysis of drugs of forensic interest
which were found to be independent of the chromatographic conditions
employed for reversed-phase chromatography [12]. The retention index was defined by the equation

$$I = \frac{\log k'_D - \log k'_N}{\log k'_{N+1} - \log k'_N} + 100N \qquad (9)$$

where k'_D was the observed capacity factor of a drug, k'_N was the
capacity factor of a 2-keto alkane standard eluting just before the
test compound and k'_{N+1} was the capacity factor of the next-higher

TABLE 10. Identification of Drugs Using Multiple Parameters

First Parameter	Second Parameter	Number of Compounds	Identifications Using First Parameter (%)	Identifications Using Both Parameters (%)
Retention time, system A	A_{254}/A_{280}	78	9	95
GLC retention time	Response index	71	41	85
Retention time, system A	GLC retention time	51	12	100
Retention time, system A	Retention time system B	35	23	83

Source: Ref. 73.

homolog of a 2-ketoalkane standard. N refers to the number of carbon atoms in the 2-keto standard eluting before the drug of interest.
As an alternative to the use of Eq. (9) for the calculation of these
indexes, semilog plots of the capacity factor and the retention indexes of ketone standard could be utilized. Upon examining a
limited number of drugs utilizing µBondapak C_{18} and µBondapak CN
columns with water, methanol, or water, acetonitrile eluants containing 0.025 M NaH_2PO_4 at pH 7, the retention indexes were found
to be fairly independent of the solvent composition, solvent type,
or column type. Thus the work of Baker and Ma attempted to solve
one of the major problems in the use of HPLC, which was the comparison of retention data from different literature sources.

Since the lipophilicity of drugs and other compounds can be
easily estimated in a linear manner, Baker combined this concept
with the use of retention indexes for the prediction of HPLC retention characteristics based on a compound's chemical structure [75].
Baker derived the following relationship between the retention
index and the partition coefficient:

$$I = \alpha \log P + \beta \qquad (10)$$

where α and β are constant and P is the partition coefficient, which
depends on the chemical structure of a compound. In the case of
2-ketoalkane standards, Eq. (10) becomes

$$I_N = 200 \log P + 343 \qquad (11)$$

Although log P values could have been estimated, it was more convenient to have used relative partition coefficients derived through
Hansch substituent constants (π values) defined as follows:

$$\pi_x = \log P_x - \log P_{ref} \qquad (12)$$

where P_x abd P_{ref} are the octanol-water partition coefficients of
the compound of interest and reference compound, respectively.
Thus the retention index of a compound x was now defined as

$$I_x = 200\pi_x + I_{ref} \qquad (13)$$

Baker calculated retention indexes for barbiturates using barbital as the reference compound. The calculated values of the retention index of barbiturates were in good agreement with the experimentally observed values even though the retention times of these compounds varied by over two orders of magnitude. For the study above, a mobile phase consisting of 6.6 g K_2HPO_4, 8.4 g KH_2PO_4, 1.6 liters of methanol, and 2.4 liters of water. The pH of the mobile phase was 7.0 before the addition of methanol. A flow rate of 2.0 ml/min was employed with dual-wavelength UV detection at 254 and 280 nm.

Baker et al. applied Eq. (13) for the estimation of retention indexes of narcotic analgesics and related compounds [13]. Mobile phase A was the same eluant used in the previous paper for barbiturates, while mobile phase B consisted of 3.3 g K_2HPO_4, 4.2 g KH_2PO_4, 2.8 liters of methanol, and 1.2 liters of water. A µBondapak C_{18} column was utilized with dual-wavelength ultraviolet detection at 254 and 280 nm. The agreement between the observed indexes and the predicted values for the narcotics and related compounds, although satisfactory, was significantly less than that observed for the barbiturates. The median error for the narcotic compounds was 100 units compared to 29 units for the barbiturates. Many of the discrepancies were possibly related to stereochemical factors such as shielding of polar substituent groups in the oxymorphone type of drugs.

D. Multiparameter Chromatographic Analysis

Wheals presented a novel technique for multiparameter chromatographic analysis of basic drugs which employed the use of a single isocratic mobile phase with three different columns [4]. The eluant consisted of methanol-2 M ammonium hydroxide-1 M ammonium nitrate (27:2:1), pH approximately 10.3. The columns utilized were silica (particle size 3-7 µm), a mercaptopropyl-modified silica, and a n-propylsulfonic acid-modified silica. As described in Chap. 3, a system employing a silica column with the mobile phase described above was successfully employed by Jane [76] for the analysis of basic drugs.

As explained earlier in this chapter, the use of an eluant with a
high methanol content and ammonia as the source of hydroxide ions
does not appear to preclude the use of silica-based columns at high-
mobile-phase pHs. This multicolumn approach is advantageous of the
use of a single column and different eluants or different columns
and different eluants since most HPLC operational problems are associated with changing eluants. Of the 162 drugs studied as indicated in Table 11, 93 had unique retention characteristics. The
ambiguity was found mainly in the area of low retention, $k' \leq 0.2$.
Thus it appears that the use of relative retention times and absorbance ratioing would be more specific for the screening of drugs of
forensic interest. The report by Wheals contained interesting discussions on the effects of various substituent groups on retention
for the various classes of drugs examined. At the eluant pH of 10.3,
most basic drugs were only partially ionized and thus Wheals postu-
lated four possible mechanisms occurring in the chromatographic
systems employed.

1. Ion exchange
2. Liquid-liquid partitioning of the free base
3. Liquid-liquid partitioning of the ion pairs between the ionized
    solute and the nitrate eluant counterion
4. Ill-defined mechanisms such as dipole interactions, hydrogen
    bonding, van der Waals forces, and so on

For the silica column the dominant separation mechanisms were believed to be ion exchange and interactions of type 4. The ion-
exchange mechanism occurred as a result of the silanol groups being
completely ionized at pH 10.3. For the mercaptopropyl-modified
silica, the dominant mechanism was probably a partition process involving the lipophilic propyl-bonded group and either the free base
or an ion pair. Finally, the dominant mechanism for the n-propyl
sulfonic acid-modified silica was ion exchange since the bonded
sulfonic acid moiety was a strong cation exchanger. Thus the three
columns employed because of the different separation mechanisms involved were well suited for a multichromatographic approach. The

TABLE 11. Retention Volume Data of Basic Drugs

Compound	pK_a	Class	k' on Si	k' on SH	k' on SCX
Buclizine	--	1	0	0	0
Meclizine	--	1	0	0	0
Benzphetamine	6.6	2	0	0	0
Chlormezanone	--	3	0	0	0
Isocarboxazid	10.4	3	0	0	0
Benzocaine	2.5	4	0	0	0.1
Phenadoxone	6.9	5	0	0.1	0
Amfepramone	--	2	0	0.1	0
Paragyline	--	3	0	0.1	0
Haloperidol	8.3	3	0	0.1	0
Pimozide	7.3	3	0	0.1	0
Trifluperidol	--	3	0	0.1	0
Anileridine	--	5	0	0.2	0
Phenazocine	--	5	0	0.2	0
Diphenoxylate	--	5	0	0.2	0
Phenbutrazate	--	2	0	0.2	0
Oxypertine	--	3	0	0.2	0
Papaverine	8.1	5	0	0.3	0
Pipamazine	8.6	6	0	0.3	0.1
Thiopropazate	7.3	6	0	0.4	0
Hydroxybuclizine	--	1	0.1	0.1	0
Nialamide	--	3	0.1	0.1	0
Phenoxypropazine	--	3	0.1	0.1	0
Nalorphine	7.8	7	0.1	0.1	0.1
Tranylcypromine	8.2	3	0.1	0.1	0.2
Caffeine	--	2	0.1	0.2	0
Trifluomeprazine	--	6	0.1	0.2	0
Flupentixol	--	8	0.1	0.2	0
Benoxinate	--	4	0.1	0.2	0.1
Fluphenazine	8.1	6	0.1	0.2	0.1
Isoproterenol	--	2	0.1	0.2	0.8
Noscapine	6.2	5	0.1	0.3	0
Pericyazine	--	6	0.1	0.3	0.1
Trimethoprim	7.2	9	0.1	0.3	0.2
Methylphenidate	--	2	0.1	0.3	0.2
Phendimetrazine	7.6	2	0.1	0.3	0.2
Benzoctamine	7.6	8	0.1	0.3	0.5
Acetophenazine	--	6	0.1	0.4	0.1
Carphenazine	--	6	0.1	0.4	0.1
Metopimazine	--	6	0.1	0.4	0.1
Perphenazine	7.8	6	0.1	0.4	0.1
Clopenthixol	--	8	0.1	0.4	0.1
Opipramol	--	8	0.1	0.5	0.1
Fonazine	--	6	0.1	0.5	0.1
Propiomazine	--	6	0.1	0.5	0.2
Emetine	7,4,8.3	--	0.2	0.1	0.2
Scopolamine	7.6	--	0.2	0.1	0.2
Tetracaine	8.5	4	0.2	0.2	0.2

TABLE 11 (cont.)

Compound	pK$_a$	Class	k' on		
			Si	SH	SCX
Cocaine	8.6	4	0.2	0.2	0.2
Procaine	9.0	4	0.2	0.2	0.2
Cyclizine	8.2	1	0.2	0.3	0.2
Phenmetrazine	8.4	2	0.2	0.3	0.2
Dibenzepin	--	8	0.2	0.4	0
Trimipramine	--	8	0.2	0.4	0.2
Tripolidine	--	1	0.6	0.7	0.4
Amphetamine	9.9	2	0.6	0.7	1.5
Amopyroquine	--	9	0.6	0.8	0.1
Thebaine	8.2	7	0.6	0.8	0.4
Methoxypromazine	--	6	0.6	0.8	0.4
Imipramine	9.5	8	0.6	0.8	0.4
Aminopromazine	--	6	0.6	0.9	0.8
Methadone	8.3	5	0.6	1.0	0.8
Narceine	9.3	5	0.7	0.5	0.4
Amydricaine	--	4	0.7	0.6	0.9
Hydroxyamphetamine	9.3	2	0.7	0.6	1.8
Codeine	8.2	7	0.7	0.8	0.5
Promazine	9.4	6	0.7	0.9	0.5
Prothipendyl	--	1,8	0.7	0.9	0.5
Tofenacin	--	3	0.7	0.9	1.0
Diphenylpryraline	--	1	0.8	0.8	0.4
Benzylmorphine	--	7	0.8	0.8	0.3
Procyclidine	--	1	0.8	0.8	0.8
Nicomorphine	--	7	0.8	1.3	0.3
Dimethoxanate	--	6	0.9	0.9	0.5
Morphine	9.9	7	1.0	0.5	0.6
Pheniramine	9.3	1	1.0	0.7	0.5
Chlorpheniramine	9.2	1	1.0	0.8	0.5
Pipazethate	--	8	1.0	1.2	0.8
Ethoheptazine	--	5	1.0	1.4	0.8
Nortriptyline	10.0	8	1.0	1.5	1.6
Phenylephrine	10.1	2	1.1	0.7	2.9
Amodiaquin	--	9	0.2	0.4	0.1
Chlorcyclizine	8.1	1	0.2	0.4	0.1
Piperacetazine	--	6	0.2	0.4	0.2
Isomethadone	--	5	0.2	0.4	0.2
Ethopropazine	--	6	0.2	0.4	0.3
Pentazocine	8.5	5	0.2	0.4	0.4
Noxiptiline	--	8	0.2	0.5	0.1
Norpipanone	--	5	0.2	0.5	0.2
Methotrimeprazine	9.2	6	0.2	0.5	0.2
Trimeprazine	--	6	0.2	0.5	0.2
Butriptyline	--	8	0.2	0.5	0.2
Chlorprothixene	--	8	0.2	0.5	0.2
Diethazine	9.1	6	0.2	0.5	0.3
Lobeline	--	--	0.2	0.5	0.5

TABLE 11 (cont.)

Compound	pK_a	Class	k' on Si	k' on SH	k' on SCX
Dipipanone	8.5	5	0.2	0.8	0.6
Meperidine	8.7	5	0.3	0.3	0.2
Metaproterenol	--	2	0.3	0.3	0.8
Triflupromazine	--	6	0.3	0.4	0.2
Orphenadrine	--	1	0.3	0.4	0.2
Thonzylamine	8.9	1	0.3	0.4	0.2
Chlorothen	8.4	1	0.3	0.4	0.2
Normethadone	--	5	0.3	0.5	0.2
Tripelennamine	9.0	1	0.3	0.5	0.3
Trifluoperazine	8.1	6	0.3	0.6	0.1
Cyproheptadine	--	1,8	0.3	0.6	0.2
Promethazine	9.1	1,6	0.3	0.6	0.2
Isothipendyl	--	8	0.3	0.7	0.2
Clomipramine	--	8	0.3	0.7	0.3
Thiethylperazine	--	6	0.3	0.9	0.2
Butaperazine	--	6	0.3	0.9	0.3
Viloxazine	--	3	0.4	0.4	0.2
Diphenhydramine	9.0	1	0.4	0.4	0.3
Alphaprodine	8.7	5	0.4	0.4	0.2
Norpseudophedrine	--	2	0.4	0.4	1.1
Mepyramine	8.9	1	0.4	0.5	0.3
Diacetylmorphine	7.6	7	0.4	0.5	0.2
Piperocaine	--	4	0.4	0.5	0.5
Propranolol	9.5	--	0.4	0.5	0.9
Mepazine	9.7	6	0.4	0.6	0.3
Amitriptyline	9.4	8	0.4	0.6	0.2
Quinine	8.5	9	0.4	0.7	0.3
Chlorpromazine	9.3	6	0.4	0.7	0.3
Dothiepin	--	8	0.4	0.7	0.3
Pipradrol	--	2	0.4	0.7	0.9
Prochlorperazine	8.1	6	0.4	0.8	0.2
Thiothixene	--	8	0.4	0.9	0.1
Thioproperazine	--	6	0.4	0.9	0.2
Fenfluramine	9.1	2	0.5	0.5	0.8
Chlorphentermine	9.6	2	0.5	0.6	1.2
Phentermine	10.1	2	0.5	0.6	1.4
Iprindole	--	8	0.5	0.7	0.3
Perazine	--	6	0.5	0.8	0.2
Thioridazine	9.5	6	0.5	1.1	0.5
Monoacetyl morphine	--	7	0.6	0.5	0.3
Methyl ephedrine	9.3	2	1.1	0.9	1.0
Pentaquine	--	9	1.1	2.0	2.2
Pholcodine	9.3	7	1.1	1.1	0.6
Ephedrine	9.6	2	1.2	1.0	1.9
Mesoridazine	--	6	1.2	1.4	0.7
Antazoline	10.0	1	1.3	1.6	2.6
Primaquine	--	9	1.3	1.9	2.7

TABLE 11 (cont.)

Compound	pK_a	Class	k' on Si	k' on SH	k' on SCX
Methyl amphetamine	10.1	2	1.4	1.2	1.9
Methdilazine	--	6	1.4	1.8	1.6
Quinacrine	10.3	9	1.4	1.8	1.6
Desipramine	10.2	8	1.5	1.8	2.1
Mephentermine	10.4	2	1.6	1.4	2.3
Protiptyline	--	8	1.6	2.2	2.6
Dihydrocodine	8.8	7	1.8	1.1	0.9
Hydromorphone	8.2	7	1.8	1.0	0.9
Strychnine	8.0	--	1.9	2.1	0.8
Maprotiline	--	8	1.9	2.4	2.8
Dihydromorphine	8.6	7	2.0	1.1	1.1
Levorphanol	8.2	7	2.0	1.4	1.3
Levomethorphan	--	7	2.0	1.9	1.3
Racemorphan	--	7	2.1	1.6	1.4
Norcodeine	--	7	2.4	1.7	2.5
Normorphine	--	7	2.7	1.6	3.2
Chloroquine	10.8	9	2.7	2.2	2.5
Norlevorphanol	--	7	3.0	2.7	4.6
Atropine	9.9	--	3.4	1.8	1.9

Note: The eluant used to achieve the retention shown below was methanol-2 M ammonium hydroxide-1 M ammonium nitrate (27:2:1) with 50 mg of sodium sulfite added to each liter of solvent. The class of compound is designated as follows: 1 - antihistamines and anti-nauseants, 2 = stimulants and anorexics, 3 = antidepressants and tranquilisers (not tricyclic compounds), 4 = local anaesthetics, 5 = narcotic analgesics (not morphinelike in structure), 6 = phenothiazines, 7 = narcotic analgesics (morphinelike compounds), 8 = tricyclic antidepressants and other related compounds, 9 = antimalarial agents. The pK_a values shown in the table were taken from *The Extra Pharmacopoeia*, Martindale, 27th ed., The Pharmaceutical Press, London.

Source: Ref. 4.

use of columns mounted in parallel as illustrated in Fig. 14 can preclude the necessity of changing columns. In this system solvent originating from a single pump was split to pass through each column and then recombined before entering the detector. Disadvantages for such a setup included doubling the solvent usage and sample dilution after the separation results in half the detection sensitivity.

PUMP

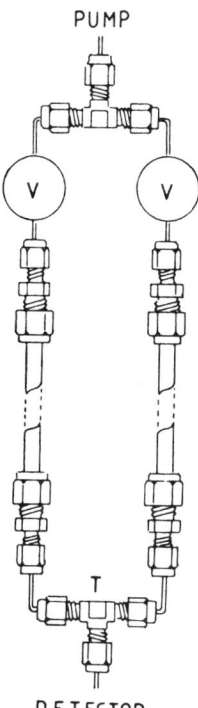

DETECTOR

FIG. 14. Diagram of an isocratic dual-column assembly. V, injection valve; T, stainless steel 1/16 in.-union tee drilled out for zero dead volume and joined to each column with a length of 1/16 in.-OD, 0.010 in.-ID stainless steel tubing. The tubes are virtually butted together in the tee. The outlet line to the detector was PTFE 1/16 in.-OD, 0.006 in.-ID; this was also butted onto the two steel tubes, leaving minimal space between. (From Ref. 4.)

E. Reversed-Phase Ion-Pairing Chromatography

As described earlier in this chapter, reversed-phased ion-pairing chromatography would be an excellent technique for the analysis of drugs of forensic interest since acidic, neutral, and basic compounds would exhibit good chromatographic properties. An ideal situation would be a single chromatographic system that could be used for a wide range of drugs. A system was developed by Lurie which approached this goal, which consisted of a μBondapak C_{18}

column and a mobile phase consisting of 40% methanol, 59% water, 1%
acetic acid, and 0.005 M heptanesulfonic acid at a pH of 3.5 [49].
Under these conditions acidic drugs such as barbiturates were
analyzed by ion suppression, while moderate to strong bases were
assayed by ion pairing. This system gave good separations for barbiturates, ergot alkaloids, phenethylamines, opium alkaloids, and
local anesthetics. Chromatograms for standard mixtures of local
anesthetics, opium alkaloids, and barbiturates are presented in
Figs. 15-17. However, certain drawbacks existed in the use of this
system. The phenethylamines, amphetamine, and methamphetamine were
poorly resolved. Two opium alkaloids heroin and acetylcodeine coeluted. Although good resolution was obtained for compounds related
to cocaine, their retention times were longer than optimum. The
same could be said for the analysis of LSD and PCP. Therefore, it
was desired to optimized these various separations for resolution
and speed. A study was undertaken to determine the effect of column
type, water-methanol ratio, counterion size, counterion concentration,
and basicity of drugs chromatographed on the capacity and selectivity

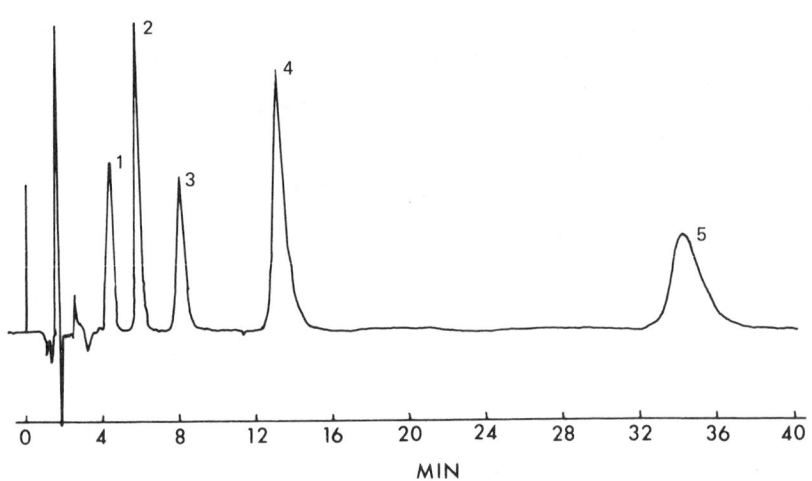

FIG. 15. Chromatogram of a standard mixture of local anesthetics:
1, procaine; 2, benzocaine; 3, lidocaine; 4, cocaine; 5, tetracaine.
(From Ref. 49.)

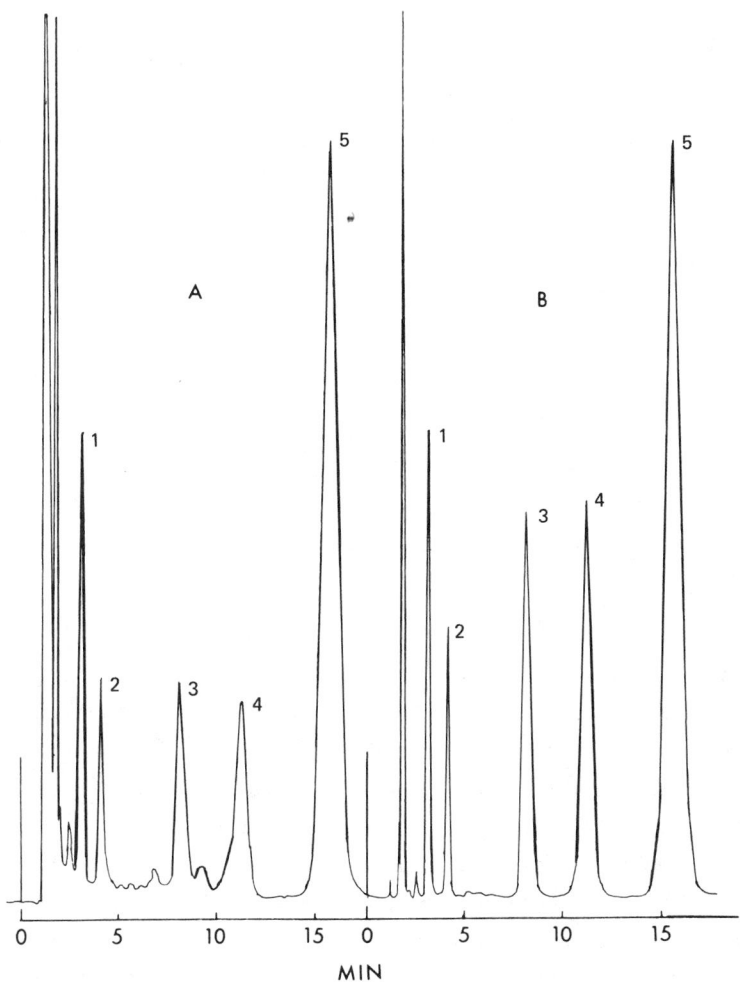

FIG. 16.   Chromatograms of A, an opium sample: 1, morphine;
2, codeine; 3, thebaine; 4, narcotine; 5, papaverine; and B, a
synthetic opium alkaloid mixture of the above.   (From Ref. 49.)

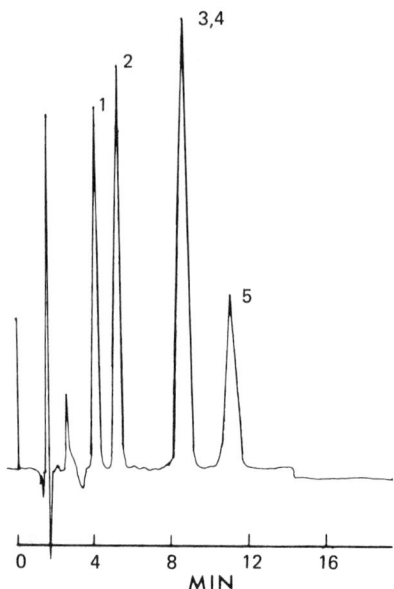

FIG. 17. Chromatogram of a standard mixture of barbiturates:
1, phenobarbital; 2, butabarbital; 3, amobarbital; 4, pentobarbital;
5, secobarbital. (From Ref. 49.)

factors for a reversed-phase ion-pairing separation [16, 50]. Resolution and speed as explained in Chap. 1 depend on the chromatographic variables k' and α. Approximately 50 drugs of forensic interest, including barbiturates, phenethylamines, opium alkaloids,
local anesthetics, and ergot alkaloids, were included in this study.
The columns employed consisted of μBondapak C_{18}, μBondapak alkylphenyl, and μBondapak cyanide. Mobile phases consisting of water,
methanol, 1% acetic acid, and 0.005 M alkylsulfonate counterion at
pH 3.5 were utilized with all three columns. Methanesulfonate,
butanesulfonate, and heptanesulfonate were used as counterions for
methanol concentrations of 40 and 40%. Only methanesulfonate was
used at 20% methanol because of excessive retention of many compounds with a more lipophilic counterion.
     A study by Lurie and Demchuk illustrated the following factors
effecting k' [16]:

1. For a given stationary phase and water-methanol ratio, the k' of barbiturates were independent of counterion.

2. For most bases at a given mobile phase, the k' increased with increasing size of counterion with both the C_{18} and alkylphenyl columns.

3. For most of the bases on the C_{18} and alkylphenyl columns and a given mobile phase, the ratio of k's for any given set of counterions were fairly constant. These ratios were independent of eluate surface area but varied with the charge of the ionized base. Quinine and quinidine, which form divalent cations, had ratios of k' values that were approximately double of other bases.

4. In general, higher variations of k' with counterions were observed on the C_{18} column than the alkylphenyl column.

5. The variation of k' with counterion size on the C_{18} and alkylphenyl columns appeared to be fairly independent of the water-methanol ratio employed.

6. Compounds which were weakly basic, such as antipyrene, benzocaine, caffeine, and methaqualone, exhibited no significant variation of k' with counterion size.

7. For the cyanide column no significant variation of k' with counterion size was observed for any of the drugs.

8. For any eluant employed, the retention order of barbiturates was C_{18} > alkylphenyl > CN.

9. When heptanesulfonate was used as a counterion, the retention of most basic drugs increased in the order C_{18} > alkylphenyl > cyanide. However, a much smaller variation of k' with stationary phase was observed when butanesulfonate or methanesulfonate were used as counterions. Retention was actually greater on the cyanide or alkylphenyl column than the C_{18} column in many instances when the latter counterions were employed.

10. In most instances increasing the ratio of water to methanol increased retention on all three columns for all the drugs studied.

11. In a study utilizing 11 drugs of the original 50, the k' of the various solutes did not vary with a change of the methanesulfonate concentration from 0.005 M to 0.02 M on the C_{18} column. On the alkylphenyl column the k' of the various solutes except for butabarbital increased with concentration of methanesulfonate.

12. On the C_{18} and alkylphenyl columns, the k' of all solutes except butabarbital increased with an increase of butanesulfonate and heptanesulfonate concentration from 0.005 to 0.02 M.

For a detailed explanation of these trends, the reader is
referred to the manuscript by Lurie and Demchuk [16]. However, a
few salient points are worth mentioning. Barbiturates which are
weak acids with pK_a values greater than 7 would exist as the free
acid at pH 3.5 and would not be expected to ion pair with anionic
counterions. The proposed mechanisms for ion-pairing chromatography
were presented earlier in this chapter. Moderate to strong bases
with pK_a values $\geq$ 4.5 would be fully ionized at pH 3.5 and would be
expected to ion pair with an anionic counterion. Weak bases with
pK_a values of less than 4.5 may not be ionized at pH 3.5 to an extent that would allow ion pairing to occur.[*] Horvath's solvophic
theory is a useful chromatographic model for explaining many of the
trends [8]. For retention in reversed-phase ion-pairing chromatography if we assume that ion pairing occurs in the mobile phase
followed by adsorption of the ion pair onto the stationary phase,
Horvath derived the following relationship based on his solvophic
theory [17]:

$$\ln K = a - b + c \Delta A \tag{14}$$

where K is the equilibrium constant for the binding of the neutral
ion pair and a, b, and c are constants depending on the solvent
and column properties. ΔA is the contact area, which is the difference between the molecular surface area of the ion-pairing
stationary-phase complex and the surface areas of the stationary-
phase ligand and the ion pair. This contact area depends on the
molecular surface area of the complex between the ion pair and the
hydrocarboneous ligand.

As a study by Lurie and Demchuk indicated, there were various
factors affecting α [50]:

1. For barbiturates that have closely related structures differing
    in aliphatic character, selectivity decreased in the order
    C_{18} > alkylphenyl > cyanide.

---

[*]pK_a values are in general determined in water. The values for
bases tend to be lower when alcohol greater than 20% is present in
the mobile phase.

2. For similar solutes which differed in aliphatic character, increasing the amount of water in the mobile phase increased selectivity.

3. Increasing the size of the alkylsulfonate counterion in mixtures of acids and bases selectively increased the retention of the moderate and strong bases on the C_{18} and alkylphenyl columns.

4. For certain pairs of bases on the C_{18} and alkyphenyl columns, the selectivity increased with decreasing size of the counterion.

5. In most instances, the selectivity of bases was invariant to changes in counterion concentration.

The paper by Lurie and Demchuk discusses in detail the trends listed above [50]. Retention data for the various drugs examined in the reports by Lurie and Demchuk are presented in Tables 12-15. These studies demonstrated that most drug exhibits could be analyzed with superior resolution and speed over Lurie's original methodology [49] using only two isocratic systems [77]. The first of these systems consisted of a μBondapak C_{18} column with a mobile phase consisting of 40% methanol, 59% water, 1% acetic acid, and 0.02 M methanesulfonic acid at pH 3.5. This mobile phase was used for the analysis of barbiturates, local anesthetics, LSD and related compounds, PCP, and methaqualone. A variation in retention time of bases with sample can occur in reversed-phase ion-pairing chromatography, especially at higher concentrations [49]. Therefore, 0.02 M methanesulfonic acid was used instead of 0.005 M methanesulfonic acid because the change in retention time of a basic drug was significantly diminished in the normal sample concentration range which was less than 1.0 mg/ml. The retention times of basic drugs such as cocaine, LSD, and PCP are significantly reduced by changing the counterion from heptanesulfonate which was used in Lurie's original methodology [49] to methanesulfonate. However, the selectivity factors for those compounds that ion paired were not significantly affected by changing the size of the counterion. Thus for local anesthetics, PCP, LSD, and related compounds, good separations were obtained in significantly less time by using methanesulfonate instead of heptanesulfonate as a counterion. The second system

TABLE 12. Capacity Factors k_i' and Selectivity Factors α_{ji} for the column µBondapak C_{18} and Mobile-Phase Methanol, Water, 1% Acetic Acid, and 0.005 M Alkylsulfonate

Drug	MSA k_i'	α_{ji}	BSA k_i'	α_{ji}	HSA k_i'	α_{ji}
			40% Methanol			
Phenobarbital	1.85		1.87		1.97	
Butabarbital	2.83	1.53	2.81	1.50	2.98	1.51
Pentobarbital	5.76	2.03	5.61	2.00	6.11	2.05
Amobarbital	5.80	1.01	5.71	1.02	6.24	1.02
Secobarbital	8.02	1.38	7.99	1.40	8.56	1.37
Benzocaine	2.85		2.74		3.18	
Procaine	0.237		0.431		2.03	
Lidocaine	0.819	3.46	1.27	2.95	5.68	2.80
Cocaine	1.77	2.17	2.66	2.09	9.29	1.64
Tetracaine	5.04	2.85	7.49	2.80	28.8	3.10
Acetominophen	0.262		0.260		0.236	
Theophylline	0.440	1.68	0.441	1.70	0.417	1.77
Caffeine	0.744	1.69	0.754	1.71	0.726	1.74
Phenylpropanolamine	0.263		0.492		2.59	
Ephedrine	0.328	1.25	0.577	1.17	2.92	1.13
Amphetamine	0.581	1.77	0.935	1.62	4.13	1.41
Methamphetamine	0.581	1.00	0.935	1.00	4.30	1.04
Phentermine	0.808	1.39	1.26	1.35	5.85	1.36
Methylphenidate	1.34	1.66	2.05	1.63	8.47	1.45
Antipyrene	1.42		1.41		1.52	
Morphine	0		0.163		1.09	
Codeine	0.225		0.419	2.57	1.99	1.83
Acetylmorphine	0.269	1.20	0.500	1.19	2.31	1.16
Aminopyrene	0.387	1.44	0.834	1.67	2.04	
Strychnine	0.732	1.89	1.19	1.43	4.73	2.32
Acetylcodeine	1.05	1.43	1.58	1.33	6.19	1.31
Heroin	1.08	1.03	1.56	1.01^a	6.11	1.01^a
Thebaine	1.07	1.01^a	1.61	1.03	6.24	1.02
Quinidine	1.02	1.05^a	1.75	1.09	15.1	2.42
Quinine	1.38	1.35	2.30	1.31	18.6	1.23
Methapyrilene	2.12	1.54	3.05	1.33	14.0	1.33^a
Narcotine	2.16	1.02	3.03	1.01^a	11.3	1.24^a
Papaverine	2.71	1.25	3.80	1.25	13.8	1.22
Mescaline	0.307		0.557		0.270	
DMT	0.501	1.63	0.799	1.43	4.30	1.59
LSD	2.53		3.67		15.3	
Lysergic acid methylpropylamide (LAMPA)	2.83	1.12	4.10	1.12	16.6	1.08
Iso-LSD	4.17	1.47	5.88	1.43	20.7	1.25
TCP	4.08	1.21	5.81	1.18	22.2	1.19
PCP	3.36		4.93		18.7	
Glutethimide	5.21		5.13		5.46	
Methaqualone	9.82		9.82		10.8	
Mecloqualone	12.1	1.23	12.3	1.25	13.0	1.20
Diazepam	21.6		20.9		23.0	
Phenmetrazine	0.581		0.820		3.64	
Phendimetrazine	0.581	1.00	0.874	1.07	3.87	1.06
MDA	0.581		0.935		4.13	
Diethylpropion	0.819		1.24		4.73	

TABLE 12 (cont.)

Drug	MSA k'_i	α_{ji}	BSA k'_i	α_{ji}	HSA k'_i	α_{ji}
			30% Methanol			
Phenobarbital	3.43		3.39	1.47	3.93	
Butabarbital	5.10	1.49	4.99	2.19	5.90	
Pentobarbital	12.2	2.39	10.9	1.05	13.4	1.50
Amobarbital	12.3	1.01	11.5	1.40	13.8	2.27
Secobarbital	17.9	1.45	16.1		19.9	1.03
Benzocaine	5.73		5.76		6.84	1.44
Procaine	0.845		1.21		4.89	
Lidocaine	1.90	2.25	2.56	2.11	13.5	2.76
Cocaine	6.09	3.20	6.91	2.70	31.0 b	2.30
Tetracaine	17.0	2.79	20.3	2.94		3.44
Acetominophen	0.443		0.424		0.492	
Theophylline	0.833	1.88	0.860	2.03	0.858	1.74
Caffeine	1.52	1.82	1.48	1.72	1.58	1.84
Phenylpropanolamine	0.492		0.796		4.50	
Ephedrine	0.679	1.38	1.09	1.37	5.52	1.23
Amphetamine	1.01	1.49	1.47	1.35	8.28	1.50
Methamphetamine	1.25	1.24	1.72	1.17	9.44	1.14
Phentermine	1.65	1.32	2.29	1.33	13.3	1.41
Methylphenidate	3.65	2.21	4.68	2.04	26.1	1.96
Antipyrene	3.05		2.79		3.07	
Morphine	0.225		0.384		2.14	
Codeine	0.753	3.35	1.04	2.71	4.87	2.28
Acetylmorphine	0.906	1.20	1.25	1.20	6.17	1.26
Aminopyrene	1.25	1.38	1.64	1.31	4.97	1.24
Strychnine	2.48	1.98	3.32	2.02	15.1	3.04
Acetylcodeine	3.40	1.37	4.45	1.34	19.9	1.32
Heroin	3.66	1.08	4.73	1.06	20.9	1.05
Thebaine	3.59	1.02[a]	4.70	1.01[a]	21.0	1.00
Quinidine	4.56	1.27	8.34	1.77	62.0	2.95
Quinine	6.41	1.41	11.5	1.38	85.1	1.38
Methapyrilene	7.38	1.15	10.6	1.08	44.9	1.90[a]
Narcotine	8.62	1.17	11.0	1.04	46.5	1.04
Papaverine	12.3	1.43	15.4	1.40	59.5	1.28
Mescaline	0.850		1.19		6.34	
DMT	1.25	1.47	1.72	1.44	8.82	1.39
LSD	9.69		12.5		62.1	
LAMPA	10.4	1.07	13.7	1.10	65.5	1.05
Iso-LSD	18.8	1.81	20.8	1.37	89.3 b	1.37
TCP	15.8	1.32	19.3	1.28		
PCP	12.0		15.1		65.6	
Glutethimide	11.3		10.8		13.4	
Methaqualone	28.1		24.8		30.2	
Mecloqualone	35.9 b	1.28	30.1 b	1.21	38.7 b	1.28
Diazepam						
Phenmetrazine	1.25		1.46		8.43	
Phendimetrazine	1.25	1.00	1.72	1.18	8.50	1.01
MDA	1.25		1.72		9.25	
Diethylpropion	2.32		2.92		12.8	

[a]Successive capacity ratios are reversed.

[b]Retention greater than 2 h.

Note: MSA, methanesulfonate; BSA, butanesulfonate; HSA, heptanesulfonate.

Source: Ref. 50.

TABLE 13   Capacity Factors k'_i and Selectivity Factors α_{ji}, for the Column μBondapak Alkylphenyl and Mobile-Phase Methanol, Water 1% Acetic Acid, and 0.005 M Alkylsulfonate.

Drug	MSA k'_i	MSA α_{ji}	BSA k'_i	BSA α_{ji}	HSA k'_i	HSA α_{ji}
			40% Methanol			
Phenobarbital	1.51		1.61		1.59	
Butabarbital	1.73	1.15	1.80	1.12	1.80	1.13
Pentobarbital	3.01	1.73	3.16	1.75	3.18	1.77
Amobarbital	2.96	1.02[a]	3.15	1.00	3.13	1.02[a]
Secobarbital	3.93	1.33	4.21	1.33	4.22	1.35
Benzocaine	2.16	—	2.37		2.27	
Procaine	0.525		0.628		1.63	
Lidocaine	0.781	1.49	1.02	1.62	2.72	1.67
Cocaine	2.04	2.61	2.72	2.67	5.97	2.19
Tetracaine	3.68	1.80	5.09	1.87	11.4	1.91
Acetominophen	0.294		0.252		0.282	
Theophylline	0.620	2.11	0.589	2.34	0.575	2.04
Caffeine	1.14	1.69	1.13	1.71	1.02	1.74
Phenylpropanolamine	0.310		0.419		1.35	
Ephedrine	0.395	1.27	0.544	1.30	1.59	1.18
Amphetamine	0.574	1.46	0.737	1.35	2.03	1.27
Methamphetamine	0.620	1.08	0.848	1.15	2.27	1.12
Phentermine	0.787	1.27	1.04	1.23	2.77	1.22
Methylphenidate	1.72	2.19	2.28	2.19	5.52	1.99
Antipyrene	1.46		1.51		1.46	
Morphine	0.163		0.242		0.857	
Codeine	0.456	2.80	0.628	2.60	1.67	1.95
Aminopyrene	0.500	1.10	0.737	1.17	1.45	1.15[a]
Acetylmorphine	0.554	1.11	0.750	1.02	1.92	1.32
Strychnine	1.67	3.01	2.28	3.04	5.31	2.77
Heroin	1.76	1.05	2.33	1.02	5.31	1.00
Acetylcodeine	1.80	1.02	2.38	1.02	5.45	1.02
Thebaine	1.85	1.03	2.52	1.06	5.61	1.03
Methapyrilene	2.43	1.31	3.71	1.47	7.49	1.34
Quinine	2.66	1.09	5.32	1.43	10.4	1.39
Quinidine	2.80	1.05	5.32	1.00	10.1	1.03[a]
Narcotine	4.08	1.46	5.42	1.02	11.4	1.13[a]
Papaverine	4.25	1.04	5.55	1.02	10.6	1.08[a]
Mescaline	0.388		0.562		1.54	
DMT	0.620	1.60	0.848	1.51	2.27	1.47
LSD	3.52		4.80		10.5	
LAMPA	3.93	1.12	5.26	1.12	11.5	1.10
Iso-LSD	4.72	1.20	6.15	1.16	12.1	1.05
TCP	4.09		5.32		11.8	
PCP	4.69	1.15	6.08	1.14	13.5	1.14
Glutethimide	4.25		4.45		4.34	
Methaqualone	9.67		9.92		9.22	
Mecloqualone	11.3	1.17	11.2	1.13	10.5	1.14
Diazepam	17.2		19.3		17.4	
Phenmetrazine	0.620		0.848		2.14	
Phendimetrazine	0.620	1.00	0.848	1.00	2.14	1.06
MDA	0.688		0.950		2.38	
Diethylpropion	0.856		1.12		2.68	

TABLE 13 (cont.)

Drug	MSA k_i'	α_{ji}	BSA k_i'	α_{ji}	HSA k_i'	α_{ji}
			30% Methanol			
Phenobarbital	2.75		2.87		3.23	
Butabarbital	3.02	1.10	3.16	1.10	3.53	1.09
Pentobarbital	5.80	1.92	6.10	1.93	6.99	1.98
Amobarbital	5.80	1.00	6.04	1.01[a]	5.80	1.20[a]
Secobarbital	7.64	1.32	8.11	1.35	9.76	1.40
Benzocaine	4.28		4.52		5.02	
Procaine	1.00		1.38		4.19	
Lidocaine	1.26	1.26	1.72	1.25	6.00	1.43
Cocaine	4.82	5.02	6.32	3.66	19.2	3.2
Tetracaine	9.62	2.03	12.8	2.03	43.2	2.25
Acetominophen	0.392		0.468		0.478	
Theophylline	1.00	2.55	1.07	2.32	1.11	2.32
Caffeine	1.95	1.82	2.10	1.72	2.21	1.84
Phenylpropanolamine	0.398		0.603		2.25	
Ephedrine	0.599	1.50	0.841	1.39	2.91	1.32
Amphetamine	0.794	1.33	1.10	1.31	3.81	1.31
Methamphetamine	1.00	1.26	1.38	1.25	4.69	1.23
Phentermine	1.22	1.22	1.68	1.22	5.67	1.21
Methylphenidate	3.51	2.88	4.65	2.77	15.6	2.75
Antipyrene	2.65		2.69		3.09	
Morphine	0.305		0.498		1.79	
Codeine	1.00	3.28	1.38	2.77	4.26	2.38
Aminopyrene	1.00	1.00	1.23	1.12[a]	3.26	1.31[a]
Acetylmorphine	1.23	1.23	1.63	1.32	5.21	1.60
Strychnine	3.92	3.19	5.19	3.18	17.6	3.38
Heroin	4.58	1.17	5.96	1.15	18.8	1.07
Acetylcodeine	4.65	1.02	5.78	1.03[a]	19.0	1.01
Thebaine	4.96	1.07	6.43	1.11	19.8	1.04
Methapyrilene	5.31	1.07	7.89	1.23	27.1	1.37
Quinine	7.00	1.32	12.1	1.53	56.3	2.08
Quinidine	6.71	1.04[a]	11.9	1.02[a]	55.3	1.02[a]
Narcotine	13.1	1.95	16.7	1.40	53.5	1.03[a]
Papaverine	14.6	1.11	18.2	1.09	51.6	1.04[a]
Mescaline	0.714		1.02		3.37	
DMT	1.19	1.67	1.66	1.63	5.25	1.56
LSD	10.2		15.5		45.8	
LAMPA	11.2	1.10	17.2	1.11	49.7	1.09
Iso-LSD	15.5	1.38	21.4	1.24	54.9	1.10
TCP	9.91		14.6		47.8	
PCP	11.2	1.13	17.0	1.16	51.6	1.08
Glutethimide	8.92		9.02		10.0	
Methaqualone	23.8		25.2		26.2	
Mecloqualone	27.4	1.15	30.4	1.21	32.8	1.17
Diazepam	50.6		64.2		60.0	
Phenmetrazine	0.962		1.38		4.47	
Phendimetrazine	1.00	1.04	1.38	1.00	4.69	1.05
MDA	1.14		1.56		5.25	
Diethylpropion	1.58		2.06		6.38	

[a]Successive capacity ratios are reversed.

Note: MSA, methanesulfonate; BSA, butanesulfonate; HSA, heptane-sulfonate.

Source: Ref. 50.

TABLE 14   Capacity Factors k'_i and Selectivity Factors α_{ji}, for the
Column μBondapak CN and Mobile-Phase Methanol, Water, 1% Acetic
Acid, and 0.005 M Alkylsulfonate

Drug	MSA k'_i	MSA α_{ji}	BSA k'_i	BSA α_{ji}	HSA k'_i	HSA α_{ji}
			40% Methanol			
Butabarbital	0.513		0.495		0.618	
Phenobarbital	0.704	1.37	0.614	1.24	0.804	1.30
Amobarbital	0.744	1.10	0.817	1.33	0.981	1.22$_a$
Pentobarbital	0.774	1.04	0.817	1.00	0.933	1.05^a
Secobarbital	0.944	1.22	0.997	1.22	1.19	1.27
Benzocaine	0.929		0.915		1.03	
Procaine	0.774		0.817		0.804	
Lidocaine	0.913	1.18	0.956	1.17	0.957	1.19
Cocaine	1.64	1.80	1.76	1.84	1.54	1.61
Tetracaine	2.80	1.71	2.93	1.67	3.06	1.98
Theophylline	0.134		0.131		0.122	
Caffeine	0.177	1.32	0.183	1.40	0.179	1.47
Acetominophen	0.199	1.12	0.188	1.03	0.203	1.13
Phenylpropanolamine	0.493		0.514		0.522	
Ephedrine	0.561	1.14	0.584	1.14	0.587	1.12
Amphetamine	0.774	1.38	0.817	1.39	0.804	1.38
Methamphetamine	0.774	1.00	0.817	1.00	0.804	1.00
Phentermine	0.774	1.00	0.817	1.00	0.804	1.00
Methylphenidate	1.07	1.38	1.14	1.39	1.10	1.37
Antipyrene	0.288		0.280		0.260	
Morphine	0.410		0.419		0.412	
Aminopyrene	0.416	1.01	0.430	1.03	0.412	1.00
Codeine	0.561	1.35	0.583	1.36	0.562	1.36
Acetylmorphine	0.774	1.38	0.756	1.29	0.758	1.35
Heroin	1.08	1.39	1.15	1.52	1.09	1.44
Acetylcodeine	1.11	1.03	1.17	1.02	1.13	1.03
Strychnine	1.29	1.16	1.32	1.13	1.33	1.18
Thebaine	1.44	1.12	1.42	1.08	1.36	1.02
Papaverine	1.50	1.04	1.58	1.11	1.41	1.04
Methapyrilene	1.88	1.25	1.85	1.17	1.87	1.33
Narcotine	1.89	1.00	1.98	1.07	1.95	1.04$_a$
Quinidine	1.95	1.03	2.13	1.08	1.83	1.07^a
Quinine	2.07	1.06	2.24	1.05	1.91	1.04
Mescaline	0.506		0.506		0.540	
DMT	1.22	2.41	1.23	2.20	1.20	2.22
LSD	2.26		2.46		2.29	
LAMPA	2.49	1.10	2.70	1.10	2.48	1.08
Iso-LSD	2.69	1.08	2.92	1.08	2.55	1.03
TCP	1.52		1.63		1.53	
PCP	1.73	1.14	1.83	1.11	1.70	1.11
Glutethimide	1.01		0.980		1.14	
Methaqualone	1.39		1.36		1.41	
Mecloqualone	1.76	1.27	1.74	1.28	1.86	1.32
Diazepam	2.35		2.34		2.56	
Phenmetrazine	0.774		0.817		0.804	
Phendimetrazine	0.774	1.00	0.817	1.00	0.804	1.00
Diethylpropion	0.774		0.817		0.804	
MDA	0.851		0.907		0.949	

TABLE 14 (Cont.)

Drug	MSA k'_i	MSA α_{ji}	BSA k'_i	BSA α_{ji}	HSA k'_i	HSA α_{ji}
			30% Methanol			
Butabarbital	0.859		0.870		0.833	
Phenobarbital	1.18	1.37	1.08	1.24	1.10	1.32
Amobarbital	1.45	1.23	1.33	1.23	1.36	1.24
Pentobarbital	1.43	1.01[a]	1.31	1.01[a]	1.31	1.04[a]
Secobarbital	1.82	1.27	1.65	1.26	1.69	1.29
Benzocaine	1.86		1.63		2.56	
Procaine	0.859		0.870		1.10	
Lidocaine	1.06	1.23	1.03	1.18	1.34	1.22
Cocaine	2.45	2.32	2.35	2.28	2.78	2.07
Tetracaine	5.36	2.19	4.91	2.09	6.34	2.28
Theophylline	0.198		0.174		0.187	
Caffeine	0.312	1.58	0.283	1.58	0.285	1.52
Acetominophen	0.318	1.02	0.286	1.01	0.290	1.02
Phenylpropanolamine	0.445		0.487		0.614	
Ephedrine	0.537	1.21	0.572	1.19	0.738	1.20
Amphetamine	0.810	1.50	0.791	1.38	1.01	1.37
Methamphetamine	0.859	1.06	0.870	1.10	1.10	1.09
Phentermine	0.859	1.00	0.870	1.00	1.10	1.00
Methylphenidate	1.42	1.65	1.42	1.63	1.75	1.59
Antipyrene	0.480		0.451		0.466	
Morphine	0.407		0.414		0.534	
Aminopyrene	0.407	1.00	0.431	1.04	0.542	1.01
Codeine	0.636	1.56	0.649	1.51	0.809	1.49
Acetylmorphine	0.859	1.35	0.870	1.34	1.10	1.36
Heroin	1.64	1.91	1.54	1.77	1.87	1.70
Acetylcodeine	1.68	1.02	1.60	1.04	1.94	1.04
Strychnine	1.85	1.10	1.52	1.05[a]	2.11	1.09
Thebaine	2.28	1.23	2.15	1.41	2.58	1.22
Papaverine	2.84	1.25	2.65	1.23	3.15	1.22
Methapyrilene	2.77	1.02[a]	2.67	1.01	2.82	1.12[a]
Narcotine	3.35	1.21	3.12	1.17	3.70	1.31
Quinidine	2.85	1.18[a]	2.72	1.15[a]	3.08	1.20[a]
Quinine	3.18	1.12	3.04	1.12	3.44	1.12
Mescaline	0.558		0.565		0.719	
DMT	1.50	2.69	1.50	2.65	1.82	2.53
LSD	4.27		3.93		4.69	
LAMPA	4.80	1.12	4.38	1.11	5.15	1.10
Iso-LSD	5.41	1.13	4.90	1.13	5.62	1.09
TCP	2.25		2.18		2.74	
PCP	2.55	1.13	2.48	1.14	3.12	1.14
Glutethimide	2.07		1.85		1.86	
Methaqualone	3.21		2.80		2.86	
Mecloqualone	4.45	1.39	3.98	1.42	3.92	1.37
Diazepam	5.69		4.92		5.05	
Phenmetrazine	0.859		0.870		1.10	
Phendimetrazine	0.859	1.00	0.870	1.00	1.10	1.00
Diethylpropion	0.859		0.870		1.10	
MDA	1.06		1.04		1.29	

[a]Successive capacity factors are reversed.

Note: MSA, methanesulfonate; BSA, butanesulfonate; HSA, heptane-sulfonate.

Source: Ref. 50.

TABLE 15   Capacity Factors k'_i and Selectivity Factors α_{ji} for Mobile-Phase 20% Methanol, 79% Water, 1% Acetic Acid, and 0.005 M Methane-sulfonate Acid Using Various Columns

Drug	μBondapak C_{18} k'_i	α_{ji}	μBondapak Alkylphenyl k'_i	α_{ji}	μBondapak CN k'_i	α_{ji}
Phenobarbital	7.21	1.51	6.60		1.61	
Butabarbital	10.9	2.47	7.01	1.06	1.20	1.34[a]
Pentobarbital	26.9	1.05	15.2	2.17	2.04	1.70
Amobarbital	28.2	1.49	14.8	1.03[a]	2.11	1.03
Secobarbital	42.0		22.8	1.54	2.66	1.26
Benzocaine	12.7		10.1		2.56	
Procaine	2.17		2.65		1.09	
Lidocaine	4.10	1.90	2.73	1.03	1.29	1.18
Cocaine	17.1	4.17	14.7	5.39	3.43	2.67
Tetracaine	52.1	3.05	32.5	2.22	8.17	2.38
Acetominophen	0.829		0.863		0.395	
Theophylline	1.76	2.12	2.02	2.34	0.256	1.54[a]
Caffeine	3.58	2.03	4.34	2.03	0.386	1.51
Phenylpropanolamine	0.840		0.743		0.471	
Ephedrine	1.23	1.46	1.10	1.48	0.587	1.25
Amphetamine	1.87	1.53	1.50	1.36	0.860	1.46
Methamphetamine	2.47	1.32	2.02	1.36	0.980	1.14
Phentermine	3.26	1.32	2.46	1.22	0.980	1.00
Methylphenidate	9.36	2.92	8.95	3.64	1.82	1.86
Antipyrene	7.73		6.52		0.726	
Morphine	0.536		0.849		0.464	
Codeine	2.04	3.81	2.79	3.29	0.778	1.68
Acetylmorphine	2.59	1.27	3.58	1.28	1.14	1.46
Aminopyrene	2.94	1.14	2.38	1.50[a]	0.469	2.43[a]
Strychnine	7.43	2.53	12.2	5.13	2.40	5.12
Acetylcodeine	12.3	1.66	16.4	1.34	2.40	1.00[a]
Heroin	13.9	1.13[a]	16.7	1.02	2.31	1.04[a]
Thebaine	13.0	1.07[a]	17.2	1.03	3.45	1.49
Quinidine	18.0	1.38	30.9	1.80	3.20	1.08[a]
Quinine	24.6	1.37	29.9	1.03[a]	3.64	1.14
Methapyrilene	21.7	1.13[a]	17.6	1.70[a]	3.57	1.02[a]
Narcotine	41.5	1.91	52.7	2.99	5.17	1.45
Papaverine	64.0	1.54	65.6	1.24	5.03	1.03[a]
Mescaline	2.47		1.77		0.681	
DMT	2.77	1.12	2.67	1.51	1.85	2.72
LSD	37.1		37.2		7.10	
LAMPA	40.0	1.08	42.6	1.15	8.01	1.13
Iso-LSD	74.1	1.85	58.0	1.36	9.33	1.13
TCP	41.5		31.1		3.27	
PCP	44.5	1.07	38.8	1.25	3.81	1.16
Glutethimide	29.6 b		24.2 b		3.15	
Methaqualone	b		b	b	5.51	
Mecloqualone	b		b		8.50	1.54
Diazepam					9.56	
Phenmetrazine	2.02		1.92		0.980	
Phendimetrazine	2.47	1.22	2.26	1.18	0.980	1.00
MDA	2.38		2.30		1.18	
Diethylpropion	5.80		3.68		1.04	

[a]Successive capacity factors are reversed.

[b]Retention greater than 2 h.

Source: Ref. 50.

consisted of a μBondapak C_{18} column with a mobile phase which consisted of 20% methanol, 79% water, 1% acetic acid, and 0.02 M
methanesulfonic acid at pH 3.5. This mobile phase was excellent
for the analysis of phenethylamines plus caffeine. A chromatogram
for these compounds is presented in Fig. 18. When a mobile phase of
40% methanol, 1% acetic acid, and 0.005 M heptanesulfonic acid was
utilized for the analysis of phenethylamines, inadequate resolution
was obtained between amphetamine and methamphetamine [49]. It was
shown that for the μBondapak C_{18} column and a given counterion, the
selectivity factor between amphetamine and methamphetamine two similar solutes differing in aliphatic character increased significantly
in going from 59% water to 79% water [50]. Although increasing
the proportion of water in the mobile phase increased retention,
this effect was offset by using the smallest lipophilic counterion.
This system is excellent for the analysis of the amphetamine "look-
alike" capsules, which contain combinations of ephedrine, pseudoephedrine, phenylpropanolamine, and caffeine. Ephedrine has a different retention time than pseudoephedrine by approximately 13 s.
In a previous paper by Lurie [49] it was shown that heroin and
acetylcodeine coeluted with a mobile phase containing 40% methanol
and 0.005 M heptanesulfonic acid. By using 20% methanol and a
methanesulfonate counterion, heroin and acetylcodeine, two similar
compounds differing in aliphatic character, were separated by a
resolution of 1. A good approach for improving the resolution for
closely related compounds differing in aliphatic character is to
increase the amount of water and lower the size of the counterion
in the mobile phase if possible. Various assays for the compounds
above agreed well with other techniques, such as ultraviolet and
fluorescence spectroscopy and gas chromatography.

The alkylphenyl column, together with the two mobile phases
described above, could be used for the analysis of most of the drugs
analyzed on the C_{18} column. Since the μBondapak cyanide column was
found to lack selectivity for many of the drugs chromatographed,
it is not recommended for the general analysis of drugs of forensic
interest. In general, samples were dissolved in methanol prior to

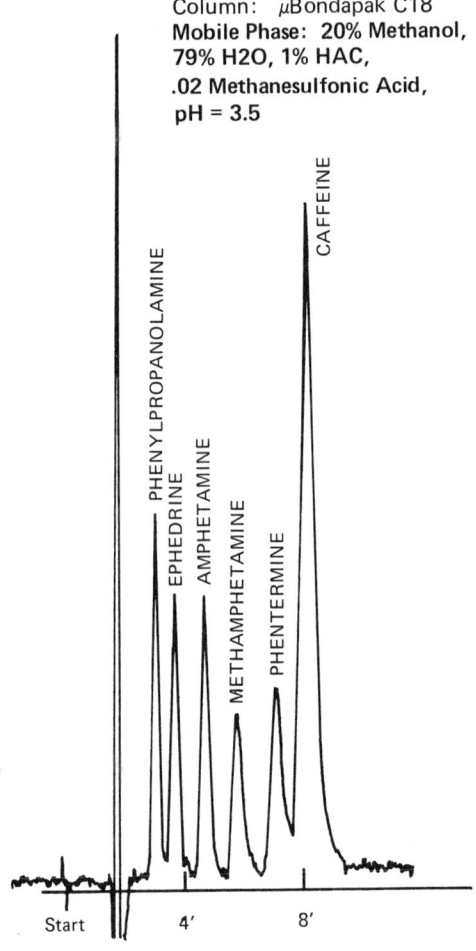

Column: μBondapak C18
Mobile Phase: 20% Methanol,
79% H2O, 1% HAC,
.02 Methanesulfonic Acid,
pH = 3.5

FIG. 18. Chromatogram of synthetic mixture of phenethylamines and related compounds. Column: μBondapak C$_{18}$; variable wavelength detector sensitivity: 0.02 AUFS; mobile phase: Water-methanol-acetic acid (79:20:1) with 0.02 M methanesulfonic acid adjusted to pH 3.5. (From Ref. 77.)

injecting onto the liquid chromatograph with the active ingredient
at a concentration of approximately 0.5 mg/ml. Starting mobile
phase is recommended to dissolve the sample for drugs slightly soluble in methanol (e.g., amphetamine sulfate). For LSD samples,
starting mobile phase is also employed with the LSD concentration
between 60 and 100 μg/ml. The use of mobile phase is advantageous
to dissolve LSD samples for two reasons. First LSD is better extracted from preparations such as blotter acids using the latter
solvent. In addition, large injections of this "quant." solution
(e.g., 2 ml) could be injected onto the liquid chromatograph, using
the identical chromatographic conditions employed for quantitative
analysis, in order to isolate the LSD peak for further spectrographic techniques such as mass spectrometry [84].

Recently, an improved reversed-phase ion-pairing chromatographic
technique for the analysis of illicit heroin exhibits has been developed by Lurie et al. [51]. The conditions employed gave approximately baseline separation between heroin and acetylcodeine.
Adulterants and heroin synthetic by-products were identified on the
basis of relative retention times and 220 nm/254 nm absorbance
ratios. As shown earlier in this chapter, the use of relative retention times and absorbance ratios are highly specific for the
identification of drugs [73]. A Partisil 10 ODS 3 column was employed with a mobile phase consisting of 12% acetonitrile, 87% water,
1% phosphoric acid, and 0.02 M methanesulfonic acid adjusted to
pH 2.2 with 2 N NaOH. A μBondapak C_{18} column which gave similar
relative retention values with the eluant above for compounds present in heroin exhibits could also be employed. The use of the
Partisil 10 ODS 3 column was preferred because of the higher efficiencies and better peak symmetrys for the various compounds in
question. Relative retention values for both columns and 220 nm/
254 nm absorbance ratios for the various compounds in heroin exhibits are given in Table 16. Since all adulterants and by-products
showed good UV responses at 220 nm, this wavelength would be more
suitable for use in absorbance ratioing than 280 nm, which was used

TABLE 16 Relative Retention Data and 220 nm/254 nm Absorbance
Ratios for Heroin and its Adulterants and By-Products

Drug	μBondapak C_{18} RRT (Heroin)	Partisil 10 ODS 3 RRT (Heroin)	220/254
L-Ascorbic acid	--	0.05	c
Isonicotinamide	0.05	0.05	c
Morphine	0.09	0.08	12.1
Aminopyrine	0.15	0.12	0.9
Procaine	0.12	0.12	4.1
Ephedrine	0.14	0.12	3.5
Acetaminophen	0.11	0.12	0.6
Theophylline	0.12	0.13	1.4
Methapyrilene	0.15	0.13	2.0
Tripelennamine	0.16	0.13	2.4
Codeine	0.17	0.16	8.9
Pyrilamine	0.18	0.17	3.2
Quinidine	0.19	0.19	0.7
Barbital	0.23	0.21	62.0
Quinine	--	0.22	0.7
Acetylmorphine	0.23	0.22	14.7
Caffeine	0.19	0.23	2.2
Phentermine	0.28	0.25	5.1
Lidocaine	0.30	0.30	16.6
Quinine (second peak)	--	0.30	0.7
Acetylprocaine	0.28	0.30	0.6
Prilocaine	--	0.31	3.5
Salicylamide	--	0.31	6.0
Antipyrine	0.36	0.36	1.4
Hyoscyamine	0.41	0.39	21.7
Strichnine	0.38	0.41	0.56
Benzocaine	0.50	0.58	4.9
Aspirin	0.60	0.58	10.7
Sodium salicylate	--	0.75	7.3
Tropacocaine	0.71	0.76	7.7
Phenobarbital	0.81	0.85	11.0
Benzoyltropeine	0.82	0.87	8.1
Acetylcodeine	0.85	0.88	11.2
Thebaine	0.88	0.92	4.6
Phenacetin	0.81	0.95	0.5
Meperidine	--	1.00	25.5
Heroin	1.00	1.00 (19 min)	24.0
Cocaine	1.05	1.11	6.0
Amylocaine	--	1.37	5.9
Phencyclidine	2.12	2.15	8.5
Noscapine	2.30	2.36	10.8
Tetracaine	2.28	2.38	10.3
Papaverine	2.80	3.17	0.4
Tartaric acid	--	(2)	a
Diphenhydramine	--	(1)	b
Methadone	--	(1)	b
Phenylbutazone	--	(1)	b

Source: Ref. 51.

[a] Exhibits no UV at 254 mm.

[b] Retention time greater than 1 hour.

[c] Elutes near solvent front.

by Baker et al. [73]. At 280 nm negligible extinction coefficients were obtained for phenethylamines and barbiturates. Peak area was utilized for absorbance ratioing for measurement convenience. Amylocaine hydrochloride was employed as an internal standard. A typical chromatogram is given in Fig. 19. Good agreement was obtained between GC and HPLC for the quantitative determination of heroin hydrochloride. In addition, compounds identified by relative retention times and absorbance ratios agreed well with the determination of these compounds by a combination of GC, TLC, GC-MS, and IR. The accuracy and precision for the quantitation of heroin was

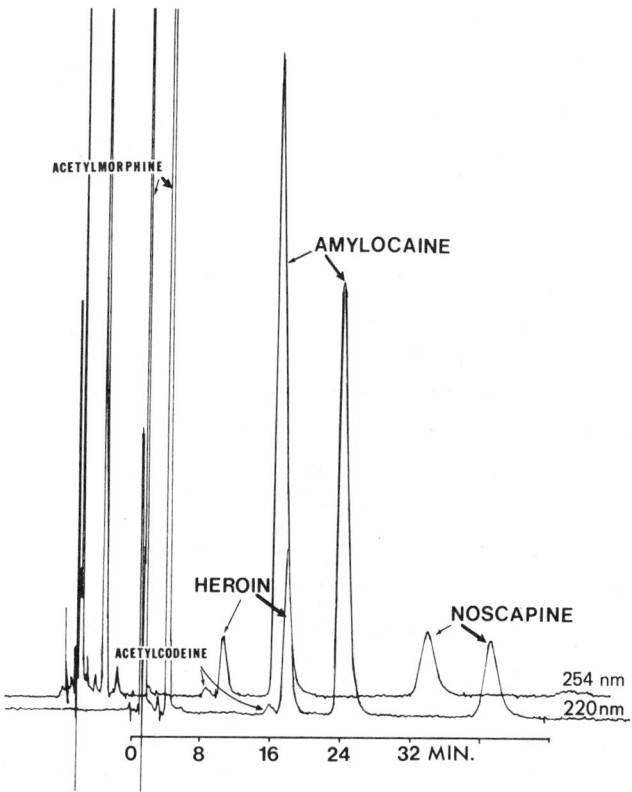

FIG. 19. Chromatogram of a heroin exhibit containing acetylmorphine, acetylcodeine, heroin, amylocaine (internal standard), and noscapine. (From Ref. 51.)

studied by analyzing five separate weighings of synthetic heroin
exhibits. Using the 220-nm UV detector, the percent error was zero
with a coefficient of variation of 1.7%. For the 254-nm fixed wavelength detector, the error was 1.2% with a 2.8% coefficient of
variation. The 220-nm UV wavelength was preferred for the quantitation of heroin because of its greater precision, accuracy, and
sensitivity.

The following sample preparation procedures were employed:

1. Less than 10% heroin: weighed approximately 100 mg of powdered
   sample into a 5-ml volumetric flask to which 1 ml of a 2.5-mg/ml
   methanolic amylocaine hydrochloride internal standard solution
   was added. The flask was then brought to volume with methanol
   prior to injecting 10 μl of sample onto the liquid chromatograph.

2. Greater than 10% heroin: weighed approximately 100 mg of powdered sample into a 100-ml volumetric flask to which 10 ml of a
   5.0-mg/ml methanolic amylocaine hydrochloride internal standard
   solution was added. The flask was then brought to volume with
   methanol prior to injecting 10 μl of sample onto the liquid
   chromatograph.

The use of surfactants such as dodecyl sodium sulfate has
proven to be valuable for forensic drug analysis [64]. The chromatographic conditions which Lurie recommended for the analysis of PCP
samples [77] was not useful for exhibits consisting of PCP on mint
leaves or parsley because of interferences from the compounds
present in plant material. The retention of PCP, the only base
present in the sample, could be selectively increased by increasing
the size and/or concentration of the counterion. By using the
technique above and increasing the amount of methanol in the mobile-
phase PCP elutes in approximately 12 min free of interferences from
the compounds present in the plant material. This mobile phase
consisted of 65% methanol, 39% water, and 1% acetic acid with 0.02
M dodecyl sodium sulfate at a pH of 3.5. These same conditions
has been utilized for the analysis of PCP samples seized in clandestine labs.

Another useful reversed-phase ion-pairing isocratic system for the analysis of drugs of forensic interest consists of a µBondapak C_{18} column and a mobile phase consisting of 57% methanol, 42% water, and 1% acetic acid with 0.02 M methanesulfonic acid at pH 3.5 [64]. This system has been used for the analysis of liquid methadone exhibits, diazapam samples, and preparations containing aspirin, phenacetin, caffeine, and propoxyphene.

F. Evaluation of Micro Cation-Exchange Column

Twitchett et al. evaluated a microparticulate cation-exchange column, the Partisil 10 SCX for the HPLC analysis of drugs [78]. The column employed was of limited value since its lifetime was only a few months. However, some interesting chromatographic trends were presented. Aqueous organic mobile phases containing ammonium dihydrogen phosphate buffers of varying ionic strength and various pH values were utilized. Methanol or acetonitrile was used as the organic modifier. For basic solutes retention was inversely proportional to the mobile-phase ionic strength, which is typical in ion-exchange chromatography. For acidic drugs there was an increase in retention with ionic strength due to salting-out effects. In general, for basic drugs, as is depicted in Fig. 20, retention increases with eluant pH, at pH values greater than 5. Twitchett et al. attributed this effect to the basic drugs becoming less ionic and therefore less retention by an ion-exchange mechanism and more retention by a partition mechanism. It is possible that a major cause for the trend in Fig. 20 is an ion exchange and adsorption with the unbonded silanol sites. For acidic drugs there was a decrease in retention with pH as the compound became more ionized. With a cation-exchange column the major mechanism for acidic and neutral drugs would be partition. The chromatographic efficiency of all types of drugs was considerably greater when an organic modifier was added to the mobile phase.

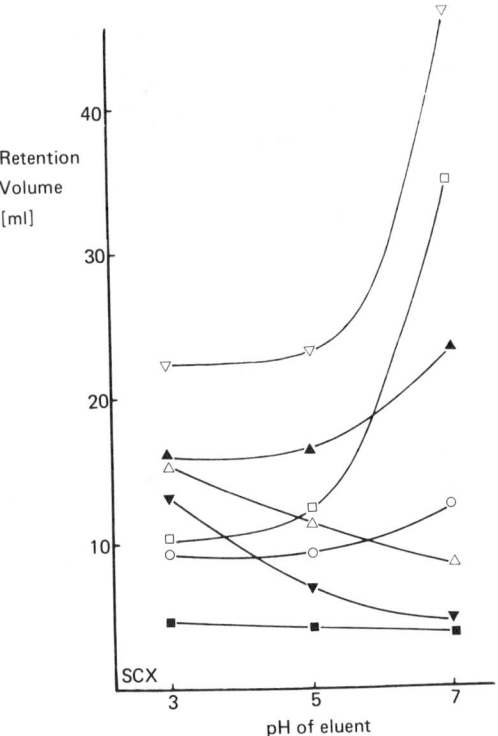

FIG. 20.   Effect of variation of the eluant pH on retention volume.
Ionic strength: 0.1 M; methanol concentration: 40% for tubocurarine; 0% for other drugs. ■, Acetylsalicylic acid; ▼, phenobarbitone; Δ, nitrazepam; O, amphetamine; ▲, methylamphetamine; □,
morphine; ▽, tubocurarine.   (From Ref. 78.)

V. NONDRUG APPLICATIONS

In this chapter we have dealt exclusively with the role of bonded-
phase columns in the analysis of drugs of forensic interest in bulk
quantities.   There are, however, excellent applications of these
columns in the analysis of excipients such as dyes, sugars, and
polyhydric alcohols that can appear in forensic drug exhibits.

A. Dye Analysis

Clark and Miller described an interesting method for the identification of dyes in illicit heroin samples [79].   Reversed-phase ion-
pairing chromatography was employed with a mobile phase at a pH

between 7 and 8 consisting of 58% methanol, 42% water, and 0.005 M tetrabutylammonium hydroxide ca. 0.03% acetic acid. In this instance since the dyes to be separated were acidic and existed as an anionic species at pH 7-8, a cationic counterion such as tetrabutylammonium hydroxide was employed for ion-pairing chromatography. A chromatogram of the various dyes encountered in heroin exhibits is presented in Fig. 21. Useful intelligence information was obtained for determining common origins from a comparison of chromatograms showing the relative ratios of dyes present in samples exhibiting similar dye patterns. In general, sample preparation consisted of dissolving the sample of interest in starting mobile phase and filtering the solution through a 0.5-μm membrane filter. In certain instances when interference was encountered from procaine and minor alkloidal constituents, a wool dye extraction procedure was employed to isolate the dyes prior to analysis.

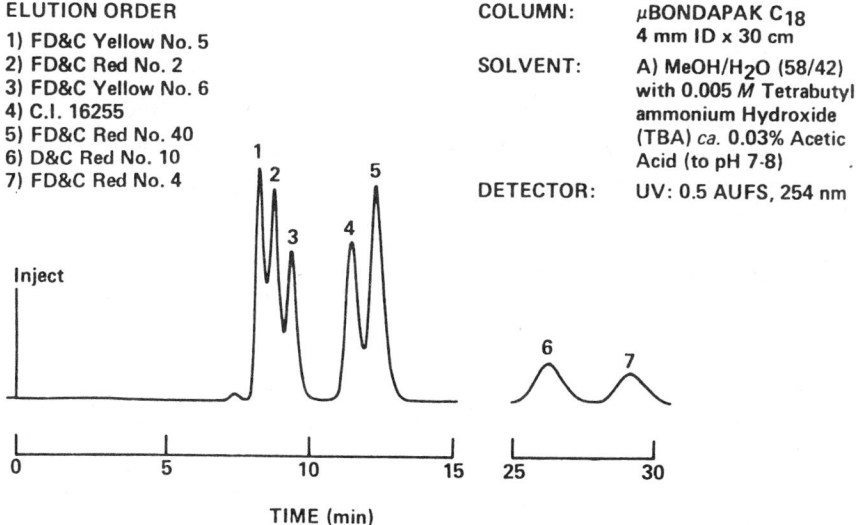

ELUTION ORDER
1) FD&C Yellow No. 5
2) FD&C Red No. 2
3) FD&C Yellow No. 6
4) C.I. 16255
5) FD&C Red No. 40
6) D&C Red No. 10
7) FD&C Red No. 4

COLUMN: μBONDAPAK C$_{18}$
4 mm ID x 30 cm

SOLVENT: A) MeOH/H$_2$O (58/42)
with 0.005 M Tetrabutyl-
ammonium Hydroxide
(TBA) ca. 0.03% Acetic
Acid (to pH 7-8)

DETECTOR: UV: 0.5 AUFS, 254 nm

TIME (min)

FIG. 21. Mixed dye standards. (From Ref. 79.)

B. Sugars and Polyhydric Alcohols

Lurie and Henderson described a procedure for the analysis of sugars
and polyhydric alcohols in heroin and cocaine samples [80]. A
Partisil 10 PAC column which has bonded cyano and amino functional
groups was employed. The mobile phase consisted of 88% acetonitrile
and 12% water. A chromatogram of the five most commonly encountered
sugars and polyhydric alcohols, consisting of dextrose, mannitol,
sucrose, inositol, and lactose, is portrayed in Fig. 22. As a
screening procedure for sugars and polyhydric alcohols, HPLC is
superior to GC, IR, x-ray diffraction, TLC, and optical crystallo-
graphic analysis. The run time of less than 20 min is significantly less than that required for the gas chromatographic analysis
of carbohydrates. Unlike GC, the HPLC methodology does not require
silylation procedures. Because the use of HPLC does not distinguish between anomers as would GC, quantitation is easier. Infrared
analysis and x-ray diffraction are less reliable when dealing with
mixtures. The visualization of polyhydric alcohols is difficult
via TLC and the use of optical crystallography is highly subject to
operator skill.

Since HPLC is a chromatographic technique and as such lacks
specificity, optical crystallography should be employed for confirmation of compounds initially identified by HPLC. Since increasing the amount of water in the eluant decreases retention for the
compounds of interest, a normal-phase mechanism is occurring. A
refractive index detector was utilized in series with a fixed ultraviolet detector at 254 nm. The sugars and polyhydric alcohols
which are non-UV adsorbing are identified by the RI detector while
the UV detector would monitor any interferences from UV adsorbing
compounds. Although the RI detector is not a very sensitive detector, with typical limits of detection in the low-microgram range,
satisfactory levels of detection are obtained for the carbohydrates
of interest. Using peak area the detection limits are 1.5% for
dextrose, mannitol, sucrose, and inositol, while lactose has a detection limit of 3%. The sample preparation for screening consists

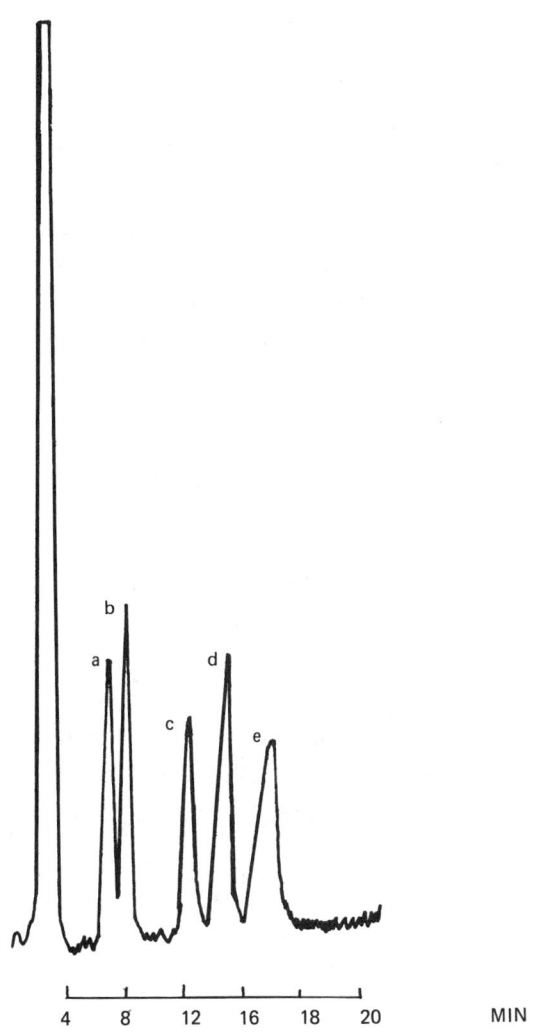

FIG. 22. Chromatogram of a standard mixture of sugars and polyhydric alcohols: a, dextrose; b, mannitol; c, sucrose; d, inositol;
e, lactose. Column: Partisil 10 PAC; mobile phase: water-
acetonitrile (12:88); flow rate: 2.0 ml/min.

of weighing approximately 100 mg of sample onto a filter paper
which is washed three times with chloroform and methanol (4:1).
The residue on the filter paper is allowed to air dry prior to
being diluted to volume in a 10-ml volumetric flask with acetonitrile and water (1:1). The resultant solution was placed on a
sonic bath for 10 min. Fifty microliters of a standard solution
consisting of dextrose, mannitol, sucrose, and inositol at a concentration of 2.0 mg/ml/ml was injected onto the liquid chromatograph followed by a 50-µl injection of sample. This methodology
is not quantitative because of some solubility of the sugars and
polyhydric alcohols in the methanol-chloroform solution. In addition, these compounds can physically adhere to the filter paper.
Preliminary work indicates the feasibility of quantitating carbohydrates via HPLC, and this avenue is presently being pursued.

VI. ISOLATION AND IDENTIFICATION OF DRUGS

Up to this point this chapter has dealt almost exclusively with
forensic analytical applications of HPLC, that is, the quantitation
of controlled substances and the identification of separated constituents on the basis of absorbance ratioing and/or retention
characteristics. Since HPLC with the use of certain detectors such
as ultraviolet or refractive index is a nondestructive technique,
peaks can be collected for further analysis by chromatographic or
spectrometric techniques. Alternatively, a liquid chromatograph
can be coupled with a mass spectrometer for specific compound
identification.

     In this chapter preparative liquid chromatography will refer
to any separation in which peaks are collected for further analysis
regardless of the amount of sample involved. For an excellent
discussion on preparative liquid chromatography, the reader is referred to the review articles by De Stefano and Kirkland [81].
For legal and intelligence purposes, the identification of collected
fractions is an extremely useful tool for the forensic chemist.
For example, if a sample contained a mixture of methamphetamine and

phentermine, two phenethylamines similar in structure, the separation of these compounds by wet chemical methods in order to obtain infared spectra is very difficult if not impossible. HPLC would offer a feasible means of isolating methamphetamine and phentermine in sufficient purity for infared analysis. In many instances a submitted sample will contain an unusual adulterant. Preparative HPLC affords a fairly simple means for the isolation and identification of this unknown compound. Let us begin by examining examples of preparative HPLC utilizing reversed-phase chromatography pertaining to forensic drug analysis. Then we will discuss the promising technique of LC-MS as it relates to forensic drug analysis.

A. Preparative Liquid Chromatography

Trinler and Reuland reported on the use of preparative liquid chromatography in order to obtain infared spectra of local anesthetics such as cocaine, procaine, tetracaine, lidocaine, and benzocaine in mixtures [82]. A column 3.2 mm in diameter by 60 cm in length containing Bondapak phenyl/Porasil B was employed with a mobile phase consisting of 85:15 acetonitrile/water by volume with 0.1% by weight ammonium carbonate. Utilized were 0.25-ml injections of 1.0 mg/ml of sample in mobile phase. Two injections were necessary in order to obtain sufficient material for infared analysis by conventional techniques. A refractive index (RI) detector was used in series with a UV detector since the former detector, being less sensitive, will keep peaks of interest on scale. After the peaks were collected from the liquid chromatograph, the eluant was evaporated to dryness in a warm bath at 50° C under the reduced pressure of a water aspirator. The resulting residue was then taken up in 0.5 ml of absolute ethyl alcohol and added to 170 mg of potassium bromide. This slurry was dried in a vacuum disiccator and the resultant powder was ground in an agate mortar and mixed on a Wig-L-Bug. A KBR pellet was then formed in a press prior to IR analysis.

A method for the unequivocal determination of heroin in street drugs by preparative HPLC was presented by Reuland and Trinler [83].

Unlike in their previous paper, where large particle size (30-75 μm)
porous packing material was employed, they utilized Partisil 10
ODS (10 μm particle size) packing material, which was slurry packed.
The less expensive columns employing hand-packed 37 to 75-μm porous
material were not employed because they lacked the resolution necessary to separate heroin from its common contaminants and excipients via an isocratic separation. The mobile phase utilized consisted of acetonitrile-water (60:40) with 0.1% ammonium carbonate.
Under these chromatographic conditions morphine, acetylcodeine,
and quinine were well resolved from heroin while acetylmorphine
and procaine were not baseline separated from heroin. Ten- to 20-μl
injections of simulated street samples of heroin containing by
weight 2% heroin, 5% each of the adulterants procaine and quinine,
and 88% of the diluent lactose dissolved in an extraction solvent
at a heroin concentration of 5.0 mg/ml. In addition, a sample consisting of heroin 5.0 mg/ml and 1.0 mg/ml of acetylmorphine, acetylcodeine, and morphine in eluting solvent was prepared. The solvent
consisted of 50:50 acetonitrile-water by volume with 0.1% ammonium
carbonate. The heroin fractions collected were treated as described previously [82] except that 20 mg of potassium bromide was
employed for the preparation of a micropellet. IR analysis via
the use of a beam condenser was necessary for the smaller sample
examined. Larger injection volumes, which would result in more
sample being collected, may not have been employed because of resultant loss of resolution between heroin and procaine and/or
acetylmorphine. Other possible approaches that were not discussed
in this publication for increasing sample throughput are injecting
larger volumes of more dilute solutions or increasing the resolution between heroin and procaine and acetylmorphine followed by
larger injections of the concentrated sample.

    Since reversed-phase ion-pairing chromatography has been shown
to be a most versatile technique for the analysis of drugs of forensic interest, it would be desirable to be able to perform prepara-

tive separations using this technique. Lurie and Weber presented
methodology for the isolation and identification of drugs of
forensic interest by reversed-phase ion-pairing chromatography [84].
Analytical separations were first developed using a 4.4 mm ID by
30 cm μBondapak C_{18} column with mobile phases containing water,
methanol, acetic acid, and alkylsulfonic acids or alkylsulfonic acid
salts at a concentration of 0.005 M. For the isolation of milligram quantities of the compounds separated on the analytical system,
a larger-diameter 9.4 mm ID by 25 cm Magnum 9 Partisil 10 ODS column
was utilized. The flow rate was increased from 2.0 ml/min, which
was used on the analytical column to 5.0 ml/min. The mobile phases
were identical to those employed on the analytical system except
that the counterion concentration was scaled up to 0.04 M. For
the preparative separations a concentrated extract of sample was
taken up in 2 ml of mobile phase and clarified through a 0.45-μm
filter prior to an injection of approximately 1.4 ml. Typically,
5-15 mg of compounds of interest would be injected onto the liquid
chromatograph. Subsequent to collection from the chromatographic
system, acidic and neutral drugs could be extracted with chloroform
two to three times and evaporated to dryness on a steam bath. Infared analysis was then performed either by a film on a KBR window
or a KBR pellet. Collected effluents containing basic drugs were
adjusted with 1 N NaOH to at least 1.5 pH units above the pK_a value
of the drug if known, or otherwise a pH of 11.5 was utilized. The
resulting effluent was then extracted two to three times with chloroform and evaporated to dryness on a steam bath prior to IR analysis. In certain instances a few drops of concentrated hydrochloric
acid was added to the chloroform to make the hydrochloride salt of
the basic drug. Examples of various samples analyzed include a
legitimate capsule containing amobarbital and secobarbital, a mixture of LSD and iso-LSD, and a mixture of ephedrine hydrochloride
and methamphetamine hydrochloride. The analytical and preparative
separation of an LSD sample is depicted in Fig. 23. For the preparative separation a RI detector was used in series with a UV

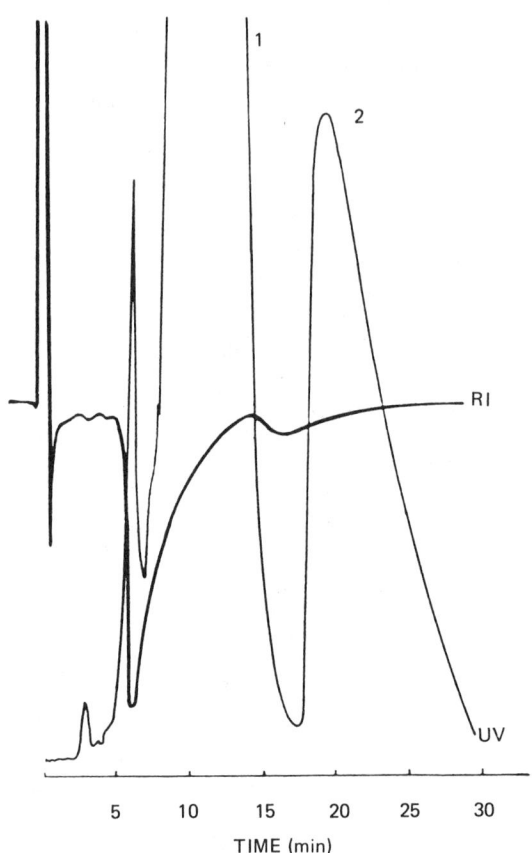

FIG. 23. Semipreparative chromatogram of a clandestine LSD sample:
1, LSD; 2, iso-LSD. Column: magnum 9 ODS; fixed wavelength detector sensitivity: 2.0 AUFS, RI: 16X; mobile phase: water-
methanol-acetic acid (59:40:1) with 0.04 M methanesulfonic acid adjusted to pH 3.5; flow rate: 5.0 ml/min. (From Ref. 8.)

detector for reasons mentioned earlier. Since milligram quantities
are readily isolated by the technique above, separated components
could be identified by nuclear magnetic resonance analysis. If
only the isolation of microgram quantities is required for analysis
by mass spectrometry, the analytical column could be utilized, as
described in Sec. III.C for the analysis of LSD.

Weber et al. presented methodology for the isolation and isolation and identification of trace impurities in clandestine methaqualone [85]. Impurities that can be present in clandestine methaqualone samples include anthranilic acid, o-toluidine, N-acetyl
anthranilic acid, N-acetyl-o-toluidine, and the carboxy homolog of
methaqualone [2-methyl-3(2'-carboxyphenyl)-4-quinazolinone]. These
impurities arise from incomplete reaction or inadequate purification of the final product. Gradient elution was necessary for the
analytical separation of methaqualone and its impurities because
an isocratic separation of these components would have required
over 2 hr. A μBondapak C_{18} column was employed with the starting
mobile phase of the gradient consisting of 10% methanol, 89% water,
and 1% acetic acid, while 80% methanol, 19% water, and a 1% acetic
acid solution was added in a linear gradient for 30 min. For the
preparative separation the same gradient conditions were employed
with a Magnum 9 Partisil 10 ODS column. To isolate sufficient
quantities of the trace impurities in clandestine methaqualone for
subsequent mass spectrometry it was necessary to inject milligram
quantities of sample onto a larger-bore column. For the preparative separation, approximately 1.4 ml of a clarified solution of
concentrated sample in starting mobile phase was utilized. All
fractions except the o-toluidine were extracted directly into
chloroform followed by evaporation to dryness. Since o-toluidine
has a pK_a value of 4.5, this compound was extracted at pH 6 to
maximize yield. The isolated fractions were identified by solid
probe mass spectrometry.

B. LC-MS

We have just examined techniques whereby the collected fractions
are extracted and evaporated to dryness prior to analysis by solid
probe mass spectrometry. An easier, though much more expensive
means of carrying out this same task would be by directly interfacing a liquid chromatograph with a mass spectrometer. As shown
in Chap. 2, this is presently being done commercially either by a

256   *Lurie*

direct liquid interface or by employing a moving belt as a mechanical interface.

Henion modified a commercially available direct liquid interface for the LC-MS analysis of drugs [86]. The advantages of LC-MS
analysis over GC-MS analysis of drugs were described. Highly polar
and thermally unstable compounds could be assayed directly by LC-MS
without the necessity of derivatization procedures. One example of
forensic interest was the analysis of dilantin using a Partisil PXS
10 PAC column with a mobile phase consisting of 25:75 methanol-
pentane. When a direct liquid interface is employed, chemical
ionization mass spectra are obtained with the mobile phase acting
as a reagent gas.

Henion reported in a paper comparing direct liquid introduction LC-MS techniques by using either microbore or conventional
packed columns that greater sensitivites could be obtained via
microbore columns for the analysis of drugs [87]. Using conventional columns, sensitivities were limited to 100 ng while by using
microbore columns, low-nanogram quantities of drugs could be detected. Using conventional columns it was necessary to split the
effluent stream so that only 1-3% of the eluant entered the MS ion
source. The use of a micro-LC at flow rates of 8 µl/min with a
microbore column allowed the introduction of the total effluent
into the mass spectrometer. Employing a SC-01 ODS C_{18} micro-LC
column with a mobile phase consisting of 90:10 acetonitrile-water,
the LC-MS analysis of 35 ng of 2-hyroxypromazine, 20 ng of acepromazine, and 30 ng of chlorpromazine was carried out as illustrated in Fig. 24. Certain drugs did not yield satisfactory LC-MS
results using the microbore columns because of precondensation in
the capillary. Using a similar system as depicted above, Henion
reported on the LC-MS analysis of 10 ng of the narcotic analgesic
fentanyl [88].

Arpino and Krien presented methodology for the LC-MS analysis
of cannabis leaves using a prototype DLI interface "Carole" [89].
A µBondapak C_{18} column was employed with an effluent consisting of

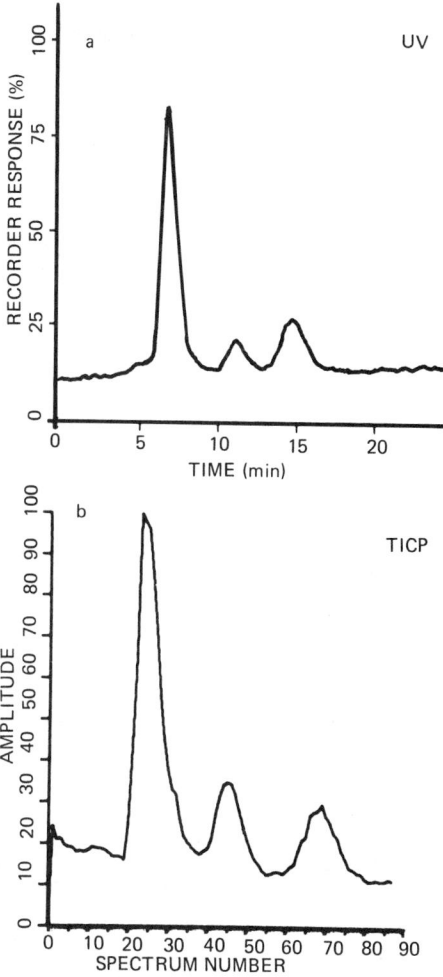

FIG. 24.   (a) Liquid chromatogram at 254 nm for the micro-LC separation of 2-hydroxypromazine (35 ng), acepromazine (20 ng), and
chlorpromazine (30 ng), using an eluant of 90:10 CH_3CN/H_2O containing 0.1% TMA at 8 µl/min on a 0.5 mm ID × 7 cm JASCO SC-01 ODS-C_{18}
micro-LC column.   (b) TICP plot from the micro LC-MS analysis of
the promazine mixture in (a) using capillary LC-MS interface and
a Finnigan 3200 CI mass spectrometer.   (From Ref. 87.   Reproduced
from the *Journal of Chromatographic Science*, by permission of
Preston Publications, Inc.)

acetonitrile. Dried cannabis leaves were extracted with pure benzene prior to a 1-µl injection onto the LC-MS system. Chemical
ionization mass spectra were obtained for cannabinol and a unresolved mixture of cannabidiol and tetrahydrocannabionol. Although
these latter compounds overlapped on the reconstructed liquid
chromatogram, they could be separately displayed by plotting limited mass chromatograms corresponding to the M + 1 ions.

Kenyon and Malera reported on the use of the direct liquid
inlet LC-MS system for the analysis of LSD in urine [90]. A RP-8
column was utilized and the solvent consisted of a linear gradient
of 10-90% acetonitrile in water for 5.5 min. This system would be
applicable to the analysis of LSD from street samples or clandestine laboratories.

Karger et al. discussed the technique of reversed-phase LC-MS
utilizing a continuous extraction interface [91]. The use of LC-MS
in the reversed-phase mode has been severely limited because of the
involatility of many of the inorganic buffers or chemical modifiers
employed in reversed-phase chromatography. By the use of an additional interface whereby solutes of interest are continuously extracted into a volatile organic solvent such as methylene chloride,
the role of reversed-phase chromatography in LC-MS could be greatly
expanded. This interface is based on segmented AutoAnalyzer technology. The organic solvent enters the mass spectrometer via a
commercially available moving-belt interface where the compounds
are mass analyzed via electron impact or chemical ionization. Although no applications to forensic drug analysis was reported in
this paper, this technique could afford great potential in this
area.

VII. PRACTICAL CONSIDERATIONS

In discussions with forensic chemists from other laboratories,
several problems in the use of HPLC are always raised regardless
of the mode of chromatography employed.

Problem: HPLC is too time consuming since a different mobile phase
has to be employed for each type of drug assayed.

Problem: If one mode of chromatography has been utilized, it is
time consuming to switch to an alternative mode.

Solution: Always attempt to analyze as many drugs as possible using
the minimal number of mobile phases. Although in certain instances this approach leads to separations that are not optimized for time, productive work can be accomplished while peaks
are eluting from the liquid chromatograph. In general for the
analysis of drugs a few select mobile phases can be employed
for most of the analysis performed. These eluants can be made
up in large quantities (4 liters or greater) by designated
personnel. For example, in one laboratory one chemist is
responsible for a 40% mathanol, 59% water, 1% acetic acid,
0.02 M methanesulfonic acid solution [77]; another chemist is
responsible for a 20% methanol, 79% water, 1% acetic acid,
0.02 M methanesulfonic acid solution [77]; and a third chemist
makes up mobile phase for the analysis of sugars and polyhydric alcohols [80].

Solution: If possible, dedicate instruments to one mode of HPLC.
Normally, changing chromatographic modes (e.g., adsorption to
reversed-phase chromatography) requires changing the column
employed. In addition, since reversed-phase solvents can be
immiscible in the mobile phases employed for adsorption
chromatography, washes with solvents soluable in both mobile
phases are required prior to switching from one chromatographic
mode to another. Problems can arise in changing chromatographic conditions in the same mode of chromatography. For
example, in reversed-phase ion-pairing chromatography, if the
mobile phase is changed from a long-chain counterion such as
dodecyl sodium sulfate to a short-chain counterion such as
methanesulfonic acid, long equilibration times are required.

Problem: It takes a long time to get the liquid chromatograph to
function from a down condition because it is not clear which
mobile phases were previously utilized or how the instrument
is hooked up.

Solution: Clearly label all mobile phases both as to content and
to which solvent inlet line is being utilized. Clearly label
all connecting tubing as to which components are being used.
If your instrument has dual pumps, injectors, detectors, and
so on, clearly label which components belong together.

VII. FUTURE DEVELOPMENTS

Most separations described in this chapter employed 10-μm particle
size analytical columns (25-30 cm × 3.5 mm ID). Since the use of
5-μm particle size columns of similar dimensions gives significantly
greater theoretical plates, similar separations to those which were
obtained with the former columns could be obtained in less time by
using 5-μm particle size columns with 10-cm lengths. Similarly,
even faster separations could be obtained by using 3-μm particle
sizes with 5 to 7-cm column lengths [15]. Since extra column effects become increasingly important when using these shorter columns,
the design of the injection system, detector, and connecting tubing
becomes crucial. Microbore (column diameter 0.5-1 mm), micropacked
capillary (diameter 50-200 μm), and open tubular (diameter 10-60 μm)
columns all appear to have potential because of lower solvent usage
and sharper peaks. These features could be advantageous for direct
LC-MS interfacing [15]. These columns could generate high theoretical plates at the expense of analysis time. It appears that microbore columns have the best potential for routine separations. One
of the problems in analyzing bases on silica-based columns with
reversed-phase chromatography has been the instability of these
columns at the high pH values that are required of the mobile phase
in order to analyze many of these compounds via ion suppression.
Presently, Hamilton offer a nonsilaceous polymeric support that can
be used at these high pH values, which in addition has good mass
transfer characteristics.

A universal detector for HPLC similar in scope to the flame
ionization detector in gas chromatography is yet to be developed
and would be of great utility for forensic analysis. Techniques
such as LC-FTIR and LC-FTNMR have great potential as an isolation
and identification technique [92, 93]. Along these lines the role
of LC-MS in forensic analysis will greatly increase. The present
revolution in microprocessor technology has resulted in the development of liquid chromatographs that are totally automated. An
analyst can store the chromatographic conditions for his assays on

a floppy disk and therefore carry out unattended runs requiring different mobile phases. In addition, the time required for method development would be greatly reduced. Multidimensional chromatographic techniques which involve the switching of peaks from one column to another offer great potential in forensic analysis [94], especially for samples derived from natural products or clandestine laboratories.

REFERENCES

1. C. Horvath and W. Melander, *J. Chromatogr. Sci*. 15:393-404 (1977).
2. M. C. Hennion, C. Picard, and M. Caude, *J. Chromatogr*. 166:21-35 (1978).
3. K. Karch, I. Sebestian, and I. Halasz, *J. Chromatogr*. 122:3-16 (1976).
4. B. B. Wheals, *J. Chromatogr*. 187:65-85 (1980).
5. G. B. Cox, *J. Chromatogr. Sci*. 15:385-392 (1977).
6. D. Locke, *J. Chromatogr. Sci*. 12:433-437 (1974).
7. R. B. Hermann, *J. Phys. Chem*. 75:363 (1971).
8. C. Horvath, W. Melander, and I. Molnar, *J. Chromatogr*. 125:129-156 (1976).
9. O. Sinanoglu and S. Abdulnur, *Fed. Proc*. 24:12-13 (1965).
10. B. L. Karger, J. R. Gant, A. Harthopf, and P. H. Weiner, *J. Chromatogr*. 128:65-78 (1976).
11. J. M. McCall, *J. Med. Chem*. 18:549-552 (1975).
12. J. K. Baker and C. Y. Ma, *J. Chromatogr*. 169:107-115 (1979).
13. J. K. Baker, R. E. Skelton, T. N. Riley, and J. R. Bagley, *J. Chromatogr. Sci*. 18:153-158 (1980).
14. C. Horvath, W. Melander, and I. Molnar, *Anal. Chem*. 49:142-153 (1977).
15. R. E. Majors, *J. Chromatogr. Sci*. 18:488-511 (1980).
16. I. S. Lurie and S. M. Demchuk, *J. Liq. Chromatogr*. 4:337-355 (1981).
17. C. Horvath, W. Melander, and P. Molnar, *Anal. Chem*. 49:2295-2305 (1978).
18. P. T. Kissinger, *Anal. Chem*. 49:883 (1977).
19. B. A. Bidlingmeyer, *J. Chromatogr. Sci*. 18:525-539 (1980).
20. E. Tomlinson, T. M. Jefferies, and C. M. Riley, *J. Chromatogr*. 159:315-358 (1978).
21. B. B. Wheals and R. N. Smith, *J. Chromatogr*. 105:396-400 (1975).
22. R. N. Smith, *J. Chromatogr*. 115:101-106 (1975).
23. R. N. Smith and C. G. Vaughan, *J. Chromatogr*. 129:347-354 (1976).
24. E. E. Knaus, R. T. Coults, and C. W. Sazakoff, *J. Chromatogr. Sci*. 14:525-530 (1976).

25. M. F. Balandrin, A. D. Kinghorn, S. J. Smolenski, and R. H. Dobberstein, *J. Chromatogr.* 157:365-370 (1976).
26. H. M. Fales and J. J. Pisano, *Anal. Biochem.* 3:337 (1962).
27. I. Jane and B. B. Wheals, *J. Chromatogr.* 84:181-186 (1973).
28. H. Bethke, B. Delz, and K. Stich, *J. Chromatogr.* 123:193-203 (1976).
29. F. Erni, R. W. Frei, and W. Lindner, *J. Chromatogr.* 125:265-274 (1976).
30. L. Szepesy, I. Feher, G. Szepesi, and M. Gazdag, *J. Chromatogr.* 149:271-280 (1978).
31. D. Sondack, *J. Chromatogr.* 166:615-618 (1978).
32. M. Wurst, M. Flieger, and Z. Rehacek, *J. Chromatogr.* 150:477-483 (1978).
33. J. R. Salmon and P. R. Wood, *Analyst* 101:611-615 (1976).
34. A. Hulshoff and J. H. Perrin, *J. Chromatogr.* 129:263-276 (1976).
35. J. B. Zagar, F. J. Van Lentern, and G. P. Chrekian, *J. Assoc. Off. Anal. Chem.* 61:678-682 (1978).
36. K. G. Wahlund and A. Sokolowski, *J. Chromatogr.* 151:299-310 (1978)
37. A. Sokolowski and K. G. Wahlund, *J. Chromatogr.* 189:299-316 (1980).
38. F. T. Noggle and C. R. Clark, *J. Assoc. Off. Anal. Chem.* 62:799-807 (1979).
39. F. T. Noggle and C. R. Clark, *J. Chromatogr.* 188:426-430 (1980).
40. M. Emery and J. Kowtko, *J. Pharm. Sci.* 68:1185-1187 (1979).
41. D. W. Newton and R. B. Kluza, *Drug Intell. Clin. Pharm.* 12:546-554 (1978).
42. W. Lund, M. Hannisdal, and T. Greibrokk, *J. Chromatogr.* 173:249-261 (1979).
43. H. B. Hanekamp, W. H. Boogt, P. Bos, and R. W. Frei, *J. Liq. Chromatogr.* 3:1205-1218 (1980).
44. D. Burke and H. Sokoloff, *J. Pharm. Sci.* 69:138-140 (1980).
45. D. M. Takahashi, *J. Pharm. Sci.* 69:184-187 (1980).
46. V. Das Gupta, *J. Pharm. Sci.* 65:1697-1698 (1976).
47. C. Y. Wu and J. J. Wittick, *Anal. Chem.* 49:359-363 (1977).
48. S. K. Soni and S. M. Dugar, *J. Forensic Sci.* 24:434-447 (1979).
49. I. S. Lurie, *J. Assoc. Off. Anal. Chem.* 60:1035-1040 (1977).
50. I. S. Lurie and S. M. Demchuk, *J. Liq. Chromatogr.* 4:357-374 (1981).
51. I. S. Lurie, S. M. Sottolano, and S. Blasof, *J. Forensic Sci.* 27:519-526 (July 1982).
52. V. Das Gupta, *J. Pharm. Sci.* 69:110-112 (1980).
53. N. Muhammad and J. A. Bodnar, *J. Liq. Chromatogr.* 3:113-122 (1980).
54. J. L. Love and L. K. Pannell, *J. Forensic Sci.* 25:320-326 (1980).
55. G. K. Poochikian and J. C. Cradok, *J. Chromatogr.* 171:371-376 (1979).
56. C. Olieman, L. Maat, K. Waliszewski, and H. C. Beyerman, *J. Chromatogr.* 133:382-385 (1977).
57. E. J. Kubiak and J. W. Munson, *J. Pharm. Sci.* 69:152-155 (1980).

58. Y. Nobuhara, S. Hirano, K. Namba, and M. Hashimoto, *J. Chromatogr.* 190:251-255 (1980).
59. F. T. Noggle, *J. Assoc. Off. Anal. Chem.* 63:702-706 (1980).
60. T. H. Beasley and H. W. Ziegler, *J. Pharm. Sci.* 66:1749-1750 (1977).
61. J. W. Hsieh, J. K. H. Ma, J. P. O'Donnell, and N. H. Choulis, *J. Chromatogr.* 161:366-370 (1978).
62. M. J. Walters and S. M. Walters, *J. Pharm. Sci.* 65:198-201 (1976).
63. R. W. Souter, *J. Chromatogr.* 134:187-190 (1977).
64. I. S. Lurie, *Am. Lab.* pp. 35-42 (Dec. 1980).
65. B. M. Thomson, *J. Forensic Sci.* 25:779-785 (1980).
66. K. Sugden, G. B. Cox, and C. R. Loscombe, *J. Chromatogr.* 149: 377-390 (1978).
67. A. C. Allen, D. A. Cooper, W. O. Kiser, and R. C. Cottrell, *J. Forensic Sci.* 26:12-26 (1981).
68. A. B. Clark and R. Gelsomino, personal communication, 1977.
69. C. R. Clark and J. L. Chan, *Anal. Chem.* 50:635-636 (1978).
70. P. J. Twitchett and A. C. Moffat, *J. Chromatogr.* 111:149-157 (1975).
71. W. A. Trinler and D. J. Reuland, *J. Forensic Sci. Soc.* 15:153-158 (1975).
72. W. A. Trinler, D. J. Reuland, and T. B. Hiatt, *J. Forensic Sci. Soc.* 16:133-138 (1976).
73. J. K. Baker, R. E. Skelton, and C. Y. Ma, *J. Chromatogr.* 168: 417-427 (1979).
74. R. Yost, J. Stoveken, and W. MacLean, *J. Chromatogr.* 134:73-82 (1977).
75. J. K. Baker, *Anal. Chem.* 51:1693-1697 (1979).
76. I. Jane, *J. Chromatogr.* 111:227-233 (1975).
77. I. S. Lurie, *J. Liq. Chromatogr.* 4:399-408 (1981).
78. P. J. Twitchett, A. E. P. Gorvin, and A. C. Moffat, *J. Chromatogr.* 120:359-368 (1976).
79. A. B. Clark and M. D. Miller, *J. Forensic Sci.* 23:21-28 (1978).
80. I. S. Lurie and R. A. Henderson, 31st Ann. Meet. Am. Acad. Forensic Sci., Atlanta, Feb. 1979.
81. J. J. De Stefano and J. J. Kirkland, *Anal. Chem.* 47:1103A-108A, (Pt. 1), 1193A-1204A (Pt. 2) (1975).
82. W. A. Trinler and D. J. Reuland, *J. Forensic. Sci.* 22:37-43 (1977).
83. D. J. Reuland and W. A. Trinler, *J. Forensic Sci.* 11:195-200 (1978).
84. I. S. Lurie and J. Weber, *J. Liq. Chromatogr.* 1:587-606 (1978).
85. J. M. Weber, I. S. Lurie, and S. Blasof, 26th Ann. Conf. Mass Spectrom. Allied Topics, St. Louis, May 1978.
86. J. D. Henion, *Anal. Chem.* 50:1687-1693 (1978).
87. J. D. Henion, *J. Chromatogr. Sci.* 18:101-102 (1980).
88. J. D. Henion, *J. Chromatogr. Sci.* 19:37-64 (1981).
89. P. J. Arpino and P. Krien, *J. Chromatogr. Sci.* 18:104 (1980).
90. C. N. Kenyon and A. Malera, *J. Chromatogr. Sci.* 18:103 (1980).

91. B. L. Karger, D. P. Kirby, P. Vouros, R. L. Foltz, and B. Hidy, *Anal. Chem.* 51:2324-2328 (1979).
92. D. T. Luehl and P. R. Griffiths, *Anal. Chem.* 52:1394-1398 (1980).
93. J. F. Haw, T. E. Glass, D. W. Hausler, E. Motell, and H. C. Dorn, *Anal. Chem.* 52:1135-1140 (1980).
94. R. E. Majors, *J. Chromatogr. Sci.* 18:571-579 (1980).

5 Liquid Exclusion Chromatography
A Review for the Criminalist

W. W. McGee

University of Central Florida
Orlando, Florida

I. INTRODUCTION

In the crime laboratory during the course of working a case, the
criminalist will, most likely, perform a comparative analysis of
the physical evidence submitted. The criminalist will make a comparison of the chemical and physical properties of questioned
samples of physical evidence collected from the scene of suspect
criminal activities with known or comparison samples collected
from the suspect or victim's person or personal property. The end
product (goal) of this effort will be an attempt to make a state-
ment concerning the source of the origin of the Q and K samples.
That is (for example), could the taillight lens fragment found at
the scene of the crime come from the fragmented lens found on the
suspect's car?

The process of making a comparative analysis presents a number
of problems for the criminalist. For example: (1) During the process of making a comparative analysis the criminalist must be prepared, if necessary, to "look" at the physical evidence using a
large number of chemical, physical, and instrumental tests. The
data obtained from these tests must be interpreted and then inte-

266   McGee

grated into the final decision concerning source of origin. To
make this decision, the criminalist must have experience in interpreting data from many chemical, physical, or instrumental tests.
(2) Reliable reference or comparison standards must be available.
This restriction can be particularly frustrating to the criminalist.
Without relevant K (known) standards collected at the crime scene,
the comparative analysis will, most likely, never get beyond the
initial examination stage. On the other hand, in an era of mass-
produced consumer products made in batch lots and engineered to
be indistinguishable, the value of relevant comparison standards
may seem diminished. Certainly, the time involved in making a comparative analysis involving mass-produced items will probably be
increased because of the manufactured similarities.

Modern liquid exclusion chromatography (LEC), which includes
both gel permeation chromatography (GPC) and gel filtration chromatography (GFC), can provide a means of overcoming some of the problems inherent in making a comparative analysis. Modern LEC can
provide a fingerprint chromatogram for use in qualitatively comparing chromatograms, numerical (quantitative) data for use in a comparison of the polymeric portion of Q and K samples of physical
evidence, or it can provide a means of separating the polymer from
the additives in a Q or K sample for additional chemical or instrumental testing. If the criminalist routinely encounters polymer-
based products such as fibers, tape (wrapping, masking, electrical),
paint and paint chips, grease and oil, grills, taillight lenses,
and window and door moldings, to name a few, the criminalist should
seriously consider integrating LEC into crime lab procedures for
analyzing these types of physical evidence.

To realize the potential that LEC can offer, the criminalist
must have a working knowledge of the practical aspects of theory,
technique, and of the range of applications, both realized and potential. The goal of this review is to provide information to en-
courage the criminalist to use LEC when making a comparative analysis of physical evidence.

In accordance with this goal, this review will emphasize the following:

An overview of the elements of theory. This should be directed toward a practical understanding of column performance.

An overview of LEC applications. The number of publications citing specific applications of LEC in forensic sciences is very small [1, 2]. These references [1, 2], as well as others that show promise for use in the crime laboratory, have been included in the applications section of this review. Hopefully, this will provide an accurate representation of the potential to be realized by using LEC in the crime laboratory.

Liquid exclusion chromatography (LEC). A variety of names have been used to describe what, in this review, will be referred to as LEC. Gel permeation chromatography (GPC), gel filtration chromatography (GFC), and high-performance size-exclusion chromatography (HPSEC) are just a few of the names currently used in the literature. Throughout this review, the terms and relationships used will be consistent with those recommended in ANSI/ASTM D3016 [3]. The term *LEC* will be used to describe an analytical procedure for separating molecules by differences in size in solution and for determining the molecular weight distribution (MWD) of polymers.

II. OVERVIEW OF THE PROCESS

Before discussing LEC theory, it will be useful to look at the process of generating and interpreting an LEC chromatogram. Specific elements of the process are discussed in detail later in the chapter.

The sample to be analyzed is weighed and dissolved in a solvent effecting a satisfactory dissolution of the major portion of the sample, and one that is compatible with the LEC columns to be used. In most instances, the mobile phase will be chosen as the solvent to dissolve the sample. Exclusion chromatography normally requires only a single solvent in which to dissolve and chromatograph a sample. Since concentration has a pronounced effect on the chromatogram generated, the concentration of sample solution prepared will normally be limited to 0.01-0.5% (w/v) [4]. In many instances, not all of the sample will dissolve. The types of polymer products encountered in the crime lab typically contain dyes, fillers, antioxidants, and delustrants which may not dissolve in

the solvent chosen specifically for its ability to solubilize the
polymer, colorant or antioxidant, and so on. During the dissolving
process, the sample/solvent may be rocked or gently agitated, but
apparently (and there is some controversy here) *not* placed in a
sonic bath. After having adequate time to dissolve, the sample is
filtered (normally using a 0.45-μm filter). A portion of the filtered sample (a rule of thumb concerning sample volume will be presented later) is injected into the moving stream of the mobile
phase. Prior to injection the instrumental conditions should have
been established (hopefully optimized) and the column set (typically
one to six columns of usually different, but optimized nominal pore
size) should have been calibrated using (typically) a wide range of
polystyrene standards. The chromatogram is recorded using a sensitivity level consistent with the signal-to-noise ratio generated
by the detector and measured by the recorder/computer.

The LEC chromatogram has a well-defined (predictable) beginning and ending which is a function of the LEC separation mechanism
and; therefore, column flow rate. At a flow rate of 2 ml/min, each
column adds about 6 min to the total chromatographic time. The recording of peaks, or portions thereof, prior to or after these well-
defined limits usually indicates a problem.

When viewing the finished chromatogram, it should be noted
that as a result of calibration, the volume (time) axis has been
converted to molecular weight and that the recorded peak(s) may not
represent one isolated molecular species [as in high-performance
liquid chromatography (HPLC) or gas chromatography (GC)], but a
family of oligomers having a common monomer (molecular weight) base.
Depending on the size and number of oligomers present, the peak(s)
may be sharp or (typically) very broad. Satisfactory chromatographic
resolution (peak separation) of chemically different species may
be difficult to achieve, but usually through careful manipulation
of columns in the set, some improvement can be effected.

Perhaps one of the most unusual characteristics of the LEC
chromatogram, and the one that provides insight into the LEC

separation mechanism is that large (high) molecular weight molecules elute first and small (low) molecular weight molecules elute
last. Some examples of LEC chromatograms are given in Fig. 1.

At this point in the comparison, the criminalist has several
options available:

1. The chromatograms generated from the Q and K samples may be
   overlaid on a light box to effect a comparison. Total chromatographic resolution of components of the solution may not be
   necessary at this point. For samples that were dissolved and

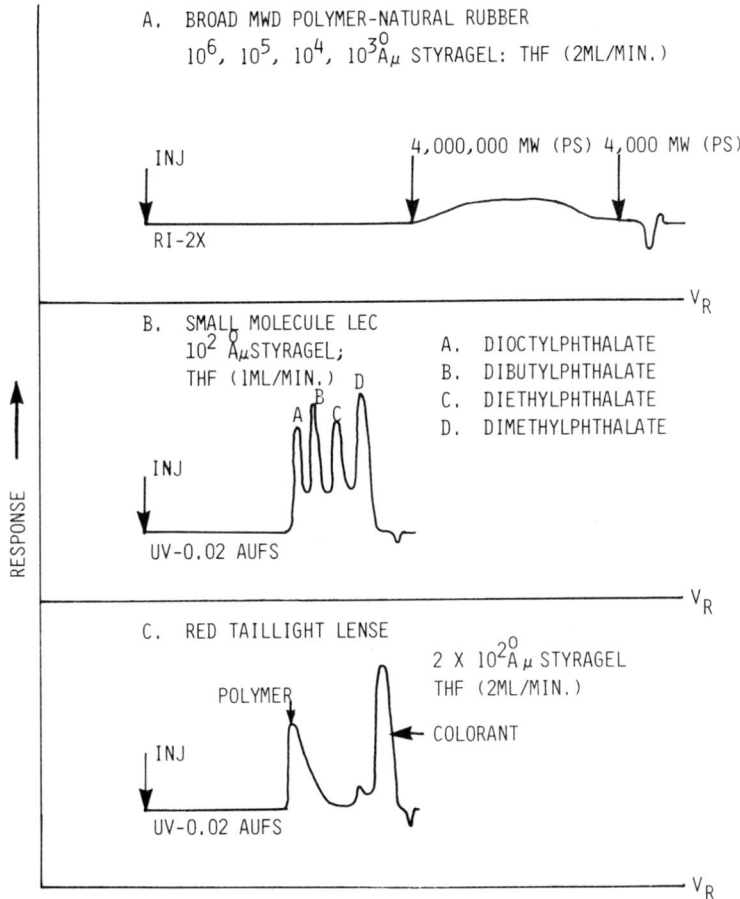

FIG. 1. Typical LEC chromatograms.

chromatographed in an identical manner, the presence of a bona
fide difference in Q and K chromatograms, in most instances,
will constitute a difference in source of origin. This is true
no matter how "good" or "bad" the technical aspects of the
resolution might be (Fig. 2).

2. If the polymer peak has been satisfactorily resolved, the
   criminalist may wish to calculate the molecular weight distribution (MWD) of the polymer to use in make the final quantitative comparison of Q and K.

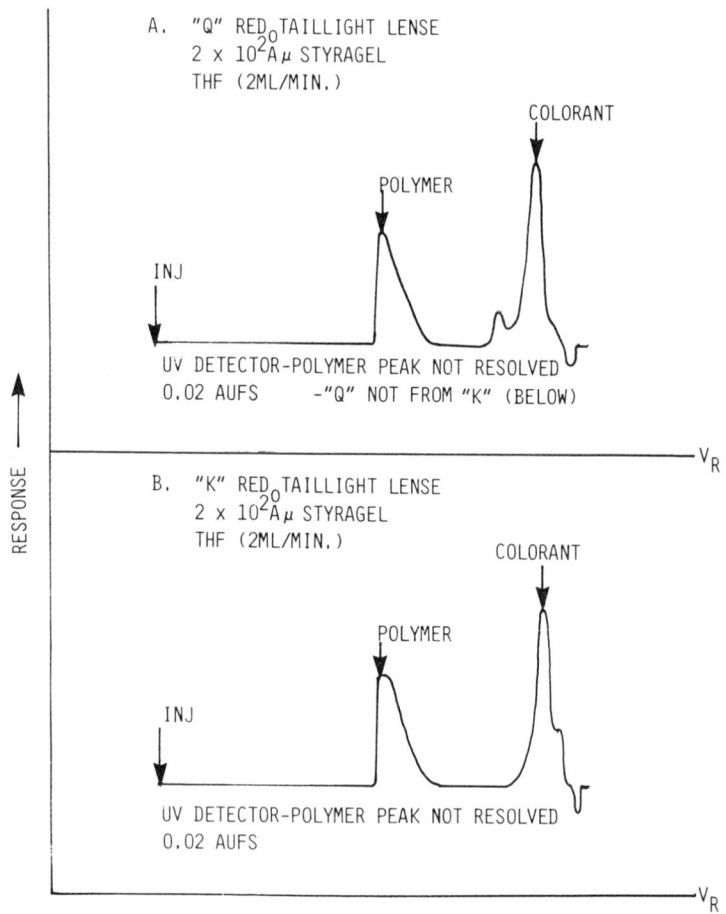

FIG. 2. Quick screen comparison chromatographs: Q and K red taillight lenses.

3. If satisfactory chromatographic resolution has *not* been achieved, then manipulation of pore size within the column set, changing the flow rate, and/or changing sample concentration (usually lowering it) may be used to improve resolution.

4. The sample may be rechromatographed and specific components (peaks) may be collected for additional chemical or instrumental testing. Using LEC, it is possible to recover the entire sample separated into its individual components.

5. The criminalist can turn to another instrumental technique better suited to providing data for making the comparative analysis.

III. SEPARATION MECHANISM AND SOLUTE RETENTION

A. Fundamental Relationships

When viewed from the outside, the microparticulate LEC column looks like any other HPLC column. Inside, the story is quite different. Each column is filled with microparticulates (less than 10 μm in size) of porous material (usually rigid styrene-divinyl benzene polymer or silica) of nominal pore size and moderately narrow pore size distribution. Columns are assembled into sets by pore size to effect a separation of molecules over the molecular weight range anticipated in the sample to be analyzed. When the set is connected to the typical HPLC instrument, the LEC capability is ready to be used.

Several theories have been proposed to explain, on the basis of molecular size in solution, the differences in solute retention volumes observed in LEC. Each theory approaches the separation mechanism, that is, the interaction of the solute and the stationary phase, in a slightly different manner. In recent years, two theories have shared the spotlight: the steric exclusion [7, 9] and restricted diffusion separation mechanisms [10, 11]. At the present time, researchers in LEC believe that the steric exclusion mechanism is most consistent with observed LEC phenomena [4]. For this reason, the steric exclusion mechanism will be used in this review of retention volume.

When the sample is injected, the solute band (zone) moves along with the mobile phase through the column and around the packing

particles. The solute molecules diffuse in and out of the pores of
the packing. According to steric exclusion [12], each molecular
species has available to it only a limited number of pores whose
size is greater than the molecular size of the solute in solution.
As a result of this pore size distribution, components of the solute mixture are separated by molecular size in solution as they
pass through the column. Large molecules (associated with high
molecular weight) elute from the column with a shorter retention
volume than small molecules (low molecular weight).

In LEC, the stationary phase is the solvent trapped in the
pores of the packing and the mobile phase is the solvent in the
interstices (between) of the packing. The mobile phase transports
the solute from the entrance to outlet of the column. Differences
in the rate at which each solute zone migrates through the column
are a function of the time each spends in the stationary phase.
This difference is reflected as a difference in the elution (retention) volume for each solute.

The volume of mobile phase in a column can be defined as

$$V_t = V_o + V_i \tag{1}$$

where V_o is the volume of mobile phase located between particles
(interstitial volume) and V_i is the volume of mobile phase in the
pores of the packing. V_t may be determined by chromatographing a
small (low molecular weight) nonretarded molecule, and V_o may be
determined by chromatographing a large (high molecular weight)
totally excluded molecule. Since V_o and V_i are fixed by the dimensions of the column and column packing, differences in retention
volume must come from variations in the distribution coefficient
(K_{LEC}) for each solute. The elution (retention) volume of each
solute (V_R) can then be described as

$$V_R = V_o + K_{LEC}V_i. \tag{2}$$

The distribution coefficient (K_{LEC}) must account for the distribution of solute between stationary phase (trapped liquid in the

packing pores) and mobile phase. In accordance with steric exclusion, it must account for the fraction of pore volume in the packing available to each molecular species and somehow be related to molecular size in solution (molecular weight).

Numerous mathematical models for describing the shape of molecules in solution and the shape of the pores in the packing have been formulated in an attempt to show the relationship between K_{LEC} and molecular weight [12, 13]. Although none of these models seems to be able to explain all possible cases (admittedly some do a better job in specific instances), they appear to reinforce the current understanding of the proposed size exclusion separation mechanism. Yau et al. [4] summarizes the current state of the size exclusion theory when he states: "The results of temperature, flow rate, and static mixing experiments clearly show that SEC retention is an equilibrium, entropy-controlled (ΔS), size-exclusion process. The mechanistic model indicates that solute diffusion in and out of the pores is fast enough with respect to (mobile phase) flow to maintain equilibrium solute distribution. Thermodynamic size exclusion is the fundamental basis common to all SEC theories. ..."

For the criminalist, the practical aspects of the retention volume relationship [Eq. (2)] should be mentioned:

1. In LEC it is assumed that the column packing is inert and that the solute does not interact with it. Under these conditions, K_{LEC} will normally never exceed unity ($0 \leq K_{LEC} \leq 1$); which means that V_R is always equal to or less than V_t. This fact accounts for the observed predictability of the LEC chromatogram.

2. The LEC calibration curve is a plot of the logarithm of polymer molecular weight versus K_{LEC} [i.e., $(V_R - V_0)/(V_i)$ or related quantity; elution (retention) volume (V_R)]. While typically sigmoid in shape, the calibration curve has a linear portion that can be expressed in the form

$$\log M = a + bV_R \qquad (3)$$

where M is the molecular weight of the polymer, and a and b are constants.

3. LEC separations are constrained to the limits imposed by V_R. Resolution is low. The criminalist must be content with chromatograms showing a few, typically broad, poorly resolved peaks. Modification of a column set (pore size) to improve a separation (resolve molecules) should be made only when the consequences of longer retention volume have been assessed.

4. When preparing polymeric samples for comparative analysis using LEC, polymer samples must be prepared in an identical fashion. Differences in sample preparation (between Q and K) could artificially create differences in molecular size in solution, thereby altering the LEC separation mechanism and subsequently distorting the chromatogram and the interpretation of any MWD data.

In light of these somewhat confining restrictions, it is reassuring to know that a great deal of "fingerprint"-type comparative information can be obtained relative to the molecular weight distribution of solutes in a sample using a single (30-cm) LEC column. Figure 3 illustrates this observation.

B. Band Broadening

In general chromatographic terms, band broadening describes the dilution of the solute zone (a single pure solute) as it passes down the column and through the chromatograph. The solute is injected as a dilute solution of small finite volume and, as a result of passing through the chromatographic system, normally elutes in a volume larger than that injected. As a result, in most chromatographic processes, peak widths generally increase (become broadened) with residence time on a column, and column efficiency (as measured by plate height) remains essentially constant. Figure 4 illustrates this effect.

Yau et al. [4] state: "All forms of band broadening are detrimental to chromatographic resolution. . . . This effect is to make peak identification and peak size analysis more difficult. . . . In the analysis of broad polymers by SEC, the effect of band broadening is to interfere with the integrity of the MWD information as displayed by the elution curve profile" -- that is, to make quantitative MWD data less reliable.

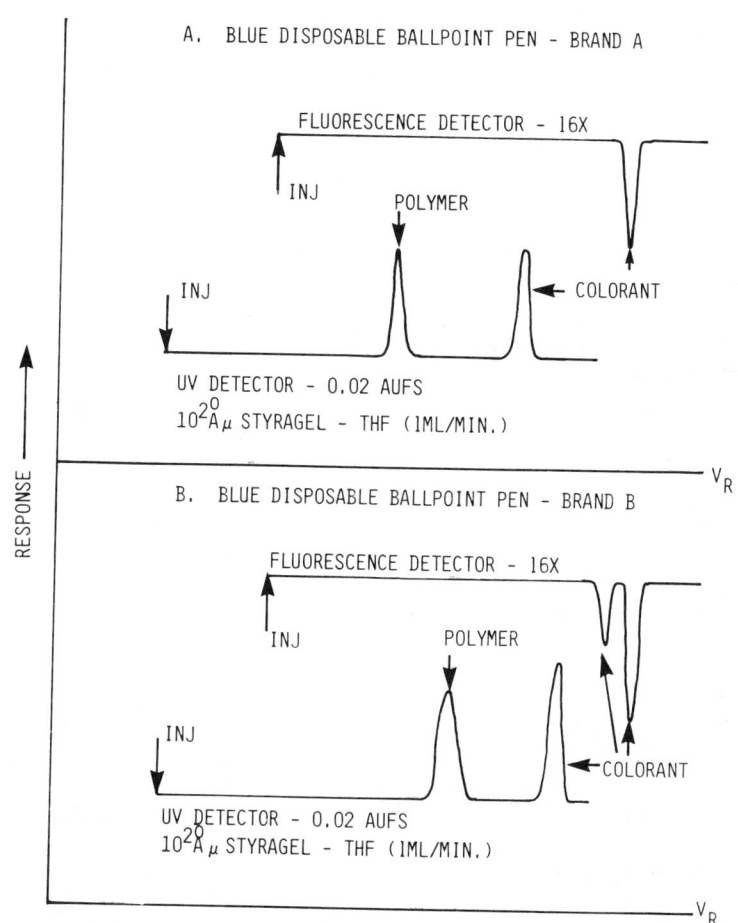

FIG. 3. Fingerprint comparison-type chromatograms: disposable ball-point ink pens.

The most compelling reason for discussing the solute broadening process in LEC is to gain a better understanding of those factors that contribute to it. From this understanding should come a pragmatic approach to manipulating LEC variables to achieve a desired separation.

The first step in this learning process is to recognize the potential sources of broadening in the chromatographic system. Hendrickson [14] and Lurie (Chap. 4 of this book) have described

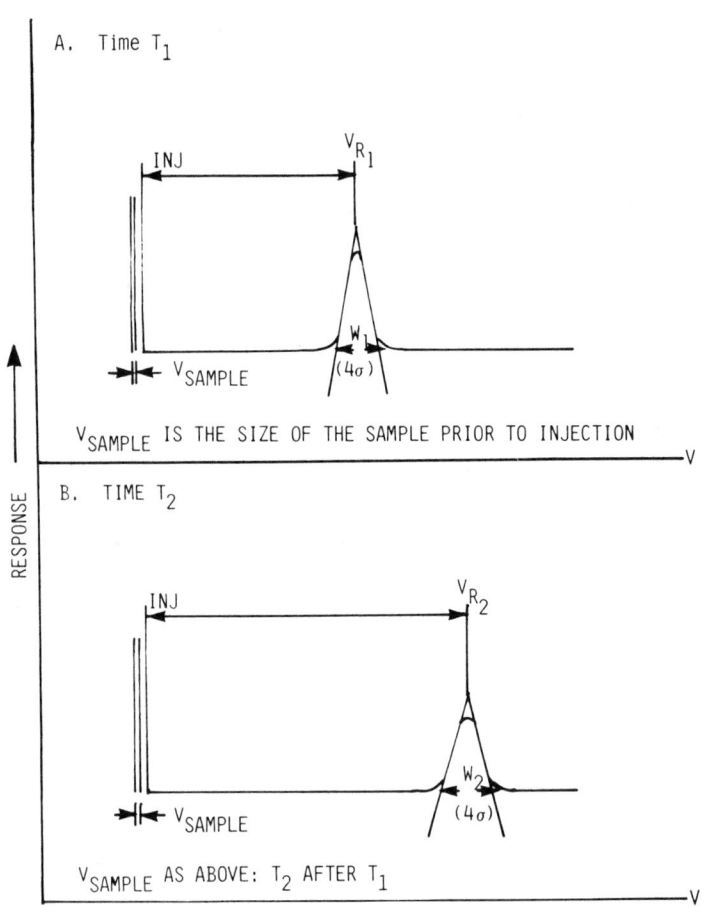

FIG. 4.   Solute dispersion-zone spreading.

the potential sources of column and extra-column (fittings, tubing,
detectors, etc.) band broadening. The next step is to focus on
those sources that contribute to broadening in the packed column.
(The reader may wish to consult Glueckauf [15], Van Deemter et al.
[16], or Lurie (Chap. 4) for a fundamental discussion of the origin
of band broadening in a packed chromatographic column and the use
of plate height as a means of measuring column performance [i.e.,
plate height -H)].) The functional dependence of H (plate height)
in HPLC has been written [14; and Chap. 4] as

$$H = H_E + H_d + H_m + H_{sm} + H_s \qquad (4)$$

where H_E, H_d, H_m, H_{sm}, and H_s are the plate height contributions due to eddy diffusion (H_E), longitudinal diffusion (H_d), mobile phase mass transfer (H_m), stagnant mobile phase mass transfer (H_{sm}), and stationary mass transfer (H_s) in HPLC processes [4].

A similar expression showing the functional dependence of H in LEC can be written from Eq. (4) by removing terms that do not apply. In Eq. (4) the longitudinal diffusion contribution (H_d) has been demonstrated to make a minor contribution in LEC [4]. The stationary phase mass transfer (H_s) has no counterpart in LEC. Giddings [17] has shown that an accurate description of H at high flow rates requires a coupling between the eddy diffusion (H_E) and mobile phase mass transfer (H_m) contributions (using H_M as the coupled term). The resultant expression describing plate height in LEC then becomes

$$H_{LEC} = H_M + H_{sm} \qquad (5)$$

Giddings [17] has expressed Eq. (5) in terms of the individual column parameters involved:

$$H_{LEC} = C_{SM} \frac{\nu d_p^2}{D_{SM}} + \frac{1}{1/ad_p + D_M/C_M \nu d_p^2} \qquad (6)$$

where

$\quad d_p$ = particle size diameter (of packing material)

$\quad D_{SM}$ = solute diffusion coefficient in stagnant mobile phase

$\quad D_M$ = solute diffusion coefficient in interparticle mobile phase

$\quad$ a and C_{SM} = coefficients of D_{SM} and D_M

The magnitude of a and C_{SM} are generally a function of the geometry of packing and pore structure. The significance of Gidding's expression lies in the fact that D_M and D_{SM} are dependent on solute molecular weight, thus making band (zone) broadening a function of sample molecular weight.

The rate theory of chromatographic band broadening [18, 19] has provided additional insight into those column parameters that influence the exclusion process. From rate theory, the stagnant mobile phase mass transfer term (H_{SM}) can be expressed in terms of experimental parameters as

$$H_{SM} = C_{SM} \frac{v d_p^2}{D_{SM}} = \frac{K_{LEC}(V_i/V_o)}{30[1 + K_{LEC}(V_i/V_o)]^2} \tag{7}$$

where V_i/V_o is the ratio of pore volume to void volume (all other terms previously defined). Combining Eqs. (6) and (7) gives the following expression for total plate height contribution in LEC:

$$H_{LEC} = \frac{K_{LEC}(V_i/V_o)}{30[1 + K_{LEC}(V_i/V_o)]^2} + \frac{1}{1/ad_p + D_M/C_M v d_p^2} \tag{8}$$

A significant amount of experimental work [4, 18, 20-22] has been conducted to examine the influence of parameters found in Eqs. (5 to 8). Some of the results of this experimental work show:

1. Column efficiency (H_{LEC}) decreases with increasing solute molecular weight because solute diffusion coefficients (D_{SM}'s) decrease with increasing molecular weight. As the molecular weight increases, the relative size of the molecules in solution increases and diffusion of the larger molecules inside the pore structure of the packing becomes more restricted. This phenomenon produces a distinctive characteristic of LEC chromatograms not observed in other forms of HPLC. The small retention volume (early-eluting high molecular weight) peaks may actually experience more band broadening than the large elution volume (late-eluting low molecular weight) peaks. In theory, column efficiency may increase slightly with elution time!

2. A decrease in support size (d_p) produces a more efficient column (H_{LEC}) because small particles permit a more homogeneous packing of particles and because movement of solute molecules in and out of the stagnant mobile phase in the pores of small molecules

is faster because penetration into the pores is not as deep in small particles as it is in large particles.

3. A decrease in mobile phase velocity (ν) normally produces a more efficient column. Small molecules (low molecular weight) do not seem to be as susceptible to changes in mobile-phase velocity as larger (high molecular weight) molecules do. Flow rate optimization for small molecules using small particle packing does not appear to be as critical as with large molecules.

4. The principal effect of a high-viscosity mobile phase is to decrease solute diffusion coefficients and to decrease column efficiency. In practice, optimization of LEC separation by using solvents of different viscosity is not a common practice. Solvents are usually selected for their demonstrated ability to dissolve a sample.

5. Solute diffusion coefficients (D_{SM}'s) and column efficiency are influenced by mobile-phase temperature. In most instances, LEC analysis will be performed at room temperature. High-temperature LEC analyses require special chromatographic equipment and special venting facilities for obnoxious/dangerous solvent gases.

6. The column parameter (V_i/V_o) is proportional to the porosity. The internal porosity is a column characteristic determined by the manufacturer of the column and one that is not widely advertised. LEC theory predicts that for optimum resolution and highest molecular weight accuracy, the internal pore volume (or specific porosity of the packing) should be as large as possible with a moderately narrow pore size distribution [4].

The purpose of this discussion was to identify the factors that contribute to band broadening. Very simply, knowledge of these factors can mean proper control of band broadening, which can result in greater column efficiency, improved resolution, and higher MWD accuracy. For the criminalist anticipating using LEC, the study of factors that influence band-broadening processes can provide a pragmatic approach to organizing and conducting comparative studies using LEC. This pragmatic approach might include the following guidelines:

1. Optimize the injection volume. Yau et al. [4] have suggested
    the following as a guide to sample injection volume: "Sample
    injection volumes should be limited to one-third or less of the
    base line volume of a monomer peak with a very small sample.
    . . ." In practical terms, injection volumes will normally be
    less than 200 μl for columns of 0.6-0.8 cm ID and 50-100 cm
    long [4].

2. Use low/no "dead volume" fittings and minimize the lengths of
    connecting tubing between columns, detectors, and so on. Use
    the appropriate internal diameter of tubing.

3. Following the suggestion given in ANSI/ASTM D3593 [23], a plate
    height (H) versus mobile-phase velocity (ν) plot should be con-
    structed when first evaluating a column or column set using both
    a small molecule (low molecular weight) and a large molecule
    (high molecular weight) standard. Construction of such a plot
    will provide an understanding of those factors that contribute
    to plate height (H_{LEC}). Such a plot can be used to monitor
    changes in column efficiency with age, and to detect undesirable
    changes in band broadening arising from sources outside the
    column (e.g., detector, fittings, etc.)

4. Although rarely used, the influence of temperature and visco-
    sity on column efficiency should be kept in mind, particularly
    when faced with a difficult analysis problem.

5. Before purchasing LEC columns, the laboratory's needs for LEC
    should be defined. Column specifications from several manu-
    facturers should be checked. Columns of appropriate particle
    size, length, diameter, solvent compatibility, and most import-
    ant, pore size and pore size distribution should be purchased
    to meet these needs.

C. Column Performance: Calibration, Resolution, and the
    Quantatitative and Qualitative Aspects of Data Interpretation

Although HPLC and LEC show many common elements of theory and equipment, this commonality does not extend to column selections, calibration, and the interpretation of data. It is in this area that
the size exclusion separation mechanism manifests itself.

Column calibration serves a somewhat different function in LEC
than in HPLC. In LEC, calibration converts elution volume (V_R) into
relative molecular weight to permit the calculation of MWD data.
The process of calibrating a set of LEC columns is described in detail in ANSI/ASTM D3536 [25]. The process consists of injecting a
series of standard polymers of known MWD and recording the peak
(apex) retention volume for each. This method of calibration is

sometimes referred to as the peak position method. Once the peak
retention volume (V_R) for each standard has been recorded, the calibration curve is constructed by plotting on semilog (at least three-
cycle) paper the retention volume (V_R) for each standard on the X
or linear axis against the corresponding value of $(M_w \cdot M_n)^{1/2}$ on the
Y or logarithmic axis. The value of M_w and M_n for each standard,
that is, the number and weight average molecular weight, are provided by the manufacturer. A typical calibration curve is shown
in Fig. 5. It consists of three distinct regions: the total permeation (low molecular weight), linear response, and total exclusion (high molecular weight) regions. The principal reason for
assembling a column set is to attempt to put the elution profile of
the polymer to be analyzed in the linear response range (elution
volume) of the calibration curve. This is one of the most important criteria for making accurate and reproducible MWD calculations.
A great deal of uncertainty is introduced into any MWD calculation
when the elution profile of a polymer unavoidably falls in either
(or both) the total exclusion or permeation regions.

   Calibration and regular recalibration of a column set are
necessitated by the following:

1. The size of any molecule in solution is a function of the chemical nature and molecular weight of the polymer, the MWD of the
   polymer, and the solvent in which it is dissolved/chromato-
   graphed. Any change in one of these factors will necessitate
   calibration.

2. Band broadening can result from improper fittings, poorly designed detectors, and too large an injection volume, in addition to column dispersion. Any change in the general overall
   assembly of an LEC instrument will necessitate column calibration.

3. The criminalist may use a number of different columns in a
   variety of combinations during the course of examining the wide
   variety of physical evidence encountered in the crime laboratory. Each assembly of columns must be calibrated before MWD
   data can be generated.

   For highest accuracy, columns should be calibrated using
standard polymers of the same chemical type as those to be investigated [e.g., polystyrene standards (PS) for PS polymers, polyvinyl

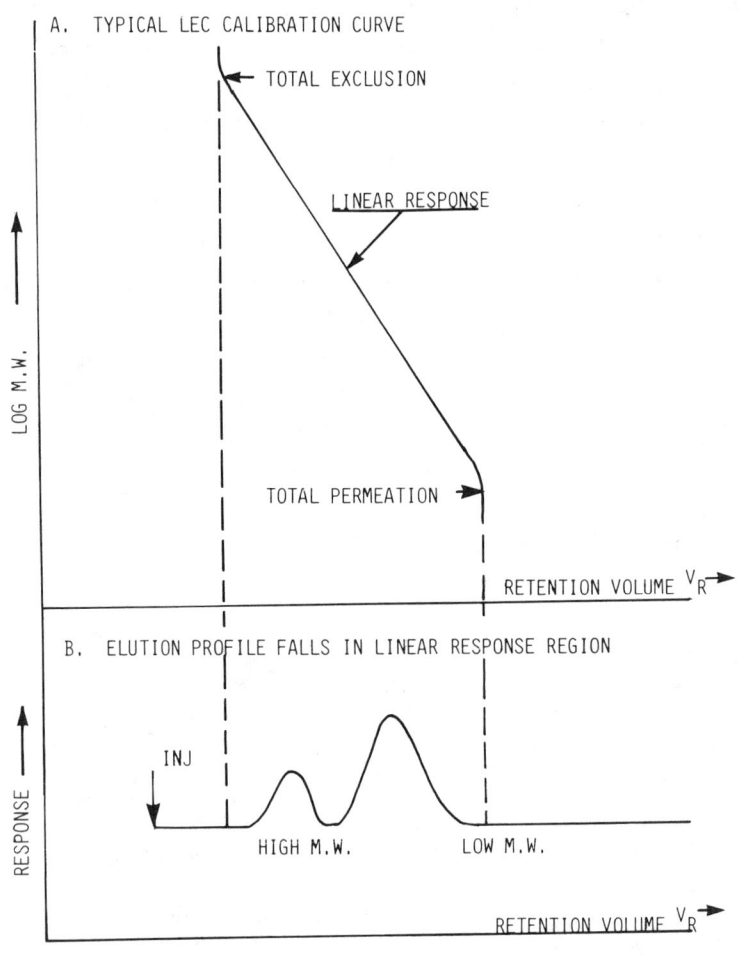

FIG. 5. Typical LEC calibration curve and elution profile.

chloride (PVC) standards for PVC polymers]. In reality, this is
almost never accomplished (except in the case of a PS polymer
analysis) because of the absence of a variety and range of "standard" polymer types of known narrow MWD (d $\leq$ 1.1). PS is about the
only polymer available in sufficiently narrow MWD range to be useful for calibrating LEC columns. The significance of this fact
should not be overlooked in subsequent MWD calculations. Once a
column set is calibrated, all MWD calculations become PS based
(i.e., linear, random coil polymer based). For polymers of different chemical type or shape in solution, the polystyrene-based MWD
data provide a numerical ranking method. The actual value of MWD
for a polymer may be very different from the one calculated based
on PS calibration data. The MWD of a polymer (regardless of shape
in solution) can be calculated from its hydro-dynamic volume, using
Benoit's [24] universal calibration method. The universal calibration method is described in ANSI/ASTM D3593 [23].

Before discussing the qualitative and quantitative aspects
of LEC, a discussion of the current rationale for assembling columns
into sets to achieve a desired separation is necessary. Again, it
must be emphasized that column selection in LEC differs considerably from HPLC. The intent of column selection in LEC is to resolve molecules based on size in solution and to place the elution
profile of the polymer on the linear portion of the calibration
curve. Little concern is given for the chemical nature of the
molecule (although fundamental exclusion theory assumes no interaction with packing material, some unwanted interaction is possible,
particularly using aqueous-based mobile-phases and silica-exclusion
columns.)

A study of column efficiency (H) can provide an understanding
of those column factors that influence the separation mechanism.
A study of resolution can provide insight into the reasons for
coupling columns to complement the separation mechanism.

The quantitative expression for degree of resolution in HPLC is

$$R_{1,2} = \frac{2(V_{R_2} - V_{R_1})}{W_1 + W_2} \qquad (9)$$

The meaning of the terms used in Eq. (9) should be clear from a study of Fig. 6.

This expression is not particularly suited to LEC. An expression describing resolution in terms of the molecular weights of the solutes involved would be more useful. For peaks eluting in the linear response region of the calibration curve, Bly [26] has proposed an expression (R_s, specific resolution) of greater functionality in LEC,

$$R_s = \frac{2(V_{R_2} - V_{R_1})}{W_1 + W_2} \left(\frac{1}{\log MW_2/MW_1} \right) \qquad (10)$$

where

$$R_s = R_{12} \left(\frac{1}{\log MW_2/MW_1} \right) \qquad (11)$$

An interesting sidelight to Eq. (11) becomes apparent when

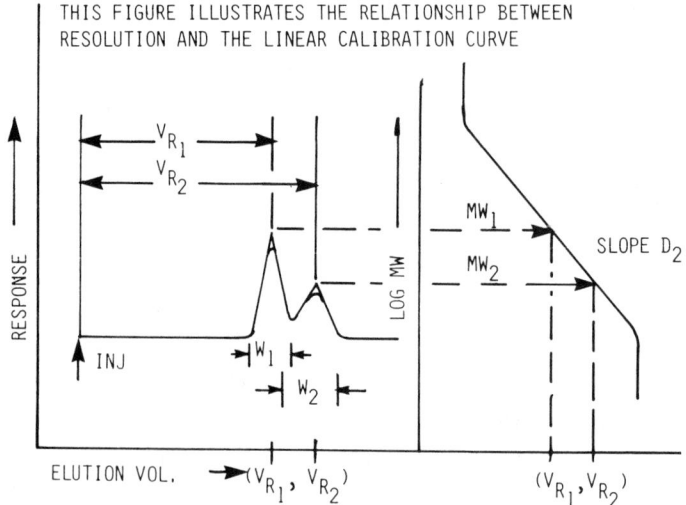

THIS FIGURE ILLUSTRATES THE RELATIONSHIP BETWEEN
RESOLUTION AND THE LINEAR CALIBRATION CURVE

FIG. 6.   Illustration of resolution relationship and linear calibration curve.

$$R_{12} = 1 \text{ (i.e., 98\% resolution between peaks)}$$

Then,

$$R_s = (R_M)^{-1}$$

where R_M is the minimum molecular weight ratio necessary to produce unit resolution.

From Eq. (11) Bly [26] has derived an expression which (from the linear portion of the calibration curve for each column) would permit a comparison of column packings (and column sets):

$$R_{sp}^* = \frac{0.567}{\sigma D_2 (L)^{1/2}} = \frac{0.144}{W D_2 (L)^{1/2}} \tag{13}$$

where L is the column length. R_{sp}^* is the usual chromatographic resolution ($R_{1,2}$) for a pair of peaks having a decade of MW difference compensated for the length of the column. Kirkland [27] has used the R_{sp}^* value to compare different LEC columns/column packing materials.

Several observations have been made as a result of using and studying Eq. (13). Some of these are:

1. W(4σ) is related to column dispersion (broadening) processes. From the prior discussion of H [Eq. (5 to 8)], it should be clear that those factors that influence column efficiency (broadening processes) also affect resolution.

2. Optimizing resolution involves manipulating those column factors that minimize the slope (i.e., a smaller value of D_2 means better resolution) of the calibration curve.

3. The combined effects of porosity (internal porosity), pore size distribution, pore size, and shape on D_2 have been studied using theoretical LEC relationships concerning the shape of molecules in solution. Of the factors studied which affect column performance, porosity has been shown to be more important than PSD, pore size, or shape [4, 28, 29]. These studies have also shown that the linear size separation range for LEC is greatly extended by optimizing PSD, pore size, and internal pore volume of particles in the column. As a result of these studies, the first set of guidelines for combining columns into sets was established.

For the criminalist, the coupling of columns into sets (by pore size) represents a compromise between two opposing factors: (1) a

wide molecular weight calibration range for convenience and versatil-
ity in handling the diversity of polymer produces encountered, and
(2) a linear response range of the calibration curve consistent
with the elution profile of the polymer to provide maximum accuracy
in calculating MWD.

In the literature, two procedures for coupling columns of different pore sizes are described; these are, the bimodel pore size
distribution (BPSD) procedure and (lacking a more accurate term)
the empirical procedure (as represented in ANSI/ASTM D3593 [23]).

The ASTM procedure recommends coupling four columns of pore
sizes 10^3, 10^4, 10^5, and 10^6 Å. Little, if any, rationale is given
for choosing columns with these pore sizes except (by inference)
that the columns will be used for generating MWD data for a wide
range of different polymers. (Universal calibration curve data for
approximately 25 different polymer types is given in ANSI/ASTM D3593
[23].)

The coupling of columns using the BPSD principle [4, 29] provides a structured approach for coupling columns of different pore
sizes to effect a desired MW resolution. Although most researchers
believe that BPSD principles work better for rigid packings (where
pore volume does not vary with the type of mobile phase used), pore
volume calculations have been made for both silica and SDVB packings. According to Bly [4], pores of one size can fractionate
polymers over nearly two decades of molecular weight range. For
separating sample components that extend over two decades of molecular weight, a column set consisting of two pore sizes (bimodal)
is optimum. Columns are combined into sets by: (1) combining
columns of nearly equal pore volume, and (2) adjacent but not over-
lapping pore size.

To make use of the BPSD selection principle, the criminalist
must prepare separate calibration curves for each column available.
From each column calibration curve, the slope of the linear portion
is calculated in terms of milliliters per decade of polymer MW resolved. Figure 7 illustrates this calculation for several μStyragel
columns.

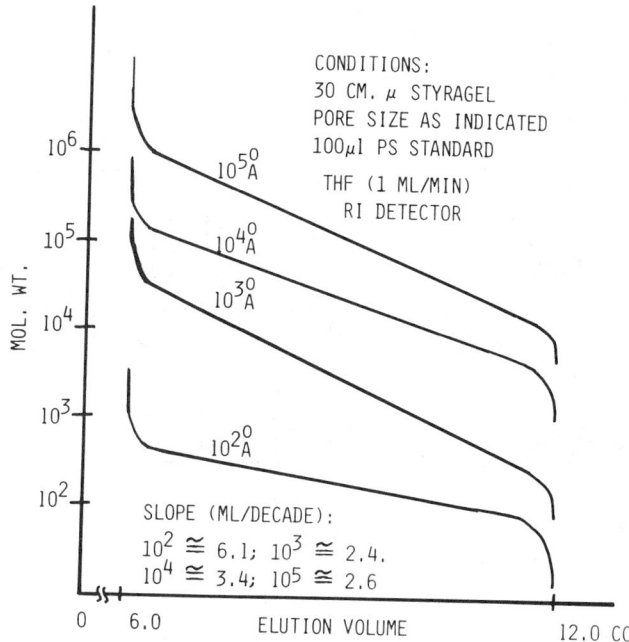

FIG. 7. Pore volume calibration for individual μStyragel columns.

The principles of BPSD can be illustrated from data taken from Fig. 7. To provide an effective linear separation range of about 10^2-10^6, a 10^3-Å μStyragel column with a pore volume of 2.4 ml/decade (and a linear molecular weight separation range of approximately 10^2-10^4) can be combined with a 10^5-Å μStyragel column (linear range 10^5-10^6) with a calculated pore volume of 2.6 ml per decade. It has been claimed [4] that through proper selection, a bimodal (two-pore-size) column set can give a wider useful MW separation range and a more even distribution of MW resolution than can be produced by the conventional (empirical) method. Within any linear separation range, once the pore sizes have been selected, resolution can be improved by coupling columns of the same pore size and volume (this has the effect of decreasing the slope of the calibration curve by doubling the pore volume). Figure 8 illustrates this observation.

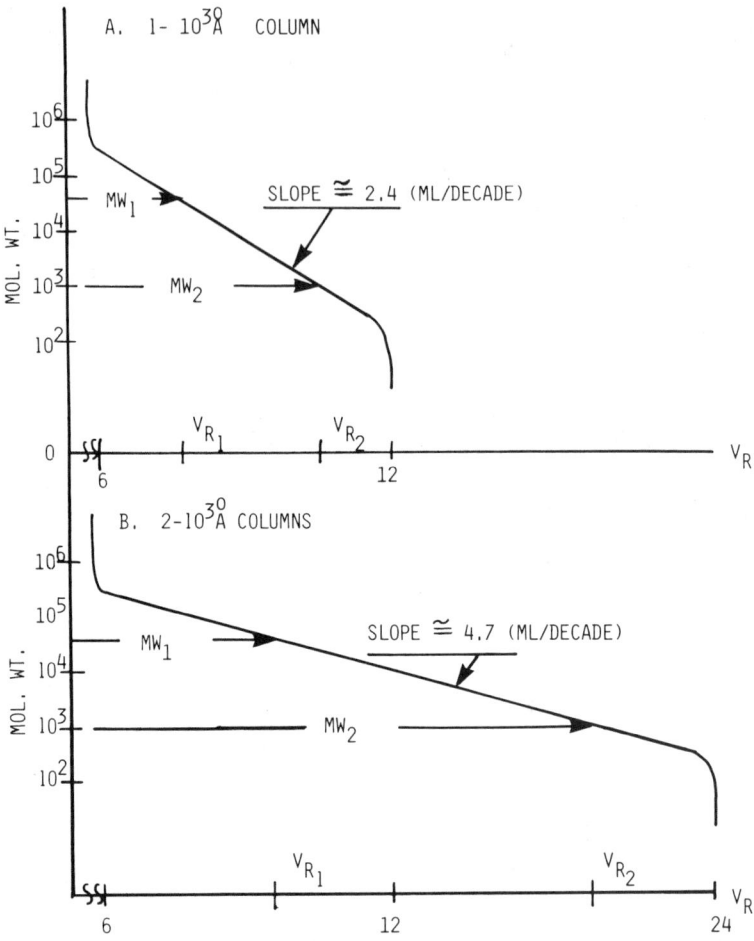

FIG. 8. Improving resolution within a pore size/pore volume range.

From a practitioner's point of view, a conscious effort should
be made to tailor columns to the separation problem at hand. The
use of columns which are, from a resolution (pore size) point of
view, not necessary can increase the elution volume (analysis time),
cause a decrease in the linear response range of the calibration
curve, and seriously affect the accuracy of the MWD data. Time
initially spent calibrating each column will be returned when the

criminalist combines the proper columns to effect a desired
separation/resolution. Using this calibration information, specific columns can be combined with authority to provide insight into
the range of molecular weight present in an unknown polymeric sample,
a "quick" screen to determine the presence of major differences in
polymer samples, or to make an "exacting" comparison of MWD data
from Q and K samples. If, at any point in the scheme for combining
columns, Q and K chromatograms show differences in peak shape, number, size, or slope, the comparison can stop!

The principles of BSPD coupling provide another (perhaps not
intended) feature which can be used in setting up specification
guidelines for purchasing columns. Since LEC columns are expensive,
a column set of two different pore sizes covering the widest linear
molecular weight range would seem like a sound initial investment.
Such a set would provide useful information in evaluating the diverse types of polymer products encountered in the crime lab. Additional column purchases would be directed toward improving resolution for specific application problems. In terms of making a comparison of physical evidence based on MWD calculations, the two-
column set may provide much of the information needed.

The interpretation of LEC chromatograms can be quantitative
and qualitative. A quantitative evaluation of an LEC chromatogram
means calculation of MWD. There are at least five terms commonly
associated with explicit MWD calculations. Three of them ($\overline{M}_w$, $\overline{M}_n$,
and d) are encountered more often than the other two ($\overline{M}_z$ and $\overline{M}_v$).
While these terms have a mystical aura about them (probably arising
from the mathematical nature in which they are calculated), they
are in fact physical properties of some significance which can be
used to describe the polymer under examination. These terms are
described as follows:

$\overline{M}_n$: *Number average molecular weight.* This is molecular species
population molecular weight average and describes the number
of molecules or equivalents per unit weight of sample. $\overline{M}_n$ is
a colligative property such as boiling point elevation, freezing point depression, or vapor and osmotic pressure. Polymer

chemists/manufacturers/fabricators are interested in this property because it can be related to the brittleness or flow properties of the formulated polymer mixture.

$\overline{M}_w$: *Weight average molecular weight.* This is the weighted average
of the mass of the individual species comprising the sample.
$\overline{M}_w$ correlates with material properties that are dependent on
mass fraction (i.e., melt viscosity and mechanical properties
such as tensile strength and hardness). It is sometimes referred to as the "light scattering" molecular weight because
the scattering intensity of any polymer is a function of the
second power of the molecular weight. Large molecules contribute most to scattering intensity.

d: *Polydispersity* ($\overline{M}_w/\overline{M}_n = d > 1$). A measure of the width of the
molecular weight distribution of the polymer. PS molecular
weight standards are manufactured such that $\overline{M}_w/\overline{M}_n = d \geq 1.1$.
(Note: $\overline{M}_n = \overline{M}_w$, when $d = 1$.)

M_z: *Z average molecular weight.* Sometimes referred to as the "centrifugation" molecular weight because when dilute polymer solutions are subjected to a centrifugal field eventually a thermodynamic equilibrium will be established where the molecules
become distributed according to their molecular sizes. Very
large molecules settle most in the gravitational field in the
centrifuge. The Z average molecular weight is weighted toward
the higher molecular weight population of MWD. It has been
correlated with material properties such as flex life and
stiffness, which are dependent on the very high molecular component of the sample.

$\overline{M}_v$: *Viscosity average molecular weight.* Dilute solution viscosities are used to measure molecular weight. Dilute solution
viscosities measure the frictional resistance to flow that results from the presence of different-size molecules of polymer
in dilute solution. The larger the molecule, the more interactions, the greater the viscosity, and thus larger molecules
contribute more to viscosity average than do small molecules.

Figure 9 shows a generalized elution profile for a polymer and
the approximate location of each molecular weight type. The molecular weight terms may be calculated from the elution profile by
using the following mathematical relationships:

$$\overline{M}_n = \frac{\sum_{i=1}^{n} h_i}{\sum_{i=1}^{n} h_i/M_i} \tag{14}$$

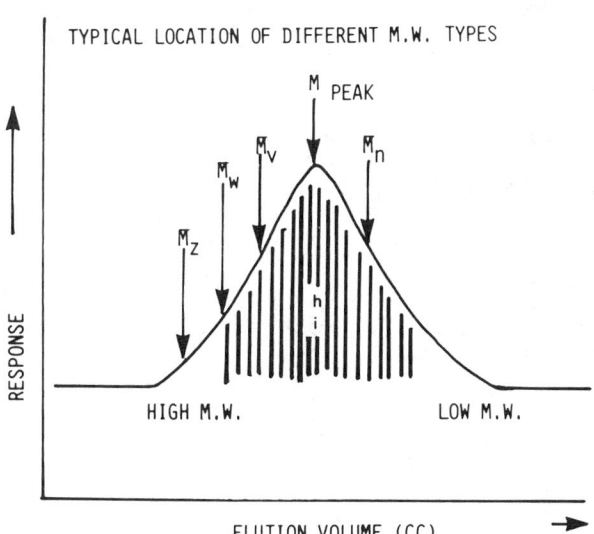

FIG. 9.   Generalized location of different molecular weight types.

$$\overline{M}_w = \frac{\sum\limits_{i=1}^{n} h_i M_i}{\sum\limits_{i=1}^{n} h_i} \qquad (15)$$

$$\overline{M}_z = \frac{\sum\limits_{i=1}^{n} h_i (M_i)^2}{\sum\limits_{i=1}^{n} h_i M_i} \qquad (16)$$

$$\overline{M}_v = \left[\frac{\sum\limits_{i=1}^{n} h_i (M_i)^a}{\sum\limits_{i=1}^{n} h_i}\right]^{1/a} \qquad (17)$$

where h_i is the peak height for molecular weight M_i, and a is the exponent constant of the Mark Houwink relationship [23].

Yau et al. [4], Cazes [30], and ANSI/ASTM D3593 [23] provide a detailed discussion of the manual digitization procedure required to calculate the different molecular weight terms from a peak elution profile.

For the criminalist interested in making MWD calculations and using the subsequent data in a comparative analysis, the following thoughts are offered:

1. The calculated MWD values are valid *only* for the *standard* polymer used to calibrate the column set (PS, in most instances). For other polymers, the PS-based MWD values are useful counting or measuring devices. The real MWD values for the actual polymers involved may be significantly different from those calculated.

2. The manual digitization of an LEC chromatogram into MWD values is a tedious, error-laden procedure best left to processor-controlled computerized data handling equipment. Instrument manufacturers are just now beginning to develop this worksaving potential.

3. The accuracy of computer-derived MWD data is a function of the accuracy of the calibration curve and range of linearity, sensitivity of the detection system to small changes in the elution profile, concentration of sample injected, and band-broadening processes (to name just a few).

4. Little information exists concerning the origin, nature, and significance of the MWD variations measured within a sample (the typical "statistical" variation of MWD values due to random error) and between samples ("batch-to-batch" variations) as measured by LEC. Whereas a 10% change in the colligative property freezing point depression can mean the difference between a frozen engine block and a smooth-running engine, a 10% variation in $\overline{M}_n$ does not appear to generate the same imagery in polymer science.

The criminalist interested in making MWD calculations and attempting to interpret the results in terms of a common source of origin will probably become disheartened with these efforts. To be able to say, based on computed MWD values differing by 3%, that the Q and K samples could have (with reasonable certainty) come from the same source of polymer is at present not possible. The state of the art for quantitative interpretation of MWD values is certainly a fertile field for future research!

For the criminalist, a qualitative evaluation of LEC data can hold some promise of reward. A qualitative evaluation of chromatograms involves overlaying the Q or K chromatograms of the polymer samples on a light or view box. As a measure of performance, chromatograms generated from consecutive injections of the same dilute polymer solution should be compared. For an LEC system of reproducible performance, the start of the chromatograms (injection mark) and the end of the chromatograms (marked by low molecular weight peaks due to solvent or impurities) should be identical. The number, size, and shpe of peaks as well as intermittent peak slopes should differ by no more than the width of the recorded line. The numerical value of the baseline (i.e., percent of scale) before the injection was made and after the low molecular weight peaks have eluted (i.e., baseline recovery) should be identical. In making a comparison of physical evidence, it would seem reasonable to expect the same performance from consecutive injections of two dilute polymer solutions (prepared in identical fashion) of a common origin or manufacturing background. The actual range of variations that one sees from chromatogram to chromatogram will be different for each LEC chromatographic system. The criminalist will have to decide (on an individual basis) what constitutes a "normal" and what constitutes a "significant" variation in elution profiles in a chromatogram.

On the other hand, for LEC chromatograms generated from dilute solutions using a reproducible chromatographic system, a significant variation in the size and number of peaks, or slopes and valleys between peaks, or the start and finish of the low molecular weight peaks would tend to indicate differences in the polymer samples and, therefore, source of origin and/or manufacturer!

IV. APPLICATIONS

A. Practical Considerations

The following is one approach to performing LEC analyses.

1. Regularly measure the mobile-phase flow rate at different flow rates over time periods consistent with the length of chromatograms typically generated. Undetected variations in flow rate constitute a serious source of error that can detract from MWD calculations.

2. Get to know your columns. Using appropriate molecular weight standards, calibrate each LEC column and construct a calibration curve for each column. From each calibration curve, determine the total exclusion and permeation limits, the linear response range, and the pore volume. Combine columns into sets through judicious choice of pore size following guidelines of the type suggested by BPSD. The criminalist may want to assembly one set of columns for screening samples (usually low V_t), another set for collecting separated portions of a sample, and a third set for making MWD calculations. A small number of columns may be used interchangeably to construct sets for different purposes. When assembling columns into sets, manufacturers do not seem to agree as to how columns should be assembled by pore size. Consult manufacturer's literature for recommendations concerning the order for assembling your columns into sets.

3. Optimizing flow rate (i.e., construct a Van Deemter plot of H versus ν) using molecular weight standards of approximately the same molecular weight expected in the physical evidence under examination. The actual flow rate used in generating a particular chromatogram should be consistent with the ultimate use of the chromatogram. Where peak resolution is the primary goal, flow optimization will be important. For a quick screen chromatogram designed to establish the future course of the comparison, flow rate optimization may not be critical.

4. Optimize the concentration of the sample to be analyzed (injected). The sample concentration must be optimized to (a) ensure reproducible operation of the LEC size separation mechanism, (b) ensure adequate detector/recorder/computer processor sensitivity, (c) ensure valid qualitative and quantitative comparison of chromatograms, and (d) provide a weight basis for comparing insoluble residues collected by filtering the samples prior to chromatographing. Small variations in peak size can be corrected by normalizing peak height prior to MWD calculations.

The final concentration of the sample used in any comparison
of physical evidence will depend on (a) the amount of sample available for analysis (as little as 1 in. of a fiber sample has been
dissolved and chromatographed), (b) the molecular weight of the
polymer in the sample (the higher the molecular weight of the polymer, the lower the concentration to be used -- 0.01-0.05%), and
(c) the volume of resultant solution. There is a physical limit to
the volume of solution that can be manipulated and filtered (5 cc
seems to be near this limit). The volume of sample solution also
becomes a factor when more than one injection will be made (e.g.,
to increase the concentration of a collected fraction).

  5. When initially setting up an instrument or when modifying
or adding components to an existing instrument, be aware of the
deleterious effects of band broadening. Use low/no dead volume
fittings. Minimize the length of connecting tubing before and after
the column set. After the column set, use the proper diameter of
tubing.

  6. A nonlinear or drifting baseline almost always signals
trouble. The baseline recovery (i.e., the characteristic return of
the recorder pen) before and after the recording of a chromatogram
should be to the same percent of scale. Poor baseline recovery and
the incomplete resolution of polymer peaks can present formidable
problems in the criminalist attempting to compare samples using MWD
data. In some instances, complete polymer peak resolution may not
be attainable. In those instances, the criminalist will have to
rely on a qualitative comparison of chromatograms. The problem of
baseline recovery and peak resolution is discussed briefly in
ANSI/ASTM D3593 [23].

  Note the peak shape. Nonsymmetrically shaped peaks can indicate poor column performance, interaction of solute and packing
materials, excessive band broadening, or adsorption of the solution
on tubing, fittings, and so on.

  Once prepared and degassed, continuously stir mobile phases.
The degassing of the mobile phase may be more critical when using

the RI detector than when using the UV detector. When using a UV detector, be aware of background changes due to the buildup of strongly absorbing material in the mobile phase [such as peroxide in tetrahydrofuran (THF)].

7. Purchase the best-quality solvents that your budget will allow. It has been our experience that filtering a single solvent mobile phase (such as THF) prior to its use introduces more particulate matter (dandruff!) than it eliminates. All mixed mobile phases (especially water or water-salt mixtures) are filtered through an appropriate 0.45-μm filter. Recognizing that no filtering process is 100% effective, an in-line 0.45-μm filter is a good investment that should be installed on your instrument to protect the columns.

Selection and use of any mobile phase [single solvent or mixture of solvents (salts)] is contingent on its compatibility with the columns and the polymeric sample under consideration. When using a solvent other than the mobile phase to dissolve a sample, check the compatibility of the resultant solution with the mobile phase.

Prior to injection *always* filter the sample solution through an appropriate 0.45-μm filter. The weight of the filter (and residue) after filtering can provide an indication of the amount of insoluble residue present in the polymer sample. Such information can be used in comparing Q and K samples. A microscopic examination of the residue can provide information about the crystalline nature of the residue.

8. Computer/processor recording and generation of MWD data should be used whenever possible for making a quantitative comparison of data. If your laboratory is contemplating the purchase of an integrator/computer/processor for HPLC and/or GC data, determine if the model(s) under consideration can make MWD calculations.

9. Place the LEC instrument in a location with adequate ventilation (to vent obnoxious or dangerous fumes), but avoid drafts which may cause ambient temperature fluctuations.

B. Current Applications

The following examples illustrate the type of physical evidence
that can be analyzed using LEC and the results to be expected. All
of the examples have been tested in the author's laboratory.

Physical Evidence Type. Taillight lenses (red, amber or yellow,
and clear). Taillight lens fragments are representative of a large
number of plastic automobile components which can be examined using
LEC. The extent of aging [i.e., ultraviolet (UV)-sun oxidation]
will not interfere with the generation of data and subsequent interpretation of results as long as representative (K) comparison standards are available.

Column type/mobile phase/detectors: SDVB/THF/RI (refractive index)
and UV.

Data generated from chromatogram: MWD for polymer/fingerprint comparison chromatograms.

Data generated from collected fractions: UV/visible spectra of
additive (e.g., dye, antioxidant),(see Fig. 10), thin-layer
chromatography (TLC) of dye fraction, infrared analysis (IR)
of polymer fraction, pyrolysis gas chromatography (GC) of
polymer fraction, weight of insoluble residue, and microscopic
examination of insoluble residue.

Physical Evidence Type: Fibers. The main problem with conducting
a successful GPC analysis of fibers is finding a satisfactory solvent for the fibers. Many, but not all (e.g., polyester, polyamide)
fibers will dissolve in THF. In some instances, useful information
can be obtained from insoluble fibers by refluxing the fiber and
analyzing the refluxed solvent for the presence of dyes, additives,
or residual monomer. Depending on fiber diameter (and overall sensitivity of detection system), as little as 1-5 mg (1-5 in.) of
fiber can produce satisfactory data.

Column type/mobile phase/detectors: SDVB/THF/RI, UV, and fluorescence.

Data generated from chromatogram: MWD for polymer/fingerprint
chromatogram showing polymer and additives, or from refluxed
sample; residual monomer and/or additives.

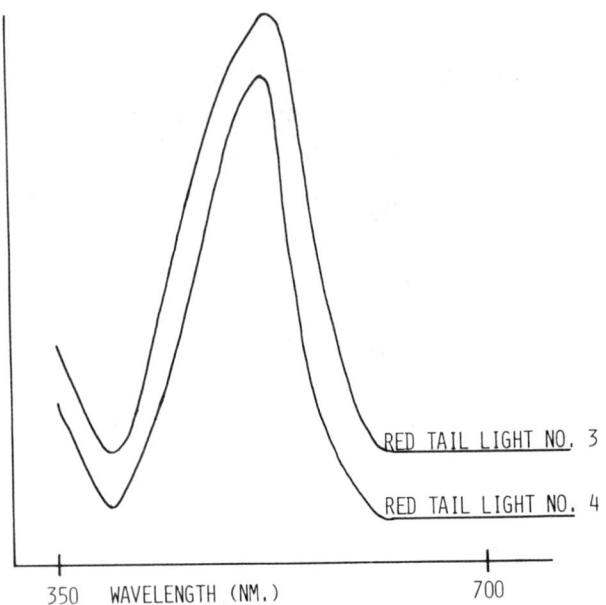

350   WAVELENGTH (NM.)   700

FIG. 10. Dye fraction collected from end taillight lenses:
visible spectra.

Data generated from collected fractions: UV/visible spectra of ad-
    ditives (particularly dyes-colorants) (see Fig. 11), IR of
    polymer fraction, pyrolysis GC of polymer fraction (see Fig.
    12), weight of insoluble residue (may contain TiO_2 delustrant),
    and microscopic examination of insoluble residue.

Physical Evidence Type: Paint/Paint Chips. Depending on the age
(or extent of aging) of the paint chip, more or less of the chip
will dissolve in THF. For those paint chips that do not readily
dissolve, the chip is refluxed and the refluxed liquid is analyzed.
Depending on the age and size of the sample, it is possible to
separate layers of a paint chip and to analyze individual layers.
As with the taillight lense fragments, the state of oxidation does
not detract from the interpretation of the resulting data as long
as representative (K) comparison standards are available.

Column type/mobile phase/detectors: SDVB/THF/RI, UV, and fluores-
    cence.

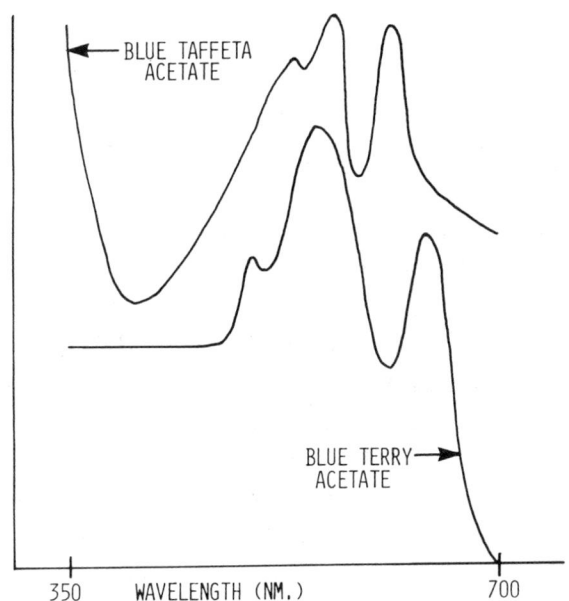

FIG. 11. Dye fraction collected from blue acetate fibers: visible spectra.

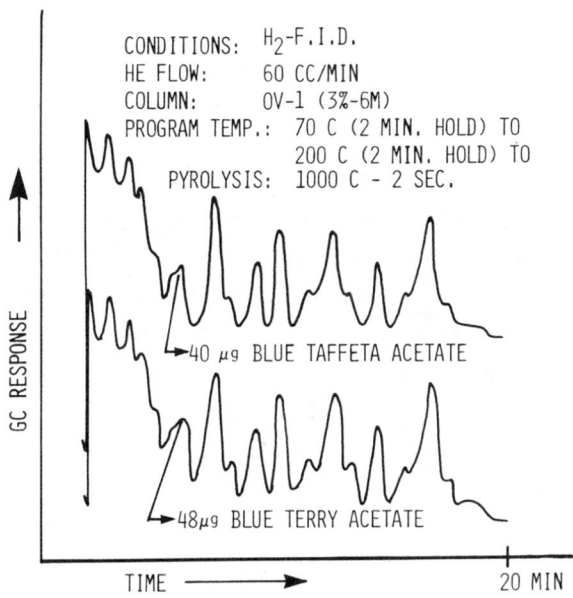

FIG. 12. Pyrolysis GC: Polymer fraction collected from blue acetate fibers.

Data generated from chromatogram: MWD for polymer/fingerprint chromatogram showing polymer and additive, or from refluxed sample; residual monomer and additives.

Data generated from collected fractions: IR of polymer fraction, microcolor and crystal tests on collected dye fraction or insoluble portion.

Physical Evidence Type: Motor Oil/Grease/Lubricants. The investigation potential of this physical evidence type seems unlimited. Physcial evidence samples are dissolved and analyzed like any other polymer sample. Upon dissolving the sample there will be a certain amount of undissolved residue which is a function of whether the physical evidence is oil/grease/lubricant. The residue from the dissolved/filtered sample may be tested for the presence of a metal (e.g., in lithium- or molybdenum-based greases) or microscopically examined for the presence of "graphite." An interesting aspect of the LEC analysis of motor oil is that preliminary studies show what appears to be characteristic decomposition of the motor oil with use (aging).

Column type/mobile phase/detectors: SDVB/THF/RI and UV.

Data generated from chromatogram: Fingerprint comparison chromatograms.

Data generated from collected fractions: None to date.

Physical Evidence Type: Copier Paper (Plastic Coated/Impregnated). The amount of plastic used in the manufacture of paper products today has increased substantially. The specific polymer and additives used in copier paper normally reflects the chemistry of the copying process (usually proprietary).

Column type/mobile phase/detectors: SDVB/THF/RI and UV.

Data generated from chromatogram: MWD and fingerprint chromatogram.

Data generated from collected fractions: IR of polymer.

In addition to these LEC applications made using SDVB columns, Wheals [1] has evaluated the use of microparticulate silica columns in a variety of forensic applications. Using a laboratory-made chromatograph and column, Wheals [1] produced LEC chromatograms of

vehicle indicator light covers (turn-signal light covers), engine
oil (used and unused), bituments, and plastic tapes. Additional
discrimination of samples was achieved by using appropriate combinations of detectors (RI, UV, or fluorescence).

C. Potential Applications

The following examples illustrate what the author believes are
future applications for LEC in the crime laboratory.

    1. *Separation of cola in soft drinks; examination of filtered
coffee.* Column type/mobile phase/detectors: One-lab-manufactured
S-100 (Merck, Darmstadt, G.F.L.) silica coated with N-(3-triethoxy-
silylpropyl) acetamide/water/RI and UV (275 nm).

    Reference: H. Englehardt and D. Mathes, Chemically bonded
stationary phases for aqueous high performance exclusion chromatography, *J. Chromatogr.* 185:305 (1979). Fingerprint chromatograms
of different cola beverages and filtered coffee are presented in
the article. Fingerprint chromatograms show brand-to-brand variations. Peaks identified in the cola chromatograms were chlorogenic
acid and caffeine. The authors used glucose, saccharose, raffinose,
and di-, tri-, and tetraethylene glycol as low molecular weight
calibration standards. Potential applications in the crime laboratory are many. The most intriguing are the comparison of common
water-based beverages and the analysis of sugars, both of which
regularly come into the drug and toxicology sections of the crime
laboratory.

    2. *Examination of gelatin. Column type/mobile phase/detector:*
One-lab-prepared Sepharose 4B/0.1 M urea-0.1 M NaCl RI.

    Reference: I. H. Coopes, Chromatography of gelatin, *J. Poly.
Sci. Symp.* 49:97 (1975). Fingerprint chromatograms of seven different gelatins are presented. One potential application in the
drug section would be the comparison of gelatin capsules used to
contain clandestinely prepared drugs/narcotics.

    3. *Examination of protein-based glues. Column type/mobile
phase/detectors:* Glyceryl-Controlled Pore Glass-10/0.05 M sodium
dihydrogen phosphate-0.1 M NaCl (pH 7)/RI/UV (254-nm).

Reference: Cormine Persiani, P. Cukor, and K. French, Aqueous GPC of water soluble polymers by HPLC using glyceryl CPG columns, *J. Chromatogr. Sci.* 14:417 (1976). Fingerprint chromatograms of protein glues are presented. An interesting series of chromatograms showing the aging of glues through bacterial degradation is also presented. This application should be of interest to the trace evidence analyst and the questioned document examiner, both of which encounter animal glues in a variety of paper-based products.

4. *Separation of human plasma proteins. Column type/mobile phase/detectors:* 4-Bondagel E columns (2-1000A, 500A, 125A)/0.05 M Trizma buffer (pH 7.4) + 1% sodium dodecyl sulfate/UV (280 nm).

Reference: R. V. Vivilecchia et al., The use of microparticulates in GPC, *J. Chromatogr. Sci.* 15:424 (1977). Albumin and immunoglobin peaks are identified in the chromatograms presented. This article is representative of a number of similar articles dealing with the LEC examination of plasma proteins. Of particular note is the resolution of characteristic patterns on an LEC chromatogram for creatine phosphokinase (CPK) and lactate dehydrogenase (LDH) isoenzymes and hemoglobin polymorphic proteins. The rapid advances in this area, spurred by research in medicine, will probably mean that aqueous LEC will become a tool in the future for the forensic serologist. The use of aqueous LEC to generate comparison chromatograms of blood and seminal stains should be of particular interest.

ACKNOWLEDGMENTS

The author would like to thank Waters Associates for their technical advice and support of numerous forensic projects. The author is particularly grateful to the many undergraduate students at the University of Central Florida whose dedicated assistance provided the data for many of the applications presented in this review.

REFERENCES

1. B. B. Wheals, *J. Liq. Chromatogr.* 2:91 (1979).
2. W. W. McGee, K. Coraine, and J. Strimaitis, *J. Liq. Chromatogr.* 2:287 (1979).
3. *1980 Annual Book of ASTM Standards*, Part 35, Procedure D3016, American Society for Testing and Materials, Philadelphia, 1980.
4. W. W. Yau, J. J. Kirkland and D. D. Bly, *Modern Size Exclusion Chromatography*, Wiley-Interscience, New York, 1979.
5. R. E. Majors, *J. Chromatogr. Sci.* 18:488 (1980).
6. H. McNaire, *J. Chromatogr. Sci.* 16:850 (1978).
7. J. F. Johnson and R. S. Porter, *Prog. Poly. Sci.* p. 201 (1970).
8. J. F. Johnson, R. S. Porter, and M. J. R. Cantow, *Rev. Macromol. Chem.* 1:343 (1966).
9. M. J. R. Cantow (Ed.), *Polymer Fractionation*, Academic Press, New York, 1967.
10. G. K. Ackes, *Biochemistry* 3:723 (1964).
11. W. W. Yau and C. P. Malone, *J. Polym. Sci. B* 5:663 (1967).
12. J. Parath, *Pure Appl. Chem.* 6:233 (1963).
13. P. G. Squire, *Arch. Biochem. Biophys.* 107:471 (1964).
14. J. G. Hendrickson, *J. Polym. Sci. A2* 6:1903 (1955).
15. E. Glueckauf, *Trans. Faraday Soc.* 51:34 (1955).
16. J. J. Van Deemter, F. G. Zuiderweg, and A. Klinkenberg, *Chem. Eng. Sci.* 5:271 (1965).
17. J. C. Giddings, *Dynamics of Chromatography*. Marcel Dekker, New York, 1965.
18. A. R. Coopes, A. R. Bruzzone, and J. F. Johnson, *J. Appl. Poly. Sci.* 13:2029 (1969).
19. L. Lapedus and N. R. Amundson, *J. Phys. Chem.* 56:984 (1952).
20. R. N. Kelley and F. W. Billmeyer, Jr., *Anal. Chem.* 41:874 (1969).
21. R. N. Kelley and F. W. Billmeyer, Jr., *Anal. Chem.* 42:399 (1970).
22. J. C. Giddings, L. M. Bowman, Jr., and M. N. Meyers, *Macromolecules* 10:443 (1977).
23. *1980 Annual Book of ASTM Standards*, Part 35, Procedure D3593, American Society for Testing and Materials, Philadelphia, 1980.
24. H. Benoit, *J. Chem. Phys.* 63:1507 (1966).
25. *1980 Annual Book of ASTM Standards*, Part 35, Procedure D3536, American Society for Testing and Materials, Philadelphia, 1980.
26. D. D. Bly, *J. Polym. Sci. C* 21:13 (1968).
27. J. J. Kirkland, *J. Chromatogr.* 125:231 (1976).
28. W. W. Yau, J. J. Kirkland, D. D. Bly, and H. J. Staklosa, *J. Chromatogr.* 125:219 (1976).
29. W. W. Yau, C. R. Ginnaid, and J. J. Kirkland, *J. Chromatogr.* 149:465 (1978).

6 High-Performance Liquid Chromatography in Forensic Toxicology

Michael A. Peat

Center for Human Toxicology
University of Utah
Salt Lake City, Utah

I. INTRODUCTION

Although chromatography, in particular gas-liquid (GLC) and thin-layer chromatography (TLC), has been used extensively in the identification and quantitation of drugs and metabolites from biological fluids, high-performance liquid chromatography (HPLC) has only recently been applied to such problems. The development of instrumentation, particularly more reliable pumping systems and more versatile and sensitive detectors, together with improvements in column technology, has contributed greatly to this increasing use of HPLC in both forensic and clinical analytical toxicology.

Analytical toxicology is concerned with the identification and quantitation of drugs and metabolites from body fluids. The clinical analytical toxicologist is concerned with the analysis of serum, plasma, and urine samples, whereas his or her counterpart in forensic toxicology must be able to assay a wide variety of body fluids and organs, most commonly blood, liver, gastric contents, urine (or kidney), and lung. The clinical toxicologist is also concerned with performing assays for therapeutic drug monitoring and is therefore interested in quantitating specific drugs and their

metabolites. In addition, he or she is often required to screen
urine and blood (or plasma) samples from suspected drug overdoses.
The forensic toxicologist invariably is called upon to screen for
the presence of drugs and metabolites and, in addition, to determine their concentrations in the biological specimens provided.
These factors influence the toxicologist in the choice of a suitable analytical procedure. For example, consider the bronchodilator
drug theophylline. The clinical toxicologist is concerned with the
determination of this drug in plasma or serum as part of a therapeutic drug monitoring program, an assay for which HPLC is particularly useful [1-5]. On the other hand, the forensic toxicologist
would prefer an analytical technique that would detect theophylline
in a screen for other acidic and neutral drugs and which would also
be applicable to a variety of biological samples. Once the drug
has been identified, the toxicologist may well use HPLC for quantitation.

   Although a large number of drugs and metabolites can now be
quantitated by HPLC, the technique has not yet been successfully
applied to the screening of biological samples. There are two
major reasons for this; one is the absence of a universal detector,
such as the flame ionization detector in gas chromatography. In
addition, the widely used ultraviolet detector is inherently insensitive for a number of drugs and metabolites (e.g., meprobamate,
the opiate narcotics, and the substituted β-phenylethylamines).
The second major problem with using HPLC for screening is the lack
of resolution for a number of structurally related drugs and the
requirement of different mobile phases to elute the commonly encountered drugs. For these reasons HPLC is much more commonly applied to either screening for a particular drug [e.g., acetaminophen (paracetamol)] or for a group of drugs (e.g., the anticonvulsant agents), rather than to more general screening.

   By far the most common application of HPLC in analytical
toxicology is the quantitation of drugs and metabolites in biological fluids. The majority of the published procedures are designed

for the analysis of plasma or serum; however, these methods are
generally applicable to the analysis of other specimens; but more
detailed extraction procedures may have to be used because of the
possible increase in background interference from forensic toxicology specimens.

It is impossible to describe in detail the applications of
HPLC to every drug or metabolite encountered in forensic toxicology. There are certain drugs and drug groups for which HPLC is
becoming the technique of choice, and these will be discussed in
more detail. It is important to realize that for the forensic
toxicologist HPLC is an analytical technique that complements other
chromatographic procedures, such as gas-liquid and thin-layer
chromatography.

II. SCREENING IN FORENSIC TOXICOLOGY

The complexity of a forensic toxicological analysis is increasing
due to a greater incidence of cases in which more than one drug is
detected and also because of the availability of more potent
therapeutic agents. The benzodiazepines illustrate this problem.
Chlordiazepoxide was first introduced in the 1950s; since then
numerous other drugs of this class have been marketed to the extent
that diazepam is the most frequently detected drug in toxicology
after ethanol. Together with the increasing number of benzodiazepines, there is a tendency toward decreasing dose due to the increased potency of the newer agents (e.g., lorazepam). It is
essential, therefore, that sensitive and reliable procedures be
available for the detection of these, and other important drugs,
in biological matrices.

In a large percentage of forensic toxicology cases, specific
drugs are suspected. It is vitally important, however, that
analytical procedures be available for screening a variety of
biological fluids for unknown drugs. Not only should these methods
be used on specimens from cases in which no drug or poison is suspected, but they should also be applied to the other cases, where

a drug is known to have been ingested, in order to confirm its
presence and to determine whether other agents are present. Such
procedures are becoming more important now that forensic toxicologists are being requested to determine drugs in samples from drivers
involved in accidents where the blood concentrations may be therapeutic or even lower and where small sample volumes are available.

It follows that a suitable screening procedure must satisfy
certain requirements, including applicability to a variety of biological fluids, specificity, and sensitivity. Although blood speci-
ments are commonly screened for drugs by forensic toxicologists,
the procedures used should also be applicable to other specimens,
such as homogenized liver samples and urine. For practical reasons
a complementary series of methods has been developed: Thin-layer
chromatography with a variety of developing reagents is widely used
for screening urine and gastric contents, whereas gas chromatographic procedures are mainly used for blood and tissue homogenates.

The methods must also be sufficiently discriminating to differentiate between a large number of drugs, their metabolites, and
naturally occurring catabolites such as cholesterol and a number of
long-chain fatty acids. If one considers the wide range of drugs
that are presently available to the public, either over the counter
(e.g., aspirin, some antihistamines and expectorants), on pre-
scription (e.g., the tricylic antidepressants, sedative hypnotics,
and nonopiate analgesics), or illicitly (e.g., cocaine, phencyclidine, marijuana, and the amphetamines), it is impractical to expect one chromatographic procedure to discriminate among them all.
For this reason, and for historical reasons, the majority of analytical toxicologists regularly perform separate extractions for
acidic and basic drugs, whereas a number isolate strongly acidic,
weakly acidic, neutral, amphoteric, and basic fractions. Following
these extractions the chromatographic procedure must be able to
discriminate only a limited number of drugs and metabolites; however, for the basic fraction there is still a wide range of drugs
and metabolites that should be detected. Without doubt, screening

procedures should be able to identify such commonly prescribed
drugs as the tricyclic antidepressants (amitriptyline, nortriptyline, imipramine, desipramine, doxepin), benzodiazepines (diazepam,
chlordiazepoxide, flurazepam), and propoxyphene.

Moffat and co-workers [6-9] have applied the concept of discriminating power (i.e., the ability, using a set of chromatographic
conditions, to discriminate between compounds) to the choice of
chromatographic techniques for the analysis of drugs. Paper and
thin-layer chromatographic systems have been compared for the analysis of the various classes of drugs (e.g., bases [6], neutrals [7]
and acids [8]), and gas-liquid chromatographic systems for the
separation of basic drugs [9]. For this work they calculated discriminating power from the formula

$$\text{Discriminating power} = 1 - \frac{2M}{N(N - 1)}$$

where M is the number of pairs of retention values that matched
within a set error factor and N the number of R values examined,
and they assumed that all the separations examined were considered
to be equally important. This, of course, may not always be so in
toxicology since some drugs occur far more frequently than others.
For example, consider the acidic drugs; then the barbiturates,
salicyclic acid, and acetaminophen are the most commonly encountered.
It is therefore advantageous to have a system that separates these
drugs from each other, remembering that it is still necessary to
separate all other drugs that may be present in a particular analysis. One possible way of reaching this compromise is to weight
the chromatographic data used to calculate the discriminating power
of a particular system. Moffat et al. [10] compared discriminating
power with weighted discriminating power for acidic drugs by thin-
layer chromatography. They used the number of fatal poisonings associated with selected acidic drugs as the weighting system. Their
findings indicate that the weighted procedure for calculating discriminating power did not result in any significant advantage over
the traditional calculations for discriminating power. For example,

using a silica gel thin-layer system with chloroform-methanol [10]
as solvent, the discriminating power using the no-weight system was
0.74, whereas that using the weighting system was 0.76.

If a chromatographic system is ideal for screening purposes,
it should have a discriminating power value approaching unity. As
an example of this approach, Moffat et al. [9] examined 62 basic
drugs using eight gas chromatographic stationary phases; 2% SE-30
and 5% OV-17 columns possessed the highest discriminating power,
although the SE-30 phase was the only one which eluted all the drugs
tested. When more than one chromatographic system is used, however,
the discriminating power of the SE-30 and OV-17 systems was only
marginally higher than that of either system alone. To date no
similar attempt has been made to define the discriminating power
of high-performance liquid chromatographic systems.

The final requirement for a satisfactory screening procedure
is that of sensitivity. Therapeutic and toxic concentrations of
drugs and metabolites vary widely, from below 0.1 μg/ml for some
basic drugs to greater than 200 μg/ml for salicylate. With this in
mind it is obvious that techniques used to detect particular drugs
must be more sensitive than those used to detect other drugs,and
metabolites. Until recently, gas chromatography with flame ionization detection was the most widely used procedure for screening.
The more specific electron-capture and nitrogen phosphorous detectors are more useful for certain drug groups and they are also more
sensitive than FID for these compounds. The electron capture detector is widely used to screen and quantitate the benzoidazepines
[11, 12], whereas Pierce and co-workers [13] have used the selectivity of nitrogen phosphorous detectors to screen for a large number of basic drugs in autopsy specimens.

With these requirements in mind what analytical procedures
are presently used in toxicological screening? Historically, both
ultraviolet and visible spectophotometry have been used, and, in
fact, ultraviolet spectrophotometry examination of the weak acid
fraction is still a commonly used procedure for barbiturate

screening. In general, these methods have been superseded by
chromatographic techniques and more recently by the development of
radioimmunoassay and enzyme multiplied immunoassay (EMIT) procedures. The latter are used for particular drugs [e.g., morphine
and related opiates, the barbiturates, and cocaine and its primary
metabolite (benzoylecgonine)], whereas thin-layer and gas-liquid
chromatographic methods have more general applications.

Where, in this general scheme for toxicological screening, does
high-pressure liquid chromatography fit? The first requirement
for a suitable chromatographic procedure is discriminating power,
and, both gas chromatography and thin-layer chromatography have been
shown to have satisfactory discriminating power when applied to
drug analyses. One of the first attempts to utilize HPLC for the
identification of basic drugs was by Jane [14]. This system uses
columns packed with silica and a methanol-2 M ammonium hydroxide-1
M ammonium nitrate mobile phase. Very similar systems have been
used by others to achieve thin-layer and HPLC separations. In the
same year Twitchett and Moffat [15] described the separation of 30
drugs on an octadecylsilane (ODS) reversed-phase HPLC system. Although these were both excellent chromatographic system, very few
drugs could be uniquely identified because of the overlap of retention volumes. For example, the system described by Jane [14]
does not resolve the commonly used analgesic codeine from the tricyclic antidepressant desipramine.

More recently Baker et al. [16] examined 101 drugs of forensic interest using three solvent systems, which were outlined in
Chap. 3. These authors found that the coefficient of variation of
HPLC relative retention volumes was 3.3%, which is typical of most
HPLC studies. It is worthwhile to note that this is considerably
larger than the coefficient of variation of gas chromatographic
retention times (0.6%). They also found that using relative retention volumes, only 9% of the drugs could be distinguished; to
increase discrimination they considered the ratio of the UV absorbances at 254 and 280 nm. When this parameter was included, the

authors claimed that 95% of the drugs could be distinguished. However, they also found that the coefficient of variation over a 2-
month period for the absorbance ratio was 21%, necessitating, in
their view, frequent calibration of the detector.

In the study of reversed-phase systems by Twitchett and Moffat
[15] it was noticed that basic drugs exhibited lower theoretical
plate counts than did acidic or neutral drugs. In the study by
Baker and co-workers [16] it was also observed that basic drugs
tended to have a lower column efficiency. However, this was only
a general trend, and numerous examples of the converse relationship
could have been cited.

Perhaps the most comprehensive attempt at developing an HPLC
system with sufficient discriminating power for basic drugs has
been that of Wheals [17]. He describes an isocratic multicolumn
approach to the identification of basic drugs using an aqueous
methanol solvent as originally described by Jane [14]. A number of
different chemically bonded packings were prepared from the same
batch of silica and examined for their ability to resolve a large
number of drugs. He found that in practice, packings with acidic
surface characteristics proved to be most useful: that is, silica,
a mercaptopropyltrimethoxy-treated silica, and an aliphatic strong
cation exchange based on an n-propylsulfonic acid modified silica.
Despite these three materials having similarity by virtue of their
ability to ionize and produce a proton, they undoubtedly differ in
their degree of ionization, and the presence of an n-propyl chain
between the siloxane bond and the ionizable groups presumably im-
parts some lipophilic properties to the mercaptopropyl and cation-
exchange packing not found with silica.

All the drugs examined eluted with capacity factors below 3.5
and many of the compounds display marked differences in capacity
factor value on the different columns. Of the 161 drugs studied,
93 have unique retention characteristics, but the remaining compounds cannot be identified conclusively. Substances in the latter
category are found mainly in the area of low retention. Wheals

also reports that using a multicolumn isocratic system, groups of
closely related compounds have differing retention characteristics
on the three columns. The differentiation is shown in Fig. 1;
Figs. 2 and 3 show examples of separation for a mixture of phenothiazines and a mixture of phenylethylamine stimulants and tricyclic
antidepressants, respectively.

Although this procedure displays unique retention characteristics for a large number of basic drugs, it has, together with the
other HPLC systems described, one major disadvantage when applied
to forensic toxicological screening: that is, the absence of any
retention data on the commonly encountered metabolites of certain
drugs. It is very likely that when these are considered, the discriminating power of the systems would be reduced. Furthermore,
the use of ultraviolet detection for a number of drug groups (e.g.,
the phenylethylamines and opiate narcotics) would not meet the
sensitivity requirements of a toxicological screening procedure.
The methods could, however, be used for those drugs that are either
present at greater than nanograms per milliliter concentrations or
have adequate absorbance (e.g., the tricyclic antidepressants at
240 nm). Indeed, HPLC is frequently used in such cases, as described in Sec. III.C.

III. HPLC ANALYSIS OF SPECIFIC DRUG GROUPS

HPLC procedures have been used extensively to monitor the plasma
concentrations of drugs and metabolites in the clinical toxicology
laboratory. Such procedures could be modified by the forensic
toxicologist to determine blood and tissue concentrations of these
agents. In the discussion that follows these individual procedures
will be referenced and greater emphasis placed on HPLC systems
that may serve as screening procedures for a particular group of
drugs or used to quantitate several drugs and metabolites.

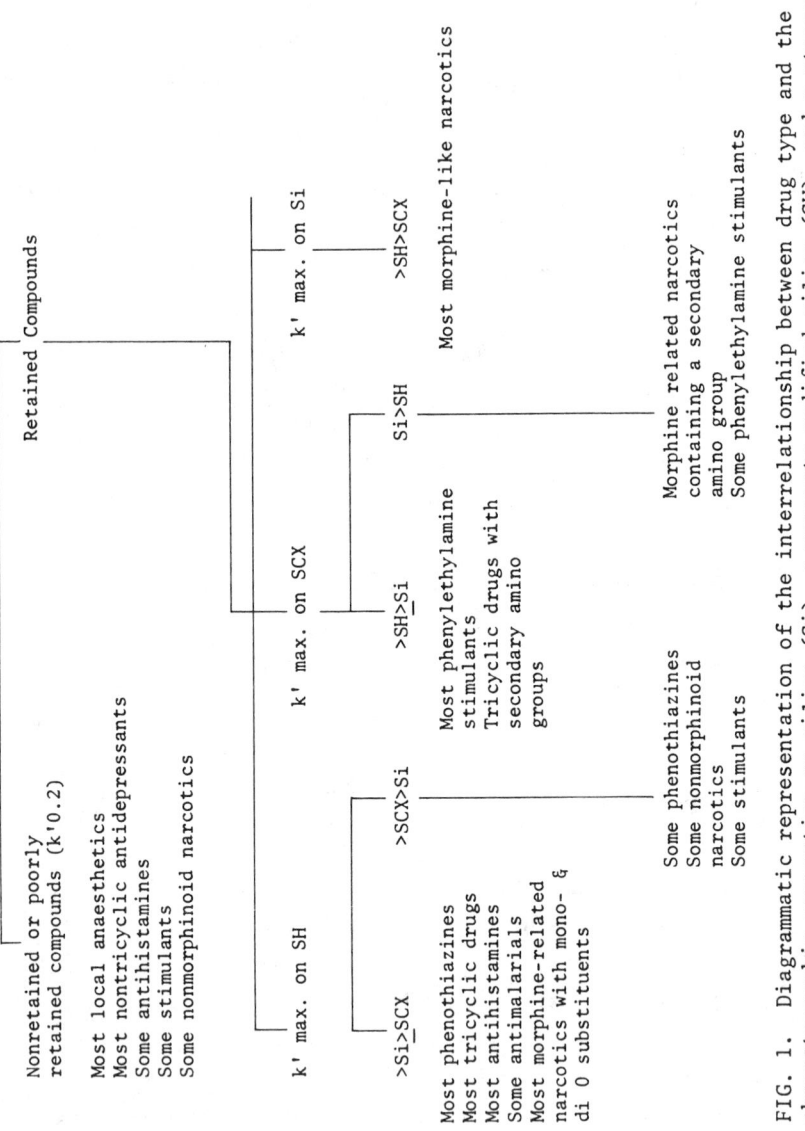

FIG. 1. Diagrammatic representation of the interrelationship between drug type and the chromatographic properties on silica (Si), a mercapto-modified silica (SH), and a strong cation exchange (SCX). (From Ref. 17.)

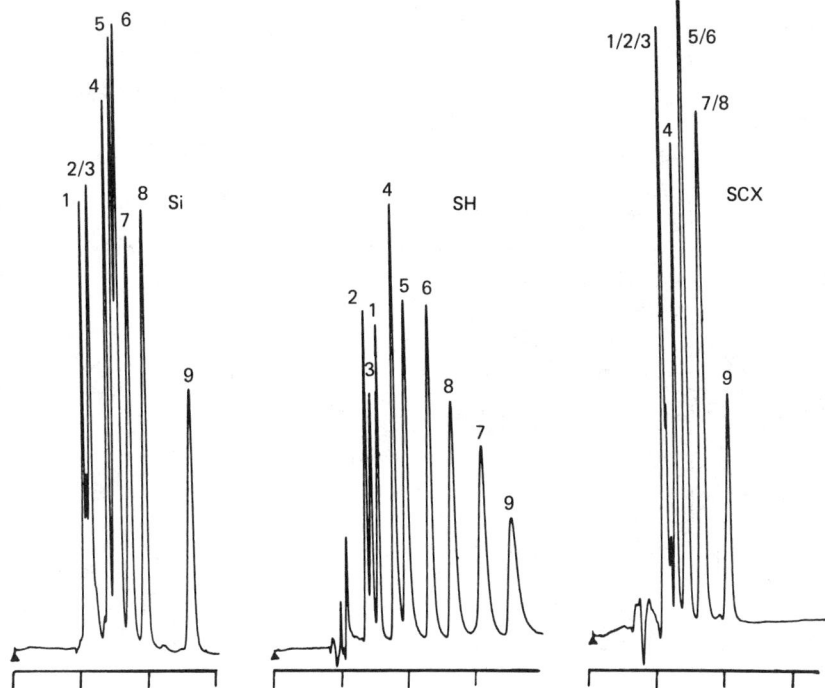

FIG. 2. Separation of 1, pipamazine; 2, fluphenazine; 3, trifluo-
meprazine; 4, trifluorpromazine; 5, mepazine; 6, chlorpromazine;
7, thioridazine; 8, promazine; 9, methdilazine. Eluant: methanol-2
M ammonium hydroxide-1 M ammonium nitrate (27:2:1) + 50 mg of sodium
sulfite per liter of solvent; columns: 25 cm × 5 mm ID packed with
silica (Si), a mercapto-modified silica (SH), and a strong cation
exchange (SCX); flow rate: 2 ml/min; detector: UV at 254 nm;
sensitivity: 0.1; time intervals on the chromatogram: 4 min.
(From Ref. 17.)

A. Benzodiazepines

Over 2000 benzodiazepines have been synthesized, and more than 100

tested for hypnotic activity, anticonvulsant activity, and treat-

ment of anxiety. Table 1 shows the structures of the benzodiaze-

pines commonly encountered by forensic toxicologists. Of these,

medazepam, nitrazepam, flunitrazepam, temazepam, bromazepam, and

triazolam are not presently available in the United States.

The first successful benzodiazepine, chlordiazepoxide, was

developed by Sternbach's group at the Roche Laboratories in the

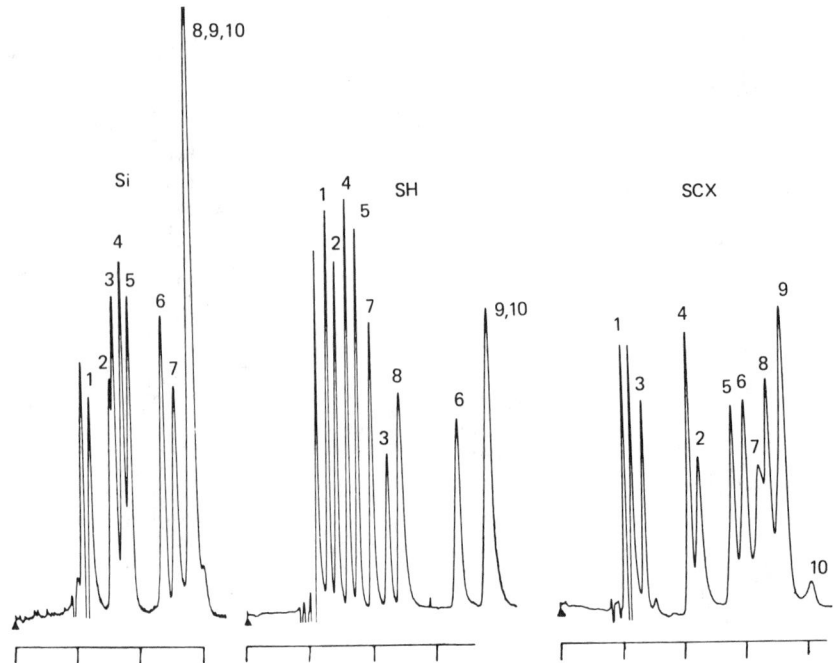

FIG. 3. Separation of 1, benzphetamine; 2, norephedrine; 3, amitriptyline; 4, fenfluramine; 5, amphetamine; 6, nortriptyline;
7, ephedrine; 8, methyl ephedrine; 9, desipramine; 10, protriptyline. Chromatographic conditions as in Fig. 2. (From Ref. 17.)

1950s, and was the first drug of this group introduced for clinical
use. Today, diazepam, chlordiazepoxide, and flurazepam are widely
prescribed in the United States. The clinical popularity of these
drugs is the result of a combination of their pharmacological actions, their relative safety, and an extraordinary demand for
agents of this type by both physicians and patients. The mechanism
of action of the benzodiazepines appears to be related to the metabolism or action of the inhibitory neurotransmiter, γ-aminobutyric
acid (GABA).

   To analytical toxicologists this group of drugs represents a
difficult problem. There are three major reasons for this:

TABLE 1. The Commonly Encountered Benzodiazepines

Name	R_1	R_2	R_3	R_4	R_5
Diazepam	CH_3	=O	H	H	Cl
Clonazepam	H	=O	H	NO_2	Cl
Clorazepate	H	$(OH)_2$	CO_2H	H	Cl
Oxazepam	H	=O	OH	H	Cl
Lorazepam	H	=O	OH	Cl	Cl
Prazepam	$-CH_2-\triangleleft$	=O	H	H	Cl
Medazepam	CH_3	H_2	H	H	Cl
Nitrazepam	H	=O	H	H	NO_2
Flurazepam	$(CH_2)_2N(C_2H_5)_2$	=O	H	F	Cl
Flunitrazepam	CH_3	=O	H	F	NO_2
Temazepam	CH_3	=O	OH	H	Cl

1. Generally, plasma concentrations resulting from therapeutic
   doses are below 1 µg/ml; and in some cases, for example lorazepam and nitrazepam, therapeutic plasma concentrations rarely
   exceed 0.1 µg/ml.

2. The presence of active metabolites that may accumulate; for
   example, flurazepam is metabolized to N-1-desalkylflurazepam
   (Table 1, R = H), which has a half-line approaching 100 h. If
   a meaningful interpretation of the data is to be made, it is
   important, therefore, that the analytical procedure used differentiate the parent drug from metabolite and also measure
   the metabolite concentrations.

3. Extensive metabolism and the presence of "common metabolites."
   With the exception of oxazepam and lorazepam, which form glucuronide conjugates, most of the other benzodiazepines available in the United States undergo phase I metabolism by hepatic
   microsomes. Table 2 lists the major metabolites of these
   drugs; from this table it is apparent that nordiazepam can be
   detected following the ingestion of diazepam, chlordizepoxide,
   and prazepam. In addition, clorazepate is converted nonenzymati-
   cally in gastric acid to nordiazepam. Those benzodiazepines,
   with a nitro group in position R_5, are metabolized to the corresponding 7-amino derivative.

A large number of papers have been published on the determination of the benzodiazepines in body fluids and tissues. Analytical
work prior to 1974 has been reviewed by Clifford and Franklin-Smythe
[18] and Hailey [19]. Since then, gas-liquid chromatography with
electron-capture detection (GC-ECD) has become the method of choice
for these drugs because it meets the necessary specificity and

TABLE 2. Metabolites of the Benzodiazepines

Parent Drug	Metabolites Detected in Plasma
Diazepam	Nordiazepam
	Oxazepam[a]
Chlordiazepoxide	Norchlordiazepoxide
	Demoxepam[a]
	Nordiazepam[a,b]
Flurazepam	N-1-Hydroxyethylflurazepam[c]
	N-1-Desalkylflurazepam
Prazepam	Nordiazepam
Clonazepam	7-Aminoclonazepam

[a]Not detected after single oral doses.

[b]Even after chronic dosing plasma concentrations rarely exceed
0.3 µg/ml.

[c]Short elimination half-life.

sensitivity requirements. Prior to the development of this method, several of the analytical procedures available to the forensic toxicologists depended on acid hydrolysis of the benzodiazepines to their corresponding benzophenones [18] followed by gas-liquid or thin-layer chromatography. Unfortunately, hydrolysis of several of the benzodiazepines results in the formation of the same benzophenone, as shown in Table 3, so that the individual parent drugs cannot be readily identified.

TABLE 3. Benzophenones and Related Benzodiazepines

Benzodiazepine	Benzophenone	R_1	R_2	R_3
Flurazepam	2-(Diethylaminoethylamino)-5-chloro-2'-fluorobenzophenone	$(CH_2)N(C_2H_5)_2$	F	Cl
N-1-Desalkylflurazepam	2-Amino-5-chloro-2'-fluorobenzophenone	H	F	Cl
Nitrazepam	2-Amino-5-nitrobenzophenone	H	H	NO_2
Clonazepam	2-Amino-5-nitro-2'-chlorobenzophenone	H	Cl	NO_2
Oxazepam	2-Amino-5-chlorobenzophenone	H	H	Cl
Chlordiazepoxide				
Nordiazepam				
Norchlordiazepoxide				
Diazepam	2-Methylamino-5-chlorobenzophenone	CH_3	H	Cl
Temazepam				
Medazepam				
Lorazepam	2-Amino-2.5-dichlorobenzophenone	H	Cl	Cl
Flunitrazepam	2-Methylamino-5-nitro-2'-fluorobenzophenone	CH_3	F	NO_2
Prazepam	2-Cyclopropylmethylamino-5-chlorobenzophenone	$-CH_2-\triangleleft$	H	Cl

During the past few years the analytical technique used most
extensively for the determination of the benzodiazepines has been
GC-ECD. De Silva et al. [20] have described a comprehensive extraction scheme using this technique for their determination in blood.
For the toxicologist, however, the procedure has major disadvantages,
such as lengthy purification steps and the necessity for relatively
large sample volumes (up to 2 ml). More recently, Rutherford [11]
and Peat and Kopjak [12] have described micro methods using GC-ECD
for their determination in plasma and blood. Although gas chromatography of a number of the intact benzodiazepines is possible, others,
and the majority of their metabolites, require derivatization into
more volatile compounds.

Scott and Bommer [21] first used HPLC for the analysis of some
benzodiazepines and applied it to animal urine. The chromatography
was limited by the state of the art at that time, but the authors
showed the potential of liquid chromatography in the analysis of
the benzodiazepines. During the past few years a number of publications [12, 22-30] have appeared on the HPLC analysis of these
drugs, in particular diazepam and its metabolites [23-27] and
chlordiazepoxide and its metabolites [22, 29, 30]. Only a limited
number of workers [12, 28, 31] have attempted to use HPLC to identify
and quantitate several benzodiazepines. Recently, Wittwer [32]
applied HPLC to the forensic analysis of several benzodiazepines.
Although the procedure was not designed for toxicological use, a
separation of 10 benzodiazepines was obtained using a 15-cm silica
column and a solvent system composed of 90 parts cyclohexane and
10 parts of the mixture ammonia:methanol:chloroform (1:200:800).
To obtain extra discriminatory power, dual-wavelength detection at
254 and 280 nm was used. Table 4 shows the retention and absorbance ratio data.

Those that have applied HPLC to the toxicological analysis of
several benzodiazepines have used reversed-phase systems to separate
the parent drugs and metabolites. Osselton et al. [28] used a pH
7.8 phosphate buffer:methanol mixture to detect and quantitate

TABLE 4. Retention Volumes and Absorbance Ratio: (254 nm/280 nm)

Benzodiazepine	Retention Volume (ml)	254 nm/280 nm Ratio
Medazepam	4.10	3.77
Prazepam	5.44	4.47
Diazepam	6.26	4.62
Flurazepam	10.26	4.97
Chlordiazepoxide	11.70	1.32
Nordiazepam	12.20	4.58
Nitrazepam	21.20	1.38
Clonazepam	26.40	1.79
Demoxepam	42.80	2.98
Oxazepam	48.40	4.20

Note: Column: 15-cm μPorasil; eluant: cyclohexane 90% and 10% ammonia:methanol:chloroform (1:300:800); detector: 254 and 280 nm.

Source: Adapted from Ref. 32.

benzodiazepines in tissues following enzymic digestion. Tissue aliquots were blended in pH 10.5 Tris base and incubated with the proteolytic enzyme subtilisin Carlsberg at 50-60°C for 1 hr with continuous stirring. After cooling and filtering through glass wool the filtrate was extracted with diethyl ether. The extract was dried with sodium sulfate, the ether evaporated, and the residue dissolved in ethanol. An aliquot was used for HPLC analysis. A 15 cm × 4.6 mm ID column packed with Spherisorb ODS (5 μm) was used for chromatographic analysis. Eluting solvents were prepared from aqueous 0.025 M disodium hydrogen phosphate and methanol and adjusted to pH 7.8 with phosphoric acid (10% w/v). In general, eluents containing 60% methanol were used; however, for medazepam and flurazepam analysis time was reduced by using 70 and 80% methanol, respectively. Benzodiazepines and their metabolites were detected by monitoring the column eluant at 254 nm, with on-column sensitivity ranging from 2.5 ng for nitrazepam to 9.0 ng for flurazepam. However, by using the λ_{max} of the individual compounds, the sensitivity range was increased to 1.5-6.0 ng on-column for the same drugs. The chromatographic data obtained by these workers, together with those obtained by Peat and Kopjak [12] and Harzer and Barchet [31], are shown in Table 5.

TABLE 5. Retention Volumes (ml) for Benzodiazepines Using Reversed-Phase Liquid Chromatography

	Reference[a]		
Benzodiazepine	12[b]	28[c]	31[d]
Chlordiazepoxide	12	6.5	6.3
Norchlordiazepoxide	9.6	NR	NR
Demoxepam	6.7	3.25	NR
Diazepam	16.1	8.1	7.5
Nordiazepam	12.7	5.9	6.53
Flurazepam	11.0 (A)	8.8 (B)	7.1 (D)
Desalkylflurazepam	10.8	NR	NR
Oxazepam	9.8	4.3	5.25
Prazepam	8.5 (A)	NR	12.8
Flunitrazepam	8.2	NR	NR
Lorazepam	NR	4.2	5.1
Medazepam	NR	10.9 (C)	3.4 (D)
Nitrazepam	NR	3.65	4.35

[a]NR, not run.

[b]Column: 30 cm × 4 mm Bondapak C_{18}; eluant: methanol:0.02 M dibasic sodium phosphate (pH 7.5), 58:42. A, 73% methanol.

[c]Column: 15 cm × 4.6 cm Spherisorb ODS (5 m); eluant: methanol: 0.025 M disodium hydrogen phosphate (pH 7.8), 60:40. B, 80% methanol; C, 70% methanol.

[d]Column: 25 cm × 4 mm LiChrosorb C_{18}; eluant: methanol:water, 70:30. D, 100% methanol.

Source: Adapted from Refs. 12, 28, and 31.

Peat and Kopjak [12] used similar HPLC conditions to those of Osselton et al. [28]. However, they used HPLC to confirm preliminary GC-ECD screening results and to quantitate chlordiazepoxide and its metabolites. Flunitrazepam was added to 1 ml of plasma or blood as internal standard and extracted at pH 9.0 with a mixed solvent system (toluene:heptane:isoamyl alcohol, 76:20:4). An aliquot of the organic layer is injected onto GC-ECD; the aqueous layer is then extracted with an additional 4.0 ml of solvent. The extract is dried and the residue examined by HPLC, using the conditions shown in Table 5. The drugs and metabolites were detected by monitoring the column eluant at 254 nm. Figure 4 shows a series of chromatograms obtained using this procedure. Although the HPLC procedure does not have the inherent sensitivity of GC-ECD, it has

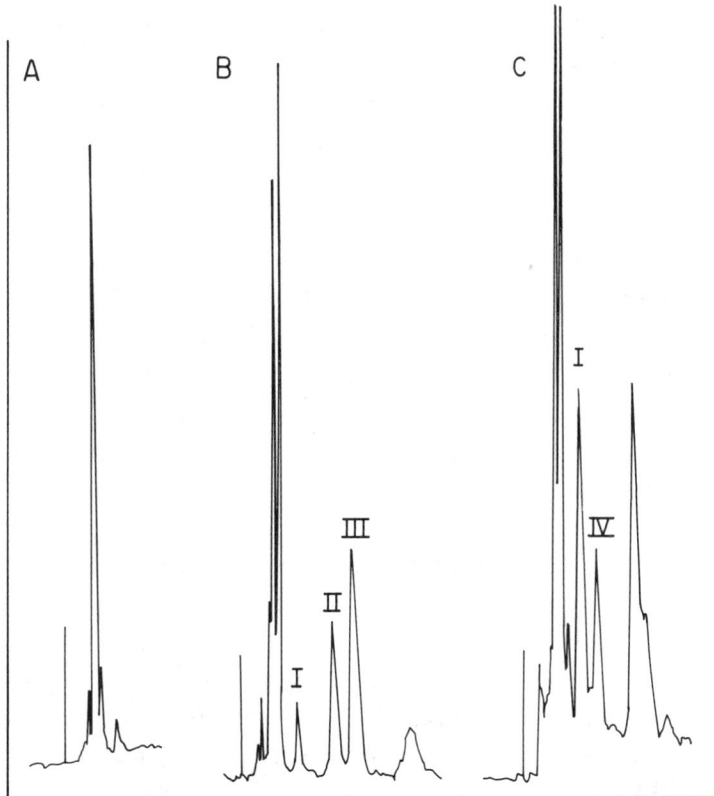

FIG. 4. (A) Extract by HPLC of blank plasma. (B) Extract by HPLC of plasma containing 1.1 µg/ml diazepam (II), 0.6 µg/ml nordizepam (III), and 0.2 µg of flunitrazepam (I). (C) Extract by HPLC of plasma containing desalkylflurazepam (IV) and 0.2 µg of flunitrazepam (I). For chromatographic conditions, see Table 5. (From Ref. 12.)

proved reliable for the confirmation of the benzodiazepines and their metabolites, and particularly useful for the quantiation of chlordiazepoxide and its metabolites.

The work described by Harzer and Barchet [31] is not as extensive as that described by the other groups; however, similar results were obtained. It is interesting to note that whereas both previous groups used a methanol:buffer mixture, these authors used a methanol:water mixture as the eluting solvent. A similar mixture

has been used by Brodie et al. [24] for the determination of diazepam and its metabolites. Although the majority of benzodiazepines
and their metabolites may elute from a reversed-phase column under
these conditions, the use of a buffer system is essential for the
detection of flurazepam and medazepam, without using 100% methanol.

In addition to using HPLC for the analysis of several benzodiazepines, Harzer and Barchet [31] also used a methanol:water
(80:20) system to separate the five benzophenones formed after the
acid hydrolsis of chlordiazepoxide, diazepam, prazepam, lorazepam,
and nitrazepam. Violon and Vercruysse [33] have also used a reversed-phase system with methanol:water (1:1) as eluant for the
HPLC determination of the benzophenones. The column used was a
LiChrosorb RP-8 (7 μm) and the hydrolysis products were monitored
at 254 nm. A separation of nine benzophenones was achieved, although the hydrolysis product of diazepam had a retention volume
of approximately 90 ml. The authors did not examine the benzophenones obtained after hydrolysis of flurazepam and its N-dealky-
lated metabolite or prazepam, which may very well have greater retention volumes. It must be stressed, however, that although
screening for the benzophenones may be a useful first step in the
analysis of benzodiazepines, it is important for the forensic toxicologist to be able to identify, with certainity, the drug and
metabolites present in the case sample.

A more unconventional HPLC approach to the determination of
the benzodiazepines and their metabolites in biological fluids has
been described by Tjaden et al. [27]. This group used a methyl-
silica (7-8 μm) packing and phosphate buffer:methanol eluants.
They measured the capacity ratios and selectivity coefficients of
the benzodiazepines as a function of the pH of the mobile phase.
If other conditions remain the same, the capacity ratios and selectivity coefficients decrease with increasing methanol content.
Hence there is no difference in the order of elution. However, on
changing the pH of the eluant, both selectivity coefficients and
capacity ratios were found to be effected, the order of elution for
several benzodiazepines changing between pH 4 and 6.

B. Sedative-Hypnotics and Anti-Convulsants

From a pharmacologist's point of view, these two drug groups should
be considered independently; however, analytically they present
similar problems to toxicologists. In addition, the majority of
the published HPLC procedures are applicable to both groups of drugs.

Of the sedative-hypnotics, the barbiturates are still the most
commonly encountered by forensic toxicologists, although the incidence of fatal poisoning with these agents has decreased over the
past decade. Other sedative-hypnotic agents that are detected, although to a far lesser degree, include glutethimide, ethchlorvynol,
meprobamate, and methyprylon. Methaqualone is also classified as
a sedative-hypnotic; however, this drug is encountered more commonly
in abuse situations than in cases arising from therapeutic use.
All these agents can produce degrees of depression of the central
nervous system (CNS), ranging from mild sedation to general ane-
sthesia. Although the exact mechanism of action of the barbiturates
is still uncertain, it is clear that at low doses they have some
GABA-like effects, suggesting similarities with the benzodiazepines.
At high doses they appear to have an effect on calcium entry into
neurons, although it is not clear if this effect has any relevance
to their CNS depressant activity.

The anticonvulsant drugs modify the ability of the brain to
respond to various seizure-evoking stimuli, reducing the spread of
excitation from the seizure foci. However, the exact mechanisms of
action of the anticonvulsant drugs are still poorly understood.
Chemically, they can be divided into distinct classes, the hydan-
toins (e.g., phenytoin), the anticonvulsant barbiturates (e.g.,
phenobarbital), the imminostilbenes (e.g., carbamazepine), the
deoxybarbiturates (e.g., primidone), the succinimides (e.g., etho-
suximide), the oxazolidinediones (e.g., trimethadione), the anticonvulsant benzodiazepines (e.g., clonazepam), and valproic acid.
HPLC has been widely used in the therapeutic monitoring of the
hydantoin, barbiturate, imminostilbene, deoxybarbiturate, and
succinimide anticonvulsants [34-41]. There have been two reports

[42, 43] in wich several anticonvulsants and their metabolites have
been assayed by HPLC simultaneously. Both procedures used a reversed-phase column at above ambient temperature. The procedure described by Kabra et al. [42] uses a 30 cm × 4 mm Bondapak C_{18} column
at a temperature of 50°C and an eluant of 21% acetonitrile in pH
4.4 phosphate buffer. The buffer is prepared by adding 0.3 ml of
1 M potassium dihydrogen phosphate to 1800 ml of distilled water,
followed by 50 µl of 0.9 M phosphoric acid. The column eluent was
monitored at 195 nm to detect the anticonvulsants. Two extraction
procedures were used; both employed 5-(4-methylphenyl)-5-phenyl-
hydantoin (MPPH) as internal standard. One method involved the
addition of 200 µl of acetonitrile, containing the internal standard, to 200 µl of plasma; after vortexing and centrifuging, 20 µl
of the supernatant is injected into the chromatograph. A modification of this procedure could be used for forensic toxicological
analysis; 300 µl of the supernatant from the procedure described
above is extracted with 8 ml of chloroform after the addition of
20 µl of glacial acetic acid. The chloroform is evaporated to
dryness and the residue dissolved in 50 µl of methanol for HPLC
analysis.

Adams et al. [43] used a similar extraction procedure, similar
HPLC conditions, and the same internal standard (MPPH). A reversed-
phase C_{18} column (25 cm × 2.6 mm) was used; the eluant consisted
of 15% acetonitrile in water at a temperature of 65°C. The UV
detector wavelength was 195 nm. Their extraction procedure consisted of the addition of 50 µl of a pH 8.0 phosphate buffer and
0.5 ml of extracting solvent. After centifuging, the organic phase
is removed and evaporated at room temperature, the residue dissolved
in methanol, and an aliquot injected into the liquid chromatography.
Table 6 lists the retention volumes of the common anticonvulsants
under the conditions used by both groups.

Both groups agreed that for adequate resolution of carbamazepine and phenytoin and for a suitable retention time it is essential to operate at a higher than ambient temperature. Adams et al.

TABLE 6. Retention Volumes (ml) of the Common Anticonvulsants

Compound	Reference 42[a]	43[b]
Ethosuximide	6.3	2.66
Trimethadione	NR	2.93
Phenylethylmalonamide	5.7	2.94
Primidone	7.2	3.28
Phenobarbital	12.3	4.17
Phenytoin	30.0	9.80
Carbamazepine	63.0	18.81

[a]Column: Bondapak C_{18}; eluant: acetonitrile:phosphate buffer pH 4.4, 21:79; temperature: 55°C. NR, not run.

[b]Column: ODS Sil-X-I; eluant: acetonitrile:water, 15:85; temperature: 65°C.

Source: Adapted from Refs. 42 and 43.

[43] point out that, as described, their procedure is optimized for the analysis of the five major anticonvulsants in a total analysis time of less than 20 min.

Kabra et al. [44, 45] have used similar HPLC conditions in developing screening procedures for sedative-hypnotics [44] and several commonly encountered drugs [45] in serum. The method described for the sedative-hypnotics uses an acetonitrile:phosphate (21.5:78.5) eluant at 50°C and extraction procedures identical to those described for the anticonvulsants, except that ethyl acetate rather than chloroform is used. The drugs examined were primidone, methyprylon, phenobarbital, butalbital, ethchlorvynol, pentobarbital, amobarbital, phenytoin, glutethimide, secobarbital, and methaqualone. Of these drugs, there was no baseline separation between pentobarbital, amobarbital, phenytoin, and gluethimide. Although the hydroxymetabolites of the barbiturates were not chromatographed, the authors do indicate that a metabolite of glutethimide interferes with butalbital.

More recent work [45] by this group has extended this screening procedure to 21 drugs, although again there is an absence of data on metabolites, particularly for the benzodiazepines. Nevertheless, this type of procedure could undoubtedly prove valuable to

the forensic toxicologist in the future. The extraction is similar
to that already described in previous work [42, 44]; however, in
this case the drugs are eluted from a reversed-phase column with a
mobile phase consisting of acetonitrile:phosphate buffer (pH 3.2),
using a two-step linear gradient. The eluted drugs are detected by
their absorption at 210 nm and a complete analysis requires approximately 45 min at a temperature of 50°C. A sensitivity of 2 µg/ml
is reported for most of the hypnotic and analgesic drugs, while
methaqualone, chlordiazepoxide, diazepam, and nordiazepam can be
detected at a concentration of 0.2 µg/ml. Figure 5 shows a chromatogram of a standard mixture of drugs. Of note is the lack of
resolution between pento- and amobarbital, phenytoin and glutethimide, and flurazepam and nitrazepam. Comparison of a Bondapak C_{18}
and a 15 cm × 4.6 mm Ultrasphere ODS (5 µm) column showed that
salicylic acid tailed and also that there was an incomplete separation of secobarbital and flurazepam on the latter column.

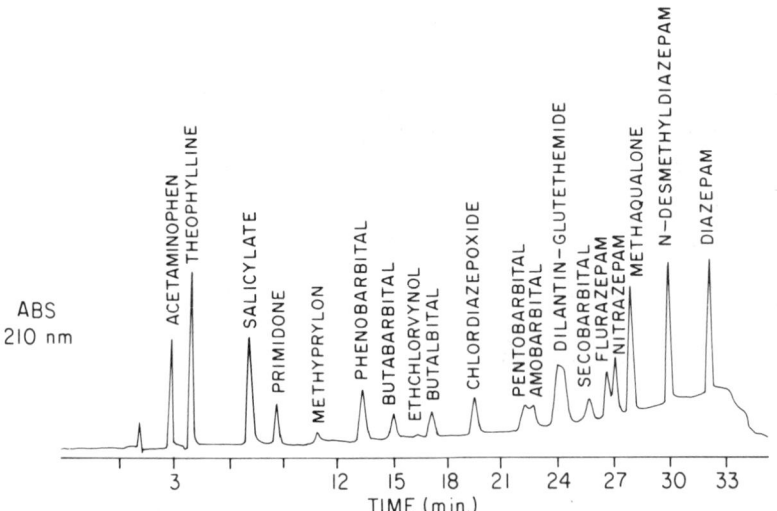

FIG. 5. Chromatogram of a standard mixture of drugs. For chromatographic conditions, see Ref. 45. (From Ref. 45.)

HPLC procedures for the analysis of barbiturates, other than
phenobarbitone, in biological fluids have been published. The
methods reported by Adams et al. [43] and Kabra et al. [44] may be
adapted to determine barbiturate concentrations, although in both
procedures amobarbital and pentobarbital are unresolved. Nevertheless, few general procedures for the identification and quantitation of a wide range of barbiturates in body fluids by HPLC have
been published. Tjaden et al. [46] used a methyl silica column
packing, similar to that used by the same group for the analysis of
the benzodiazepines [27], with methanol:water as eluant. The barbiturates were monitored at 220 nm. The methanol content of the
eluant varied depending on the particular barbiturates to be assayed. The authors applied the procedure to the analysis of these
drugs in blood and siliva, although a complicated extraction procedure was required to remove interfering peaks.

More recently, Gill [47] has developed a rapid, simple, and
sensitive procedure for the identification and quantitation of barbiturates in small volumes of blood, including hemolyzed samples.
Talbutal was used as internal standard; after its addition to blood
(100 µl), the volume was made up to 1 ml with pH 7.5 phosphate
buffer. Hexane:diethyl ether (1:1) was then added as extracting
solvent and after extraction a portion of this was evaporated to
dryness. The residue was dissolved in the HPLC eluant and an aliquot injected onto the HPLC column. The column (10 cm × 5 mm ID)
was packed with Hypersil ODS (5 µm) and the eluant consisted of
aqueous sodium dihydrogen phosphate (0.1 M) and methanol (6:4), adjusted to pH 8.5 with concentrated sodium hydroxide solution before
use. The eluant was monitored at 240 nm. Figure 6 shows an HPLC
of an extract of spiked blood containing five barbiturates under
these conditions.

C. Tricyclic Antidepressants

The structures of the tricyclic antidepressant compounds commonly
encountered by forensic toxicologists are given in Table 7. All of
these drugs block the neuronal uptake of norepinephrine, serotonin,

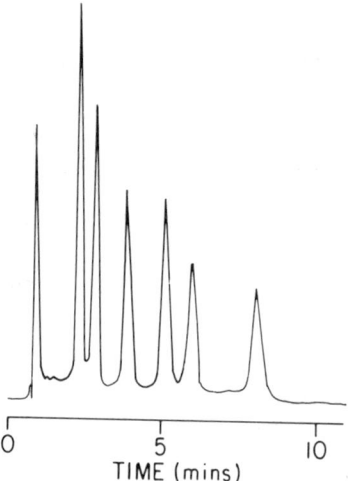

FIG. 6. HPLC of extract of spiked blood. Eluant: aqueous sodium
dihydrogen phosphate (0.1 M) and methanol (60:40 v/v) adjusted to
pH 8.5; column: 10 cm × 5 mm ID packed with 5-μm Hypersil ODS;
flow rate: 2 ml/min; detector: 240 nm; order of elution: cyclo-
barbital, butobarbital, talbutal, amobarbital, pentobarbital,
secobarbital. (From Ref. 47.)

or both. Clomipramine is a rather selective blocker of serotonin
uptake, whereas the demthylated metabolites are thought to be much
more selective in blocking the uptake of norepinephrine. At
therapeutic doses, they also have pronounced anticholinergic activity, manifested in blurred vision, dry mouth, constipation, and
urinary retention. Tachycardia is also frequently seen. At higher
doses there appears to be a direct depressant toxicity of the myo-
cardium.

Significant toxic effects of tricyclic antidepressants are
relatively common, and estimates of prevalence have run as high as
5%. Most of these reactions involve antimuscarinic effects of the
drugs and cerebral toxicity, but cardiac toxicity also represents
a serious problem. Rose [48] has described three stages to poisoning by these drugs. The vast majority of cases are in mild stage 1,
represented by anticholinergic activity. Some patients reach stage
II with major CNS effects and increasing intracardiac block. Stage

TABLE 7. Tricyclic Antidepressants

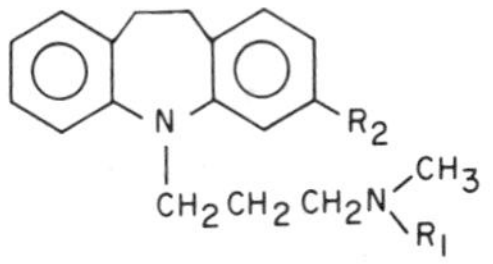

R_1 = CH_3, R_2 = H   Imipramine

R_1 = H, R_2 = H   Desipramine

R_1 = CH_3, R_2 = Cl   Clomipramine

R_1 = H, R_2 = Cl   Desmethyclomipramine

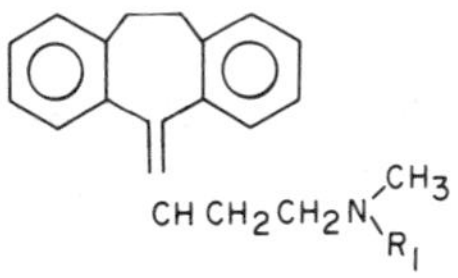

R_1 = CH_3 Amitriptyline

R_1 = H   Nortriptyline

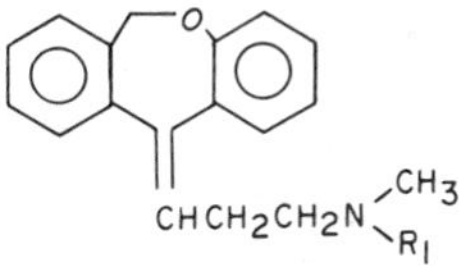

R_1 = CH_3 Doxepin

R_1 = H   Nordoxepin

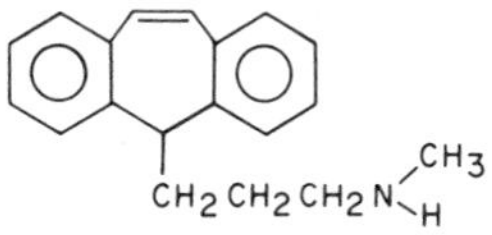

Protriptyline

III, which Rose [48] encountered in less than 5% of poisonings, is a potentially fatal situation, with respiratory arrest, convulsions, and ventricular arrhythmias.

A large number of procedures have been published for the analysis of the tricyclic antidepressants. Common to the majority of these methods is the need for extraction of the drug from the biological medium before analysis. This usually involves extraction at a basic pH with organic solvent; the extract is then commonly assayed by either gas-liquid chromatography or HPLC. GC is still the technique most widely used by forensic toxicologists. Methods [49-51] using the flame ionization detector and developed for forensic purposes have been published. Other workers [52, 53] have used derivative formation and electron capture detection for measuring the demethylated metabolites in plasma. GC with selective nitrogen phosphorous detector has become increasingly popular for quantitation of these drugs. Some investigators find that derivatization of the secondary amines is necessary [54-56], whereas others do not [57-59]. Gas chromatography-mass spectrometry (GC-MS), either in the electron impact [60] or chemical ionization

[61] mode, has also been used to monitor the tricyclic antidepressants.

The application of HPLC to the separation of tricyclic drugs was first described in 1975 [62, 63]. Since then, numerous procedures for monitoring drug and metabolite concentrations in plasma have been published; however, most of these quantitate a single drug and metabolite. Clomipramine and desmethylclomipramine have been analyzed by both ion-pair partition [64, 65] and adsorption [66] chromatography, whereas amitriptyline and metabolites have been assayed by reversed-phase [67, 68], ion-pairing partition [69], and adsorption chromatography [70]. Other groups [71-74] have attempted to determine several of the tricyclic antidepressants using a single system. A summary of their results is shown in Table 8.

Sensitivities in the lower therapeutic ranges for these drugs were obtained by all four groups, except for deZeeuw and Westenberg [73], who reported sensitivities of approximately 0.1 and 0.25 μg/ml for the tertiary and secondary amines, respectively. It must be noted that their method is intended for screening purposes in overdose cases. Multiple extraction procedures were used by all except Vandemark et al. [71], who used an extraction at pH 9 into hexane:isoamyl alcohol. Internal standards were used for quantitation by Vandemark et al. [71], Proelss et al. [72], and Wallace et al. [74]. Interferences from other tricyclic drugs were a problem in all the assays. Under the conditions described by Vandemark et al. [71], doxepin coeluted with amitriptyline and several tricyclic drugs were not examined. Protriptyline coelutes with doxepin using the system of Proelss et al. [72], although by monitoring the eluant at 290 nm, the secondary absorption maximum of protriptyline, the two drugs could be differentiated. In addition, chlordiazepoxide and some of the phenothiazines interfere with the analysis of doxepin and other phenothiazines and antihistamines with the assays of desipramine and nortriptyline. Similarly to the studies of Vandemark et al. [71] and Wallace et al. [74], a number of the tricyclic drugs were not examined. Wallace et al.

TABLE 8. Retention Volumes (ml) for Tricyclic Antidepressants

	Reference[a]				
			73[d]		
Compound	71[b]	72[c]	1	2	74[c]
Amitriptyline	6.0	20.3	2.98	6.12	2.16
Nortriptyline	11.3	15.0	13.2	4.44	16.0
Desmethylclomipramine	NR	NR	17.4	4.2	NR
Imipramine	6.75	16.3	4.06	7.39	18.8
Desipramine	15.0	13.5	23.0	4.51	14.4
Protriptyline	19.5	11.4	13.1	4.44	13.8
Trimipramine	NR	NR	2.06	5.16	NR
Doxepin	6.0	11.3	3.91	7.01	14.6
Cloimipramine	NR	NR	3.29	6.07	NR
Nordoxepin	NR	NR	NR	NR	11.0

[a]NR, not run.

[b]Column: 25 cm × 4.6 mm Silica (5 μm); eluant: acetonitrile:concentrated ammonium hydroxide (99.3:0.7); detector: 211 nm; temperature: 65°C.

[c]Column: 30 cm × 4.0 mm Bondapak C_{18}; eluant: methanol:acetonitrile:water (41:15:44), containing 5 mmol of pentanesulfonic acid per liter of phosphate buffer (0.1 M, pH 6.5); detector: 254 nm; temperature: ambient.

[d]Column: 10 cm × 4.6 mm LiChrosorb Si 60 (5 μm), eluant: 1, hexane:dichloromethane:methanol (8:1:1), hexane contained 10 ppm methylamine. 2, 0.05 M sodium bromide in methanol; detector: 250 nm; temperature: ambient.

[e]Column: Micropak MCH-10; eluant: 65-70% acetonitrile:30-35% 0.056 M phosphoric acid in 0.01 M potassium dihydrogen phosphate, pH 2.7; temperature: 40°C.

Source: Adapted from Refs. 71-74.

[74] found that doxepin, protriptyline, and desipramine coelute and that flurazepam interferes with nortriptyline.

Perhaps the most complete HPLC separation of the tricyclic antidepressants has been that obtained by deZeeuw and Westenberg [73] using an adsorption phase system with hexane:dichloromethane: methanol as eluant. Figure 7 shows a chromatogram from an extract of spiked plasma. However, protriptyline and nortriptyline coelute on both this system and that involving ion pairing with sodium bromide. The authors indicate, however, that the two drugs may be differentiated by monitoring the eluant at 290 nm or by changing

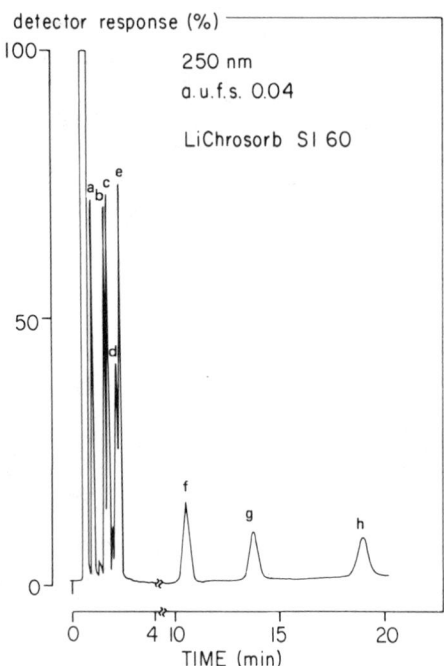

FIG. 7. Liquid chromatogram of a separation of some tricyclic
antidepressant extracted from spiked plasma. Column: LiChrosorb
SI 60, 5 μm, 10 cm × 4.6 mm ID; eluant: hexane:dichloromethane:
methanol (8:1:1), the hexane containing 10 ppm methylamine. a,
Trimipramine; b, amitriptyline; c, clomipramine; d, doxepin, e,
imipramine; f, nortriptyline; g, desmethylclomipramine; h, desipramine. (From Ref. 73.)

the mobile phase to methanol:ammonia (100:1.5) [63]. If 0.05 M
sodium bromide in methanol is used as eluant, the tricyclics are
chromatographed as bromide ion pairs, resulting in a totally different elution order, with the secondary amines eluting before the
tertiary compounds. The authors, however, did not study possible
interference from other basic drugs, so it is unclear how specific
this procedure is.

Recent work by Sonsalla [75] has also demonstrated the utility
of silica columns for tricyclic antidepressant analysis. Using a
Micropak Si 10 column, and an eluant of methanol:2 M ammonia:1 M

ammonium nitrate (95:3:2) and detection at 254 nm, a separation of
eight tricyclics was acheived in a total run time of 7 min. Figure
8 shows chromatograms of standards, drug-free plasma, and a patient
sample containing amitriptyline and nortriptyline. However, clomipramine coeluted with amitriptyline and some interference was also
noted from nontricyclic antidepressants; for example, quinidine
(and quinine) coelute with doxepin.

    Although these procedures may prove very useful to the foren-
sic toxicologist for the quantitation of tricyclic antidepressants,

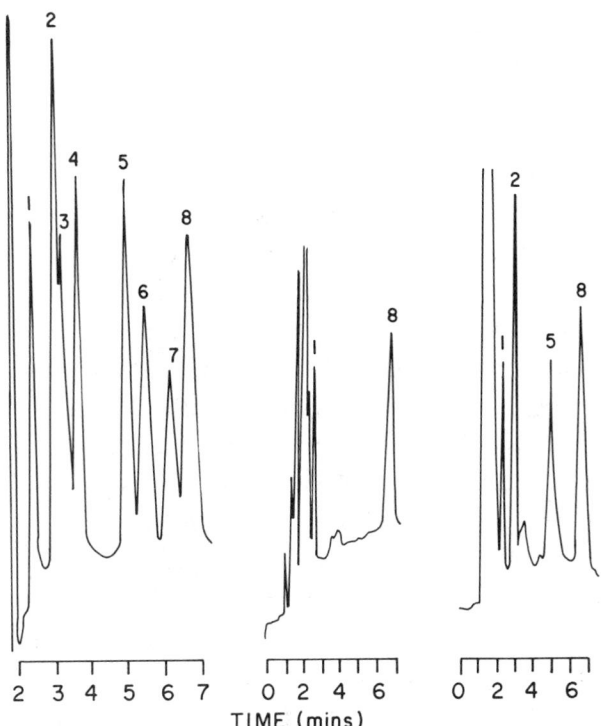

FIG. 8. Chromatograms of standards, drug-free plasma, and a sample
containing amitriptyline and nortriptyline. Eluant: methanol-2
M ammonium hydroxide-1 M ammonium nitrate (95:3:2); column: Micropak Si 10, 30 cm × 4.6 mm ID; detector: 254 nm, 1, Trimipramine;
2, amitritpyline; 3, doxepin; 4, imipramine; 5, nortriptyline;
6, nordoxepin; 7, desipramine; 8, protriptyline. (From Ref. 75.)

because of the interferences observed they should be used cautiously
for screening purposes. This problem is compounded when the hydroxylated metabolites of the tricyclics are considered. These compounds are present in high concentrations in liver and urine samples
and may even be detectable at significant concentrations in blood
after an overdose. To date, few authors have considered the chromatography of these compounds. Sonsalla [75] noted that 10-hydroxy-
amitriptyline coelutes with the parent drug, whereas 10-hydroxy-
nortriptyline is well resolved from nortriptyline but does coelute
with nordoxepin. Other groups [68, 69, 76] have used HPLC to
examine the hydroxylated metabolites of amitriptyline and nortriptyline, without testing for interference from related tricyclic
antidepressants or other drugs.

D. Antidysrhythmic Drugs

The antidysrhythimic drugs -- procainamide, N-acetylprocainamide
(NAPA), lidocaine, quinidine, disopyramide, and propranolol --
are important in the treatment of a variety of cardiac disorders.
They all have an action on the electrophysiological properties of
the heart; for example, quinidine, disopyramide, and procainamide
reduce the membrane responsiveness of Purkinje fiber. Plasma concentrations of these drugs relate more accurately with clinical
efficacy than does dosage. It is therefore important for the
toxicologist to have an understanding of the pharmacokinetics of
these drugs, and how they are affected by disease, if he is to
make a meaningful interpretation of the data. Lidocaine, for
example, must be administered by either intravenous injection or
infusion, as it undergoes extensive first-pass metabolism. Severe
hepatic disease and reduced perfusion of the liver in congestive
heart failure decrease the rate of metabolism to monoethylglyclxyli-
dide (MEGX). The clearance of lidocaine approaches the rate of
hepatic blood flow and is thus very sensitive to changes in this
parameter. The volume of distribution is also substantially reduced in patients with heart failure.

The antidysrhythmic drugs and their bioactive metabolites have been determined by several methods, including colorimetry [77], flurometry [78], gas chromatography [79, 80], gas chromatography-mass spectrometry [81], thin-layer chromatography [82], and ultra-violet spectroscopy [83]. HPLC has also been widely used to monitor plasma or serum concentrations of quinidine [84-86], procainamide and NAPA [87-89], lidocaine [89], disopyramide [90-92], and proprano-lol [93-95]. However, there are only three reports [96-98] describing the determination of more than two antidysrhythmic drugs.

Lagerstrom and Persson [96] used liquid-solid chromatography to assay six antiarrhythmic drugs: disopyramide, lidocaine, tocainide, procainamide, aprinidine and quinidine. Three different packing materials were used. LiChrosorb Si 60 (7 μm), LiChrosorb Si 100 (10 μm), and Partisil 5 (5 μm); the choice of packing material was not regarded as important by the authors. All the drugs were extracted from alkalinized plasma into dichloromethane. The chromatographic conditions are detailed in Table 9.

In liquid chromatography on silica, amines can be retained either as ion pairs or bases. Both types of separation systems were used by these authors. Aqueous perchloric acid was used as the acidifying and ion-pairing agent, dissolved in mixtures of chloroalkanes and alcohols as mobile phases. The concentration of perchloric acid in the mobile phase was found to have a limited effect on retention; however, if the alcohol concentration in the mobile phase was increased slightly, there was a much greater de-crease in the capacity factor. Ion pairing on LSC was used to monitor disopyramide and its metabolite, lidocaine, and tocainide.

The LSC of amines in the straight-phase mode usually utilizes mobile phases containing a base in order to improve the chromato-graphic characteristics of the solute. Such a separation system was used to assay procainamide and NAPA. If the ammonia was replaced with an aliphatic amine, a lower column efficiency was obtained.

Aprinidine was assayed using a very polar mobile phase with a major proportion of methanol in an aqueous buffer solution. A

TABLE 9. Liquid-Solid Chromatography of Antidysrhthymic Drugs

Drug	Mobile-Phase Component	Proportion (v/v)	Detection Wavelength (nm)
Disopyramide	Perchloric acid (1 M aq)	1	265
(N-Desisopropyl-	Methanol	9	
disopyramide)	Dichloromethane	90	
Lidocaine	Perchloric acid (1 M aq)	0.3	228
	Methanol	4	
	Dichloroethane	95.7	
Tocainide	Perchloric acid (1 M aq)	0.5	230
	Methanol	10	
	Dichloroethane	89.5	
Procainamide	Methanol	10	280
	Dichloromethane	20	
	Dichloromethane (saturated with ammonia solution)	70	
Aprinidine	Ammonia acetate (1 M aq)	1.5	254
	Ammonia (1 M aq)	3.5	
	Methanol	95	
Quinidine	1-Butanol	10	254
	Dichloromethane	70	
	n-Hexane	20	

Source: Adapted from Ref. 96.

similar system has been used by Peat and Jennison [84] to determine
quinidine concentrations in plasma. In such chromatographic systems, it may be questioned if the silica packing material is still
the most polar phase or if there is a reversed-phase separation
mechanism.

Reversed-phase systems were used by Flood et al. [97] and
Kabra et al. [98]. Both groups used extraction into dichloromethane at alkaline pH, after the addition of an internal standard.
Flood et al. [97] used a Bondapak C_{18} column with an eluant consisting of 28% acetonitrile in 30 mM potassium dihydrogen phosphate
(pH 4.45) and detection at 205 nm to determine plasma concentrations of quinidine, lidocaine, and disopyramide. The pH of the
buffer was found to be a crucial factor for good separation and to
obtain efficient chromatography for quinidine. Separation of the

drugs of interest and internal standard (p-chlorodisopyramide) was
achieved within the pH range 4-6, whereas at pHs above 4.5 the peak
shape for quinidine was unsuitable for quantitation. Other antidysthythmic drugs -- propranolol, procainamide, and NAPA -- were
tested for interference. Procainamide and NAPA eluted in the solvent front, whereas propranolol interferred with the internal standard; lowering the buffer strength and using a high temperature resolved these two compounds.

The most comprehensive work to data on tbe HPLC analysis of
the antidysrhythmics has been that of Kabra et al. [98]. To eliminate the poor resolution and peak tailing associated with the use of
octadecyl columns, they used an Ultrasphere Octyl 8 column. Table
10 summarizes the HPLC conditions used for the isochratic separation of the drugs. In addition, a gradient of 7-30% acetonitrile
in phosphate buffer (pH 3.0) separated seven antiarrhythmic drugs.
A chromatogram of a standard mixture is shown in Figure 9. Fluores-
cence (λ_{ex} 290 nm, λ_{em} 350 nm) could also be used to detect quinidine
and its metabolites, and propranolol and its metabolites. Although
procainamide fluoresces at similar wavelengths in basic solution,
it does not interfere because of the acidic eluant used. In addition to increasing the sensitivity for these drugs, fluorescent
detection will also increase the selectivity of the system if it
was to be used as a screening procedure.

Interferences from metabolites and other commonly encountered
drugs were tested, although only a limited number of drugs were
chromatographed. Diazepam and flurazepam coeluted with quinidine,
and would therefore interfere with the UV detection of this drug.
However, fluorescent detection could still be used.

HPLC has proved to be especially useful, in the clinical
toxicology laboratory, to monitor the plasma or serum concentrations
of the antidysthythmic drugs. It may very well prove to be more
useful as a screening procedure for these agents and related drugs
(e.g., beta blockers) for the forensic toxicologist.

TABLE 10. Reversed-Phase Chromatography of Antidysrhythmic Drugs

Drug or Metabolite	Eluant[a] 25.5% acetonitrile in 75 mM potassium dihydrogen phosphate (pH 3.4)	Detector 216 nm	Retention Volume (ml)	Eluant[a] 9.5% acetonitrile in 25 mM potassium dihydrogen phosphate (pH 3.0)	Detector 280 nm	Retention Volume (ml)[b]
Procainamide			3.0			5.4
NAPA						8.4
Disopyramide			9.9			ND
N-Desisopropyldisopyramide			5.7			
Lidocaine			6.15			42.0
MEGX			4.65			
Quinidine			7.35			ND
2-Quinidinone			4.5			
3-Hydroxyquinidine			3.75			
Quinidine N-oxide			7.35			
Dihydroquinidine			9.3			
Propranolol			18.0			ND
4-Hydroxypropranolol			6.75			
p-Chlorodisopyramide (IS)			20.25			
Pronethalol (IS)			10.50			
N-Propionylprocainamide (IS)						18.0

[a]Temperature 40°C.
[b]ND, not detected.

Source: Adapted from Ref. 98.

340

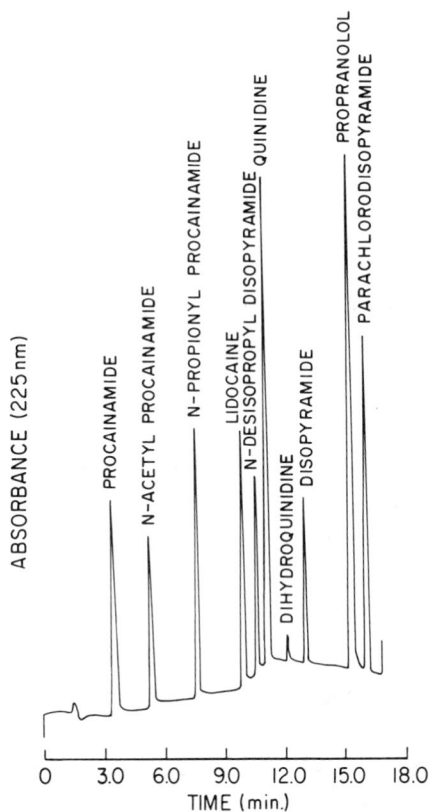

ABSORBANCE (225 nm)

PROCAINAMIDE

N-ACETYL PROCAINAMIDE

N-PROPIONYL PROCAINAMIDE

LIDOCAINE
N-DESISOPROPYL DISOPYRAMIDE
QUINIDINE

DIHYDROQUINIDINE
DISOPYRAMIDE

PROPRANOLOL
PARACHLORODISOPYRAMIDE

0   3.0   6.0   9.0   12.0   15.0   18.0
TIME (min.)

FIG. 9. Chromatogram of a standard mixture of drugs run under
gradient conditions. (From Ref. 98.)

E. Acetaminophen (Paracetamol)

Acute overdosage with acetaminophen causes fatal hepatic necrosis,
and the number of self-poisonings and suicides has grown alarmingly
in recent years. Symptoms during the first 2 days of acute poisoning by this drug do not reflect the potential seriousness of the
intoxication. Nausea, vomiting, anorexia, and abdominal pain occur
during the initial 24 h. Clinical indications of hepatic damage
manifest themselves within 2-6 days of ingestion of toxic doses.
Biopsy of the liver reveals centralobular necrosis with sparing of

the periportal area. In nonfatal cases, these lesions are reversible over a period of weeks or months.

Measurement of the plasma half-life of acetaminophen is useful
in evaluating the possibility for hepatic damage. If it is in excess of 4 hr, hepatic necrosis is likely, and hepatic coma should
be anticipated if it exceeds 12 h. The hepatotoxicity has been
attributed to an "active metabolite" which binds covalently to
cellular macromolecules. This metabolite is normally inactivated
by conjugation with glutathione; however, in overdose situations
the supply of glutathione is rapidly depleted, allowing the metabolite to bind to cellular constituents. A promising method of
treatment is the administration of N-acetylcysteine, which is ab-
sorbed orally. This compound acts by replenishing the store of
sulfhydryl groups in the liver.

A variety of methods have been used to quantitate acetaminophen in biological fluids, including spectrophotometry [99, 100],
gas chromatography [101-104], and HPLC [105-110]. A number of the
HPLC procedures can also quantitate phenacetin [105, 106] and
salicylate [108]. Acetaminophen and phenacetin will be detected in
HPLC systems similar to those used by Kabra et al. [45]. Most of
the HPLC procedures used to quantitate the drug use a reversed-
phase octadecyl column with sodium acetate (pH 4) buffer and acetonitrile as eluant. The system described by Manno et al. [110]
is typical. They use a Bondapak C_{18} column with 7% acetonitrile in
sodium acetate buffer (pH 4.0) and detection at 254 nm. A suitable
internal standard in this system would be β-hydroxyethyltheophylline.
Care must be taken, however, that the HPLC conditions used give
adequate separation of acetaminophen, phenacetin, and the xanthines
(e.g., theophylline). The majority of methods utilized in the
clinical toxicology laboratory use direct injection techniques,
whereas extraction of the drug and internal standard at a neutral
pH into chloroform, or similar solvent, is still used by the forensic toxicologist.

F. Morphine and Other Opiate Narcotics

The determination of therapeutic levels of morphine in small blood
samples is a problem commonly encountered by forensic toxicologists.
Therapeutic levels of morphine range from a few nanograms to about
1 µg/ml. Methods used must therefore be sensitive and specific.
Presently, radioimmunoassay (RIA) is widely used. However, the
morphine antibody cross-reacts with the major morphine metabolite
morphine-3-glucuronide; thus the RIA result gives an estimate of
the "total morphine concentration." Cross-reaction also occurs
with other levoratotory morphinans, such as codeine. Gas chromatography with detection by mass spectrometry [111], flame ionization
[112], and electron capture [113] has been widely used for confirmation. In addition, HPLC has been used to confirm the presence of
morphine in urine samples. Jane and Taylor [114] used the ferri-
cyanide reaction to form the dimer pseudomorphine, which was then
chromatographed on a silica column and detected by fluorescence.
More recently, Ulrich and Rügsegger [115] used a reversed-phase
system and detection at 230 nm to detect codeine and morphine in
urine samples. Unfortunately, these HPLC methods are either lacking in sensitivity or convenience of analysis.

    More recently, a number of workers [116-118] have shown that
HPLC with electrochemical detection is a suitable alternative.
The sensitivity of this method of detection for phenolic compounds
has been demonstrated by Refshange et al. [119], who designed a
detector with a carbon-paste electrode capable of detecting picogram quantities of catecholamines. Other workers [120], using
glassy carbon for the electrode, have built flow cells more suitable for use with nonaqueous solvents. Table 11 shows the HPLC
conditions and detector parameters used by three groups who have
used electroehcmical detection for morphine.

    All these authors claim sensitivities of less than 1 ng of
morphine injected, although each procedure required extraction after

TABLE 11.   Electrochemical Detection of Morphine

HPLC Conditions		Electrochemical Cell Configuration	Potential	Ref.
Column:	20 cm × 4.6 mm silica	Wall-jet and thin-layer hybrid	+0.6 V	116
Eluant:	Methanol:pH 10.2 ammonium nitrate (9:1)			
Flow:	1 ml/min			
Column:	30 cm × 4 mm LiChrosorb RP18	Metrohm 611	+1.0 V	117
Eluant:	Methanol:0.01 M potassium dihydrogen phosphate (85:15)			
Flow:	1 ml/min			
Temperature:	40°C			
Column:	30 cm × 4 mm Bondapak C18	TL-3 electrochemical cell (Bioanalytical Systems)	+0.8 V	118
Eluant:	Methanol:water (20:80) containing 50 mM tetramethyl ammonium hydroxide (pH 6.1)			
Flow:	2.0 ml/min			

Source: Adapted from Refs. 116, 117, and 118.

the addition of internal standard.   Figure 10 shows the extraction method used by White [116].

According to White [116], there are two waves to the voltage curve.   The first wave, at $E_{1/2}$ = + 0.4 V versus a silver-silver chloride reference electrode (SSCE), is probably due to a one-electron oxidation of morphine followed by dimerization of the free radical to pseudomorphine.   The second wave, at $E_{1/2}$ = + 0.7 V against SSCE, may be due to a further oxidation of pseudomorphine, or to a two-electron oxidation of morphine to an intermediate phenoxium ion.

The choice of electrode potential for electrochemical detection depends on two factors.   First, although increasing the potential increases the response, the background current and consequent noise level are also increased.   Second, increasing the potential above that required for oxidation decreases the specificity of the

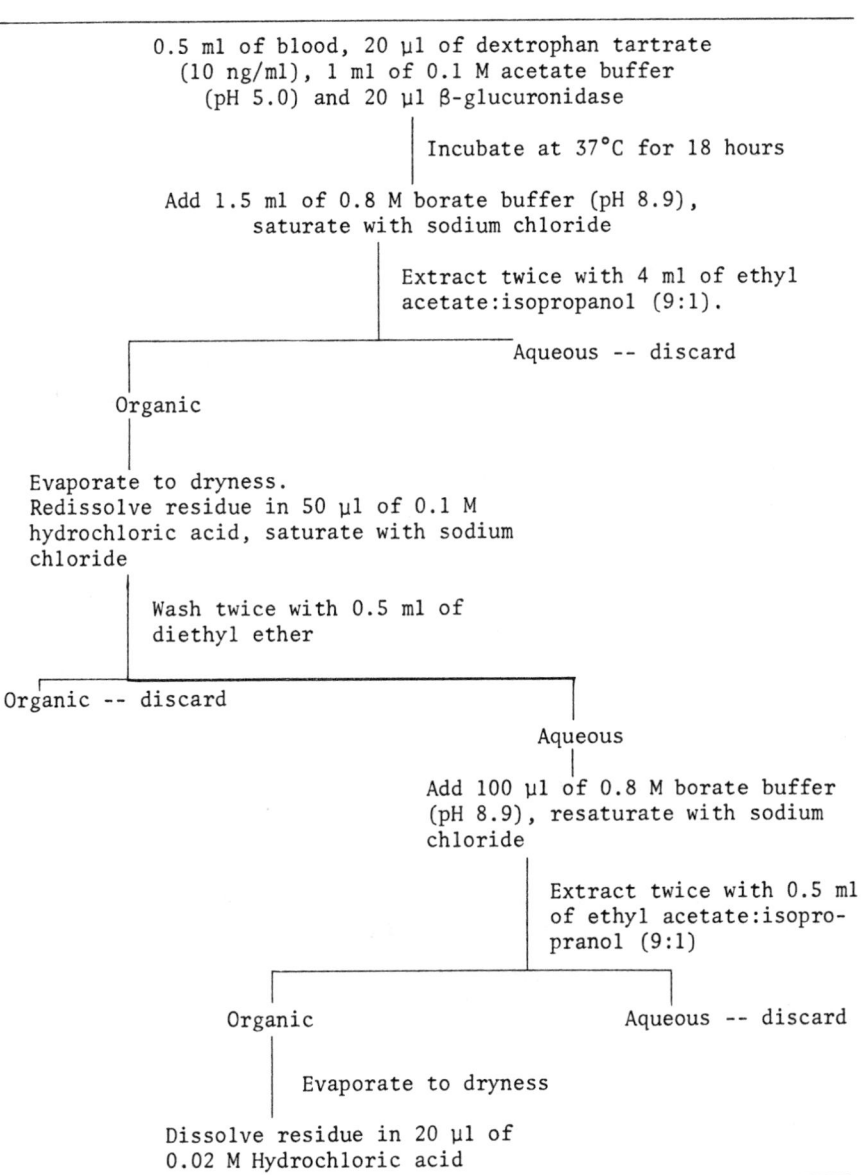

0.5 ml of blood, 20 µl of dextrophan tartrate
(10 ng/ml), 1 ml of 0.1 M acetate buffer
(pH 5.0) and 20 µl β-glucuronidase

Incubate at 37°C for 18 hours

Add 1.5 ml of 0.8 M borate buffer (pH 8.9),
saturate with sodium chloride

Extract twice with 4 ml of ethyl
acetate:isopropanol (9:1).

Aqueous -- discard

Organic

Evaporate to dryness.
Redissolve residue in 50 µl of 0.1 M
hydrochloric acid, saturate with sodium
chloride

Wash twice with 0.5 ml of
diethyl ether

Organic -- discard

Aqueous

Add 100 µl of 0.8 M borate buffer
(pH 8.9), resaturate with sodium
chloride

Extract twice with 0.5 ml
of ethyl acetate:isopro-
pranol (9:1)

Organic   Aqueous -- discard

Evaporate to dryness

Dissolve residue in 20 µl of
0.02 M Hydrochloric acid

FIG. 10.   Extraction scheme for morphine.

detector. In all three systems, other phenolic compounds (e.g.,
naloxone, nalorphine) are the most susceptible to electrochemical
oxidation, although they are separated chromatographically. Codeine
and compounds without a phenolic grouping are detected only at very
high concentrations or at higher electrode potentials.

IV. HIGH-PERFORMANCE LIQUID CHROMATOGRAPHY IMMUNOASSAY METHODS

During the last decade the most significant advances in analytical
toxicology have been the development of more selective and specific
detectors for gas chromatography (e.g., the nitrogen phosphorous
detector and the mass spectrometer when used in the single-ion
monitoring mode), and the development of immunoassay procedures
based on radiolabeled or enzyme tracers. The introduction of manufactured immunoassay tests for the screening of drugs of abuse in
urine has revolutionized the analytical procedures used in methadone treatment programs.

Radioimmunoassay kits (Roche Diagnostics, Abuscreen) are now
available for screening urine samples for a number of drugs of
abuse, including morphine and related opiate narcotics, barbiturates, cocaine and its metabolite, amphetamine, phencyclidine, and
methaqualone. These kits have also been used by a number of workers
to screen blood samples, although the amphetamine procedure does
not have the required sensitivity. In addition to these kits, which
are intended primarily for screening purposes, there are other kits
designed solely to quantitate drugs in serum, plasma, or blood
samples. An example is the quantitation of digoxin in biological
fluids by radioimmunoassay.

The enzyme-multiplied immunoassay technique (EMIT, Syva Company) uses an enzyme reaction as the marker and was originally
developed for screening urine samples for drugs of abuse. However,
the major application of this technique today is in the area of
therapeutic drug monitoring, for which a number of assays have been
developed. Using this technique, it is now possible to quantitate
the major anticonvulsant and antiarrhythmic drugs in small plasma

samples. Compared to radioimmunoassay, EMIT is much faster and demands less skill. However, it is less sensitive, and for the drugs of abuse is applicable only to urine samples.

One of the major drawbacks to HPLC as a screening procedure in toxicology is the low sensitivity for many drugs and metabolites of the widely used ultraviolet detector. To overcome this disadvantage and to improve the selectivity of HPLC, workers at the Home Office Central Research Establishment in the United Kingdom have used immunoassay procedures as the detection system for specific groups of drugs. Combined HPLC-immunoassay also serves to increase the specificity of an immunoassay procedure if there is significant cross-reactivity, since the first stage of the procedure is a chromatographic separation. Methods described in the literature suffer from the drawback that they cross-react with structurally related compounds and with metabolites of the drug under investigation. This cross-reactivity can also serve to increase sensitivity. For example, the antibody developed for morphine cross-reacts with morphine-3-glucuronide, thus increasing the sensitivity of the assay in urine. In addition, the cross-reactivity can be minimized by careful design of the drug protein conjugate used to raise antisera.

Combined HPLC immunoassay methods have been used for the analysis in biological fluids of morphine, codeine and their metabolites [121], Δ^9-THC and its metabolites [122, 123], lysergic acid diethylamide (LSD) [124], and for the cardiac glycosides [125].

Nelson and co-workers [121] used both radioimmunoassay and EMIT as the detection system for HPLC to analyze codeine, morphine, and their metabolites in biological fluids. The cross-reactivities of opiates, as determined by the authors, to the Abuscreen and EMIT morphine reagents are shown in Table 12. Changes in the morphine molecule at the 6-position and the 4-position altered reactivity less than changes at the bridge-head nitrogen; for example, normorphine showed virtually no cross-reaction in either assay.

TABLE 12. Cross-Reactivities of Other Opiates

Compound	Relative Reactivity of EMIT	Relative Reactivity of RIA
Morphine	1.00	1.00
Codeine	1.04	1.48
Dihydrocodeine	0.93	0.87
6-Monoacetylmorphine	0.50	0.12
Dihydromorphine	0.43	0.17
Morphine-3-glucuronide	0.24	0.41
Nalorphine	0.11	0.006
Norcodeine	0.08	0.04
Codeine-N-oxide	0.08	0.04
Normorphine	0.005	$<10^{-4}$
Morphine-N-oxide	9×100^{-4}	$<10^{-4}$

Note: The relative reactivity is the ratio of the amount of morphine to the amount of opiate required to give an equivalent cross-reaction to 0.5 mg of morphine for EMIT and 20 ng of morphine for RIA.

Source: Adapted from Ref. 121.

A reversed-phase HPLC system incorporating bromide ions in the eluant was found to separate efficiently a number of frequently encountered opiates, although oxycodone, a constituent of Percodan preparations, was not chromatographed. Table 13 lists the retention volumes of those narcotics examined.

Urine samples were injected without pretreatment, and blood samples from which erythrocytes and ghost cells had been removed by centifugation were diluted with distilled water to give a concentration of approximately 0.3 µg/ml of opiate cross-reactivity. A 1-ml syringe was used for injection. The procedure was used to examine blood and urine specimens from cases of suspected morphine overdose and also urine samples from volunteers who had ingested codeine. As an example, the data from one case are listed:

	Morphine-3-Glucuromide	Morphine	Unidentified	Total
Urine (HPLC-EMIT)	11.0	7.0	1.6	19.6
Blood (HPLC-RIA)	0.2	0.3	0	0.5

Note: Data are concentration values (µg/µl), expressed as morphine equivalents.

Source: Ref. 121.

TABLE 13. Retention Volumes of Opiate

Compound	Retention Volume (ml)
Morphine-3-glucronide	1.0
Normorphine	1.7
Dihydromorphine	1.7
Morphine	1.8
Morphine-N-oxide	1.8
Codeine-N-oxide	4.3
Dihydrocodeine	4.8
Nalorphine	4.9
Codeine	5.3
Norcodeine	5.7
6-Monoacetylmorphine	10.4

Note: Column: 10 cm × 4.6 mm Hypersil ODS (5 μM); eluant: 12.5% methanol in 0.01 M potassium phosphate buffer (pH 3) containing 0.1 M potassium bromide.

Source: Adapted from Ref. 121.

In all cases examined, the total cross-reacting opiates, measured by HPLC-immunoassay, were approximately half the level determined by direct immunoassay of the biological fluid. The cause of this variation is not apparent.

Examination of urine samples from volunteers who had injested codeine, using the combined HPLC-immunoassay approach, showed that RIA and EMIT have different cross-reactivities to codeine-6-glucuromide (EMIT > RIA) than they do to codeine (see Table 12).

Fletcher and co-workers [125] have used a combined HPLC-RIA approach for the identification of cardiac glycosides in human body fluids. Although in most cases of overdose with cardiac glycoside the agent is known and in the majority of cases is digoxin, for which an RIA procedure is available, occasionally other cardiac glycosides are suspected. Although commercial RIA reagents are also available for a digitoxin assay, no such kits are available for the less common therapeutic glycosides, β-methyldigoxin, lantoside C, and deslanoside. Previous work [126] had shown that the commercial digoxin RIA reagents cross-react with these glycosides.

Both normal- and reversed-phase HPLC systems were investigated for separation of the glycosides, and although the reversed-phase system gave a more efficient separation of the six therapeutic

glycosides, the normal-phase system offers better separation of the
four which cross-react to the digoxin RIA. For this reason it was
chosen for the HPLC-RIA work.

HPLC is routinely used to identify lysergic acid diethylamide
(LSD) and other ergot alkaloids in "street drugs." However, the
identification of this hallucinogen in body fluids is considerably
more difficult, owing primarily to the low dose of the drug (1
µg/kg orally) and its extensive metabolism. Although fluorescence
could be used, problems are encountered with interfering fluorescent compounds and the fact that the fluorescence spectrum of LSD
is indistinguishable from the other ergot alkaloids. Christie et al.
[127] described an HPLC-fluorescence procedure for analysing LSD in
serum on urine, thus improving specificity. However, large volumes
of sample are needed to achieve the required sensitivity of 1 ng/ml.
RIA has also been used to screen for LSD in biological fluids. Unfortunately, the majority of these procedures have significant
cross-reactivity to LSD metabolites and compounds with a similar
structure. However, that described by Ratcliffe et al. [128] is
relatively specific for LSD and has been used in an HPLC-RIA for
LSD biological fluids. This method uses HPLC-fluorescence with a
relatively specific for LSD and has been used in an HPLC-RIA for LSD
in biological fluids. This method uses HPLC-fluorescence with a
the reversed-phase HPLC system. Although the method is tedious, it
is presently the only procedure available for identifying and
quantitating LSD in biological fluids. Fortunately, this particular problem is not routinely encountered by forensic toxicologists.

No problem has been given more attention by analytical toxicologists in the last decade than the quantitation of Δ^9-tetrahydro-
cannabinol (Δ^9-THC) and its metabolites in biological fluids. Emphasis has been placed on two approaches, gas chromatography-mass
spectrometry [129] and immunoassay [130-133]. In fact, both
RIA [131] and EMIT [133] procedures are now available. Although
most groups [130, 131] have concentrated on the development of
specific antisera for Δ^9-THC and its major metabolites, 11-hydroxy-

Δ^9-THC and Δ^9-THC-11-oic acid, Teale et al. [132] developed an antisera with significant cross-reactivity to both metabolites and related compounds. It is this antisera that has been used in the combined HPLC-RIA approach for the analysis of cannabinoids in plasma and urine. The initial work reported by Williams et al. [122] used a stepped-gradient reversed-phase HPLC system, after methanol extractions of the plasma or urine cross-reacting peaks were observed at retention volumes corresponding to Δ^9-THC, Δ^9-THC-11-oic acid, and its glucuronide conjugate.

V. CONCLUSION

This chapter has not described all the HPLC procedures used in analytical toxicology; this would result in far too much detail and destroy its aim. It has, however, attempted to place HPLC into perspective, as far as the forensic toxicologist is concerned. Emphasis has been placed on procedures that may well benefit toxicologists in their attempts to design a systematic approach to identifying the toxicological unknown.

The future of HPLC in forensic toxicologist is dependent on improvements in the discriminating power of HPLC systems; these may result from the development of packing materials or through the use of more selective detectors in series. Undoubtedly, the use of postcolumn derivatization techniques, and improvements in the reliability of some of the present detectors (e.g., the electrochemical detector) will improve the sensitivity and specificity of the technique. The forensic toxicologist, interested in obtaining the maximum benefit from this technique, should become aware of these changes.

ACKNOWLEDGMENTS

The author gratefully acknowledges the secretarial assistance of Cindy Daybell and Marilyn Silcox.

REFERENCES

1. M. A. Peat, T. A. Jennison, and D. M. Chinn, J. Anal. Toxicol.
   1:204-208 (1977).
2. L. C. Franconi, G. L. Hawk, G. J. Sandman, and W. C. Harvey,
   Anal. Chem. 48:372-375 (1976).
3. R. F. Adams, F. L. Vandemark, and G. T. Schmidt, Clin. Chem.
   22:1903-1906 (1976).
4. P. J. Naish, M. Cooke, and R. E. Chambers, J. Chromatogr.
   163:363-372 (1979).
5. D. R. Clark, Clin. Chem. 25:1183 (1979).
6. A. C. Moffat and K. W. Smalldon, J. Chromatogr. 90:9-17 (1974).
7. P. Owen, A. Pendlebury, and A. C. Moffat, J. Chromatogr. 161:
   187-193 (1978).
8. P. Owen, A. Pendlebury, and A. C. Moffat, J. Chromatogr. 161:
   195-203 (1978).
9. A. C. Moffat, A. H. Stead, and K. W. Smalldon, J. Chromatogr.
   90:19-33 (1974).
10. A. C. Moffat, P. Owen, and C. Brown, J. Chromatogr. 161:179-
    185 (1978).
11. D. M. Rutherford, J. Chromatogr. 137:439-448 (1977).
12. M. A. Peat and L. Kopjak, J. Forensic Sci. 24:46-54 (1979).
13. W. O. Pierce, T. C. Lamoreaux, F. M. Urry, L. Kopjak, and
    B. S. Finkle, J. Anal. Toxicol. 2:26-31 (1978).
14. I. Jane, J. Chromatogr. 111:227-233 (1975).
15. P. J. Twitchett and A. C. Moffat, J. Chromatogr. 111:149-157
    (1975).
16. J. K. Baker, R. E. Skelton, and C. Y. Ma, J. Chromatogr. 168:
    417-427 (1979).
17. B. B. Wheals, J. Chromatogr. 187:65-85 (1980).
18. J. M. Clifford and W. Franklin-Smythe, Analyst 99:241-272
    (1974).
19. D. N. Hailey, J. Chromatogr. 98:527-568 (1974).
20. J. A. F. deSilva, I. Bekersky, C. V. Puglisi, M. A. Brooks,
    and R. E. Weinfeld, Anal. Chem. 48:10-19 (1976).
21. C. G. Scott and P. Bommer, J. Chromatogr. Sci. 8:446-448 (1970).
22. H. B. Greizerstein and C. Wojtowicz, Anal. Chem. 49:2235-2236
    (1977).
23. P. M. Kabra, G. L. Stevens, and L. J. Marton, J. Chromatogr.
    150:355-360 (1978).
24. R. R. Brodie, L. F. Chasseaud, and T. Taylor, J. Chromatogr.
    150:361-366 (1978).
25. R. J. Perchalski and B. J. Wilder, Anal. Chem. 50:554-557
    (1978).
26. A. Bugge, J. Chromatogr. 128:111-116 (1976).
27. U. R. Tjaden, M. T. H. A. Meeles, C. P. Thys, and M. van der
    Kaay, J. Chromatogr. 181:227-241 (1980).
28. M. D. Osselton, M. D. Hammond, and P. J. Twitchett, J. Pharm.
    Pharmacol. 29:460-462 (1977).
29. M. A. Peat, B. S. Finkle, and M. E. Deyman, J. Pharm. Sci.
    68:1467-1468 (1979).

30. N. Strojny, C. V. Puglisi, and J. A. F. deSilva, *Anal. Lett.* B11:135-160 (1978).
31. K. Harzer and R. Barchet, *J. Chromatogr.* 132:83-90 (1977).
32. J. D. Wittwer, Jr., *J. Liq. Chromatogr.* 3:1713-1724 (1980).
33. C. Violon and A. Vercruysse, *J. Chromatogr.* 189:94-97 (1980).
34. S. J. Soldin and J. C. Hill, *Clin. Chem.* 22:856-859 (1976).
35. R. F. Adams and F. L. Vandemark, *Clin. Chem.* 22:25-31 (1976).
36. P. M. Kabra, G. Gotelli, R. Stanfill, and L. J. Marton, *Clin. Chem.* 22:824-827 (1976).
37. G. W. Mihaly, J. A. Phillips, W. J. Louis, and F. J. Vadja, *Clin. Chem.* 23:2283-2287 (1977).
38. P. M. Kabra, P. E. Stafford, and L. J. Marton, *Clin. Chem.* 23:1284-1288 (1977).
39. J. E. Slovek, G. W. Peng, and W. L. Choiu, *J. Pharm. Sci.* 67:1462-1464 (1978).
40. R. W. Dykeman and D. J. Ecobichon, *J. Chromatogr.* 162:104-109 (1979).
41. R. Farinotti and G. Mahuzier, *J. Chromatogr.* 2:345-364 (1979).
42. P. M. Kabra, D. M. McDonald, and L. J. Marton, *J. Anal. Toxicol.* 2:127-134 (1978).
43. R. F. Adams, G. J. Schmidt, and F. L. Vandemark, *J. Chromatogr.* 145:275-284 (1978).
44. P. M. Kabra, H. Y. Koo, and L. J. Marton, *Clin. Chem.* 24:657-662 (1978).
45. P. M. Kabra, B. E. Stafford, and L. J. Marton, *Anal. Toxicol.* 5:177-182 (1981).
46. U. R. Tjaden, J. C, Kraak, and J. F. K. Huber, *J. Chromatogr.* 143:183-194 (1977).
47. R. Gill, personal communication 1980.
48. J. B. Rose, *Clin. Toxicol.* 11:391-402 (1977).
49. G. Norheim, *J. Chromatogr.* 88:403-406 (1974).
50. M. Shennan and P. Haythorn, *J. Chromatogr.* 132:237-247 (1977).
51. H. E. Hamilton, J. E. Wallace, and K. Blum, *Anal. Chem.* 47:1139-1143 (1975).
52. M. Ervik, T. Walle, and H. Ehrsson, *Pharm. Suec.* 7:623-634 (1970).
53. O. Borga and M. Garle, *J. Chromatogr.* 68:77-88 (1972).
54. D. N. Bailey and P. I. Jatlow, *Clin. Chem.* 22:1697-1701 (1976).
55. R. N. Gupta, G. Molnar, R. E. Hill, and M. L. Gupta, *Clin. Biochem.* 9:247-251 (1976).
56. M. T. Rosseel, M. G. Bogaert, and W. M. Llaeys, *J. Pharm. Sci.* 67:802-805 (1978).
57. L. A. Gifford, P. Turner, and C. M. B. Pare, *J. Chromatogr.* 105:107-113 (1975).
58. D. N. Bailey and P. I. Jatlow, *Clin. Chem.* 22:777-781 (1976).
59. S. Dawling and R. A. Braithwaite, *J. Chromatogr.* 146:449-456 (1978).
60. J. T. Biggs, W. H. Holland, S. Chang, P. P. Hipps, and W. R. Sherman, *J. Pharm. Sci.* 65:261-268 (1976).
61. D. M. Chinn, T. A. Jennison, D. J. Crouch, M. A. Peat, and G. W. Thatcher, *Clin. Chem.* 26:1201-1204 (1980).

62. J. H. Knox and J. Jurand, *J. Chromatogr.* 103:311-326 (1975).
63. I. D. Watson and M. J. Stewart, *J. Chromatogr.* 110:389-392 (1975).
64. B. Mellstrom and S. Eksborg, *J. Chromatogr.* 116:475-479 (1976).
65. B. Mellstrom and G. Tybring, *J. Chromatogr.* 143:597-606 (1977).
66. H. G. M. Westenberg, B. F. H. Drenth, R. A. deZeeuw, H. de Cuyper, H. M. van Praag, and J. Korf, *J. Chromatogr.* 142:725-733 (1977).
67. R. R. Brodie, L. F. Chasseaud, and D. R. Hawkins, *J. Chromatogr.* 143:535-539 (1977).
68. J. C. Kraak and P. Bijster, *J. Chromatogr.* 143:499-572 (1977).
69. B. Mellstrom and R. Braithwaite, *J. Chromatogr.* 157:379-385 (1978).
70. I. D. Watson and M. J. Stewart, *J. Chromatogr.* 132:155-159 (1977).
71. F. L. Vandemark, R. F. Adams, and G. J. Schmidt, *Clin. Chem.* 24:87-91 (1978).
72. H. F. Proelss, H. J. Lohmann, and D. G. Miles, *Clin. Chem.* 24:1948-1953 (1978).
73. R. A. deZeeuw and H. G. Westenberg, *J. Anal. Toxicol.* 2:229-232 (1978).
74. J. E. Wallace, E. L. Shimek, Jr., and S. C. Harris, *J. Anal. Toxicol.* 5:20-23 (1981).
75. P. K. Sonsalla, M.S. thesis, University of Utah, 1981.
76. I. D. Watson and M. J. Stewart, *J. Chromatogr.* 134:182-186 (1977).
77. J. Koch-Weser and S. W. Klein, *JAMA* 215:1454-1460 (1971).
78. G. Cramèr and B. Isaksson, *Scand. J. Clin. Lab Invest.* 15:553-556 (1963).
79. T. Walle, *J. Pharm. Sci.* 63:1885-1891 (1974).
80. J. D. Hawkins, R. R. Bridges, and T. A. Jennison, *Ther. Drug Monit.* (1981) in press.
81. D. H. Huffman and C. E. Hignite, *Clin. Chem.* 22:810-812 (1976).
82. K. Y. Lee, D. Nurok, A. Zlatkis, and A. Karmen, *Clin. Chem.* 23:636-638 (1978).
83. D. Martin, L. Burke, W. Nodin, and S. Chen, *Clin. Chem.* 24:991 (1978).
84. M. A. Peat and T. A. Jennison, *Clin. Chem.* 24:2166-2168 (1978).
85. J. T. Powers and W. Sadee, *Clin. Chem.* 24:299-302 (1978).
86. W. G. Crouthamel, B. Kowarski, and P. K. Narany, *Clin. Chem.* 23:2030-2033 (1977).
87. K. Carr, R. L. Woosley, and J. A. Oates, *J. Chromatogr.* 129:363-368 (1976).
88. L. R. Shukur, J. L. Powers, R. A. Marques, M. E. Winter, and W. Sadee, *Clin. Chem.* 23:636-638 (1977).
89. R. F. Adams, F. L. Vandermark, and G. Schmidt, *Clin. Chim. Acta* 69:515-523 (1976).
90. J. Lima, *Clin. Chem.* 25:405-408 (1979).
91. P. J. Meffin, S. F. Harapot, and D. C. Harrison, *J. Chromatogr.* 132:503-510 (1979).

92. G. Nygard, W. H. Shelver, and S. K. Khalil, *J. Pharm. Sci.* 68:1318-1320 (1979).
93. M. Lo and S. Reigelman, *J. Chromatogr.* 183:213-220 (1980).
94. R. L. Nation, G. W. Peng, and W. L. Chiou, *J. Chromatogr.* 145:429-436 (1978).
95. A. M. Taburet, A. A. Taylor, J. R. Mitchell, D. E. Rollins, and J. L. Pool, *Life Sci.* 24:209-217 (1979).
96. P.-O. Lagerstrom and B.-A. Persson, *J. Chromatogr.* 149:331-340 (1978).
97. J. G. Flood, G. N. Bowers, and R. B. McComb, *Clin. Chem.* 26:197-200 (1980).
98. P. W. Kabra, S. W. Chen, and L. J. Marton, *Ther. Drug Monit.* 3:91-101 (1981).
99. J. Epton and J. Grove, in *Methodology for Analytical Toxicology,* (Ed.). CRC Press, Cleveland, 1976, p. 16.
100. S. Kendall, G. Lloyd-Jones, and C. F. Smith, *J. Int. Med. Res.* 4(Suppl. 4):83-88 (1976).
101. H. V. Street, *J. Chromatogr.* 109:29-36 (1975).
102. J. J. Thoma, M. McCoy, T. Ewald, and N. Myers, *J. Anal. Toxicol.* 2:226-228 (1978).
103. W. A. Dechtiaruk, G. F. Johnson, and H. M. Solomon, *Clin. Chem.* 22:879-883 (1976).
104. J. Grove, *J. Chromatogr.* 59:289-295 (1971).
105. G. R. Gotelli, P. M. Kabra, and L. J. Marton, *Clin. Chem.* 23:957-959 (1977).
106. K. S. Pang, A. M. Taburet, J. A. Hinson, and J. R. Gillette, *J. Chromatogr.* 174:165-175 (1979).
107. J. H. Knox and J. Jurand, *J. Chromatogr.* 142:651-670 (1977).
108. J. N. Micell, M. K. Aravind, S. M. Cohen, and A. K. Done, *Clin. Chem.* 25:409-412 (1979).
109. C. G. Fletterick. T. H. Grove, and D. C. Hohnadel, *Clin. Chem.* 25:409-412 (1979).
110. B. R. Manno, J. E. Manno, C. A. Dempsey, and M. A. Wood, *J. Anal. Toxicol.* 5:24-28 (1981).
111. P. Clarke and R. L. Foltz, *Clin. Chem.* 20:465-469 (1974).
112. G. R. Wilkinson and E. L. Way, *Biochem. Pharmacol.* 18:1435-1439 (1969).
113. B. Dahlström and L. Paalzow, *J. Pharm. Pharmacol.* 27:172-176 (1975).
114. I. Jane and J. F. Taylor, *J. Chromatogr.* 109:37-42 (1975).
115. L. Ulrich and P. Rüegsegger, *Arch. Toxicol.* 45:241-248 (1980).
116. M. W. White, *J. Chromatogr.* 178:229-240 (1979).
117. J. E. Wallace, S. C. Harris, and M. W. Peek, *Anal. Chem.* 52:1328-1330 (1980).
118. R. G. Peterson, G. H. Rumack, J. B. Sullivan, Jr., and A. Makowski, *J. Chromatogr.* 188:420-425 (1980).
119. C. Refshange, P. T. Kissinger, R. Dreiling, C. L. Blank, R. Freeman, and R. N. Adams, *Life Sci.* 14:311-342 (1974).
120. C. Bollet, C. Olivia, and M. Caude, *J. Chromatogr.* 149:625-644 (1977).

121. P. E. Nelson, S. M. Fletcher, and A. C. Moffat, *J. Forensic Sci. Soc.* 20:195-202 (1980).
122. P. L. Williams, A. C. Moffat, and L. J. King, *J. Chromatogr.* 155:273-283 (1978).
123. P. L. Williams, A. C. Moffat, and L. J. King, *J. Chromatogr.* 186:595-603 (1979).
124. P. J. Twitchett, S. M. Fletcher, A. T. Sullivan, and A. C. Moffat, *J. Chromatogr.* 150:73-84 (1978).
125. S. M. Fletcher, G. Lawson, B. Law, and A. C. Moffat, *J. Forensic Sci. Soc.* 20:203-210 (1980).
126. S. M. Fletcher, G. Lawson, and A. C. Moffat, *J. Forensic Sci. Soc.* 19:183-188 (1979).
127. J. Christie, M. W. White, and J. M. Wiles, *J. Chromatogr.* 120:496-501 (1976).
128. W. A. Ratcliffe, S. M. Fletcher, A. C. Moffat, J. G. Ratcliffe, W. A. Harland, and T. E. Levitt, *Clin. Chem.* 23:169-174 (1977).
129. R. L. Foltz, in *GC/MS Assays for Abused Drugs in Body Fluids,* R. L. Foltz, A. F. Fentiman, Jr., and R. B. Foltz (Eds.). NIDA Res. Monogr. 32, 1980, pp. 62-89.
130. C. E. Cook, M. L. Hawes, E. W. Amerson, C. G. Pitt, and D. Williams, in *Cannabinoid Assays in Humans,* R. F. Willette (Ed.). NIDA Res. Monogr. 7, 1976, pp. 15-27.
131. S. J. Gross, J. R. Soares, S. L. Wong, and R. E. Schuster, *Nature* 252:581-582 (1974).
132. J. D. Teale, E. Forman, L. J. King, and V. Marks, *Nature* 249:154-155 (1974).
133. R. Rodgers, C. P. Crowl, W. M. Eimstad, M. W. Hu, J. K. Kam, R. C. Ronald, G. L. Rowley, and E. F. Ullman, *Clin. Chem.* 24:95-100 (1978).

7 HPLC Analysis of Explosives and Related Materials

Ira S. Krull

Northeastern University
Boston, Massachusetts

I. INTRODUCTION AND BACKGROUND

The trace analysis of organic compounds used as explosives has been
a subject of considerable interest throughout the history of these
materials. That is, ever since explosives were used for criminal,
military, or terrorist purposes, there have been forensic analysts
interested in determining the source and/or nature of the explosives
used. As each new technique, method, or instrumentation in analytical chemistry has become available, it has eventually been applied
to the trace analysis of explosives. Thus, with the current widespread interest and enthusiasm in high-performance liquid chromatography (HPLC), it comes as no surprise that this approach is now
being intensely applied to the trace determination of materials
used as explosives. Within the past few years, a number of excellent review articles have appeared, emphasizing the application of
HPLC in explosives analysis [1-7]. It has been apparent to many
investigators in forensic analysis that a simple, rapid, reliable,
accurate, precise, and inexpensive method of trace assay for a wide
variety of explosives in a single sample has never been available.
That is, it has often been necessary to undertake extensive sample

357

preparation, cleanup, extractons, and then a combination of separation-detection techniques, such as thin-layer chromatography (TLC),
gas chromatography (GC), off-line liquid chromatography-mass spectrometry, and related approaches, or combinations thereof [7]. Absolute identification and quantification of the materials of interest,
especially where these appeared at the trace or ultra-trace levels,
has often been difficult and/or questionable. Hence it has become
obvious that there is a real need for the application of HPLC to
explosives and related materials analysis. A number of commercial
HPLC manufacturers are now devoting considerable efforts, time, and
money to demonstrate the direct applicability of their products for
explosives determination [8]. It has become apparent to many, if
not most, investigators, that future, rapid progress in the analysis of trace amounts of explosives, with a high degree of accuracy
and precision, both qualitative and quantitative, can proceed only
via the application of HPLC with appropriate detection systems
[1-5]. Before we proceed to describe and critically discuss that
which has already been described in this area, it may prove worthwhile first to understand just why HPLC is so uniquely applicable
and/or adaptable to explosives analysis.

It is precisely those physical-chemical properties of explosive
materials and propellants which make them useful for such purposes
that also make them so difficult to analyze for at trace levels.
That is, materials such as 2,4,6-trinitrotoluene (TNT), 2,4,6-
trinitro-m-cresol (TNC), 1,3,5-trinitro-1,3,5-triazacyclohexane
(RDX), 2,4,6-N-tetranitro-N-methylaniline (TETR), ethylene glycol
dinitrate (EGDN), pentaerythritol tetranitrate (PETN), nitroglycerin
(NG), and related compounds are almost always thermally unstable,
shock sensitive, electrically sensitive, chemically reactive, highly
polar, and often UV sensitive. Thus they are generally difficult
to work with for long periods of time; their purity degrades rapidly
if not stored properly, and they are difficult to analyze for by
traditional methods of trace analysis. More specifically, medium-
or high-temperature gas chromatography is generally impossible to

employ with most explosives, since these will not routinely and
reproducibly survive the thermal conditions required for their
separation. Thus, in the past, most published analyses for explosives have used older chromatographic approaches for initial separations, especially TLC with a variety of sometimes nonspecific colori-
metric spray reagents for detection. Where GC has been most useful
for explosives has been in the case of polynitro aromatic compounds,
such as the dinitro or trinitro aromatics (TNT, tetryl, etc.).

Thus, for the trace analyst interested in undertaking explosives detection, until the mid-1970s, he was forced to rely heavily
on nonquantitative TLC methods, and the forensic literature is re-
plete with various references and reports in this regard. In the
early 1970s, a few investigators began to realize the potentials of
HPLC in explosives analysis, and following their initial reports,
there began to develop a more widespread interest in this field
[9-11]. HPLC was found to offer significant advantages and capabilities for trace explosives detection, not to be found in TLC,
GC, or any prior separation-detection methods. Separations and
final detection in HPLC methods can operate routinely at ambient or
subambient temperatures, thus making it very easy to work with
thermally and shock-sensitive materials. Because HPLC work operates
in the absence of light, any UV-sensitive materials will not suffer
degradation during the separation-detection schemes. HPLC provides
a wide variety of separation-detection methods, as described below,
thus enabling one to separate nonpolar, polar, and highly polar compounds within one run or a combination of two HPLC runs, often all
performed on-line with the detection system. Thus there is a minimum of sample preparation, workup, and handling required in most
HPLC applications to explosives analysis. This reduces sample requirements, time needed, energy input, personnel requirements, and
final overall costs per analysis. In a number of instances, it has
even been possible to avoid all sample workup, and to inject crude
or semicrude environmental, blast residue samples directly onto the
HPLC system. This is the ideal situation in any sort of explosives

determination, to be able to collect a sample, concentrate where
necessary, avoid all workup, preseparation steps, and inject directly onto the HPLC-detection system(s). Whereas GC allows only
for the analysis of compounds that are thermally stable, HPLC permits for the separation-identification of both thermally stable and
unstable compounds, at the same time, with a single injection, using
appropriate separation-detection techniques.

    With the rapid development of more and more separation methods
in HPLC, it is possible to separate large numbers of widely differing explosive materials, and then to detect each of these as they
elute from the chromatographic column. In addition, it is possible
to switch certain parts of the eluents from a first column directly
onto a second column for further refinements in the overall separation process, and then to have the finally resolved materials enter
a detection system [12, 13]. It is not possible here to more fully
discuss and describe the more recent advances in HPLC, but there
are a number of recent references and texts that can be referred to
[14-27]. Recent advanced microprocessor instumentation has become
available, at affordable prices, which permits for the total automated control of all aspects of HPLC instrumentation [21, 27]. This
is yet another advantage in applying HPLC methods and instrumentation to explosives analysis, since it allows for reduced time and
labor requirements for each separate analysis. Almost all commercial HPLC manufacturers now offer their own microprocessor controller for their own HPLC components, but often these are not operational with other commercial components. Some of the most modern
microprocessor-controlled HPLC equipment allows for an automated
methods development, run overnight, varying solvent composition,
flow rates, gradient elution, and other parameters in order to optimize the final separations achieved. Come morning, the analyst can
then choose the system that best suits his or her current needs,
set the necessary microprocessor controls, and allow a number of
automated analyses to be run, unattended during the same day or the
next evening, as desired. In addition, it is also possible to vary

the detector capabilities, as a function of time during each analysis, so that individual peak components of a complex mixture can be more accurately and sensitively detected during a programmed run. Final calculations, both qualitative and quantitative, can be performed automatically, again under microprocessor control, so that the necessary analytical results and quality control data are produced and processed unattended. In these ways, with appropriate microprocessor-controlled HPLC instrumentation, a large number of samples can be analyzed with a minimum of attendent personnel requirements.

The initial history of explosives analysis by HPLC suffered from a lack of selective and highly sensitive detectors. That is, the early detectors, such as fixed-wavelength UV-visible or refractive index (RI), were not generally selective enough to pick out an explosive compound in a complex environmental, postblast residue matrix. Thus because selectivity or specificity was often poor for the explosive compounds of interest, the final results were sometimes questionable, especially in the absence of additional confirmatory methods. Naturally, as the area of HPLC progressed, more and more academic and industrial researchers applied their efforts toward the development of more useful detection systems. Again, it is not possible here to describe in detail some of the more useful detectors for explosives analysis, although this is attempted somewhat, but the reader is again referred to the pertinent literature [18, 19, 28-34]. More and more detectors are being researched and developed, with specific applications for explosives analysis, such as the thermal energy analyzer (TEA), electron capture detector (ECD), mass spectrometer (MS), and electrochemical detector (EC). These are discussed further below.

There are certain fundamental requirements in trying to develop a new HPLC-detector system for explosives. The HPLC solvents and packing materials must be fully compatible with the explosives being analyzed; there can be no degradation or decomposition of the organic compounds being studied because of unwanted interactions

with either the solvent or substrate. The sample of explosives
must be soluble with the mobile phase being used for the separation,
and this is often a difficult requirement to meet with many explosives. These compounds are often very polar, highly insoluble materials, except in very polar solvents, such as acetone, ethanol,
methanol, dimethylsulfoxide, or dimethylformamide. Such solvents
for preparing the injection solutions are sometimes not compatible
with the HPLC mobile phases being used, and this must always be considered in designing the final HPLC analysis scheme. Ideally, the
final HPLC system of mobile phase, packing material, flow rate,
temperature, and so on, should provide baseline resolution of the
compounds of interest from each other and all sample interferents.
Not only should there be good baseline resolutions, but it is also
necessary to have reproducible retention times, narrow peak widths,
intense peak heights, symmetrical peak shapes, and retention times
of practical use, generally no longer than 15-20 min per analysis.
If trace analysis is the goal, the final chromatography must be
such that neither specificity nor sensitivity are compromised or
sacrificed. With regard to the detector being used, the solvent
used as the mobile phase must be fully compatible with the detector,
there should be no loss of solute signal, no loss of sensitivity,
no excessive background noise, no elevation of baseline, and no
quenching of the solute signal by the solvent(s). It is terribly
important in all trace analysis, especially that for explosives,
that the solvents needed to obtain the desired chromatography do
not sacrifice detector usefulness for the solutes of interest. Of
course, there must be no loss of column materials or coatings during the chromatography stages, possibly as a result of the mobile-
phase conditions finally chosen for the explosives of interest.

Although HPLC seems to offer some very attractive advantages
for trace explosive analysis, it is also obvious that care must be
exercised in selecting both the chromatographic system and detec-
tor(s). In addition, it is economically advantageous to select
one's HPLC instrumentation with great care, making sure that it will

be able to perform the trace analyses necessary and desirable. If a large number of such analyses are to be undertaken, additional care should be invoked to include in one's instrumentation the microprocessor controller and ancillary equipment described above, such as automatic sample injector and automatic recording integrator-data acquisition system. The available commercial instrumentation literature and appropriate references should be consulted in this regard [18, 19, 28-34].

We will now discuss much of the existing literature pertaining to the analysis of explosives via HPLC, emphasizing those detection systems and HPLC conditions that have proven most useful and advantageous. Some of this work is currently in progress, and therefore its description and discussion is somewhat limited and reserved. However, in other areas there is a great deal that has already been described, especially with regard to successful HPLC separation schemes. Where appropriate we will also describe the sample work-up methods, but in most instances this is left to the reader to determine as needed, from the existing and referenced literature. If we have inadvertently omitted certain earlier references, we ask the forgiveness of those involved, but it is felt that the more recent work in this area is often the more useful and applicable today.

II. HPLC-ULTRAVIOLET ANALYSIS AND APPLICATIONS

The vast majority of published reports on explosives analysis have employed ultraviolet (UV) detection, initially at fixed wavelengths of 254 or 280 nm. However, more recently, detection by UV has employed the variable-wavelength HPLC detector, coming with the realization that many nitro derivatives absorb more strongly at 210-230 nm than at higher wavelengths. Hence sensitivity has been greatly improved in recent years, concomitant with a lowered limit of detection for those explosives that are UV absorbing. However, it must be realized (Table 1) that only a fraction of the more commonly used and encountered explosives will absorb anywhere in the UV, and

most of the others have few if any such properties. Of interest
today are the newer vacuum UV detectors, which are capable of operating in the range 190-210 nm with no significant problems [35-37].
It is of interest that there have been few literature reports wherein
vacuum-UV HPLC detectors have been employed for trace explosives
analysis. One would expect that the alkyl nitro group -- O-nitro,
N-nitro, or C-nitro -- should have some measurable absorption in
this region of the spectrum. Of course, the problem in utilizing
the vacuum-UV region is that this again restricts the range of organic solvents that can now be utilized to perform the HPLC separations. It also means that those solvents which are used are of the
highest chemical purity possible, often requiring purification
beyond the customary distilled-in-glass step.

TABLE 1. Some of the More Commonly Encountered Explosives in
Trace Analysis

Abbreviation Used	Common and Formal Names
a. TNT	2,4,6-Trinitrotoluene, Tri, Trotyl, Trolit, sym-TNT, α-TNT
b. TNC	2,4,6-Trinitro-m-cresol
c. RDX	1,3,5-Trinitro-1,3,5-triazacyclohexane, hexogen, cyclonit, T4, 1,3,5-trinitro-s-triazin, hexahydro-1,3,5-trinitro-s-triazine
d. HMX	1,3,5,7-Tetranitro-1,3,5,7-tetrazacyclooctane, octogen, homocyclonit
e. TETR	2,4,6-N-Tetranitro-N-methylaniline, CE, tetryl
f. EGDN	Ethylene glycol dinitrate, 1,2-ethanediol dinitrate, glycoldinitrate, dinitroglycol
g. PETN	Pentaerythritol tetranitrate, 2,2-bis(nitroxy-methyl)-1,3-propanediol-1,3-dinitrate, pentrit, pentryl, penta, nitropenta
h. NG	Nitroglycerin, glycerol trinitrate, trinitroglycerin
i. NGU	1-Nitroguanidine
j. NC	Nitrocellulose
k. HNS	Hexanitrostilbene, 2,2',4,4',6,6'-hexanitrostil-bene
l. TNB	Trinitrobenzene, benzit, 1,3,5-trinitrobenzene
m. DNT	Dinitrotoluene (2,6; 2,5; 2,4; 2,3; 3,5; 3,4)
n. MNT	Mononitrotoluene (o-, m-, p-)

(continued)

TABLE 1 (Cont.)

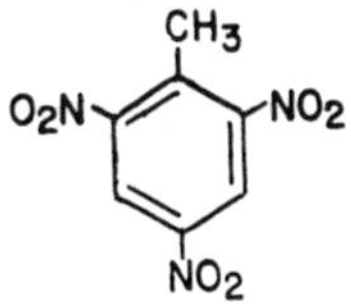

a. TNT

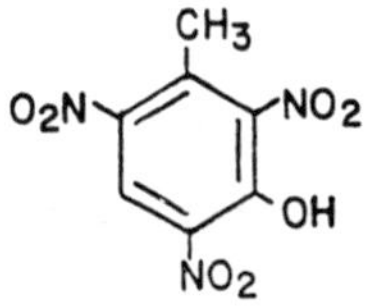

b. TNC

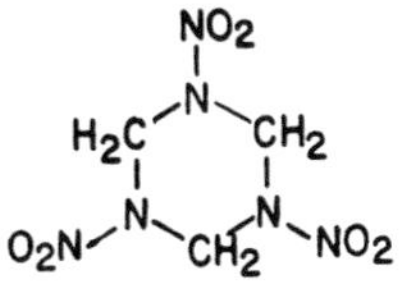

c. RDX

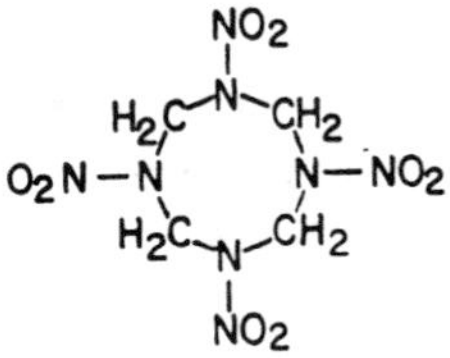

d. HMX

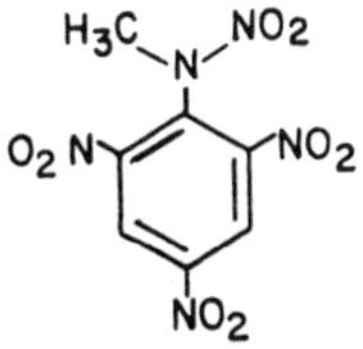

e. Tetryl

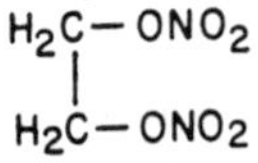

f. EGDN

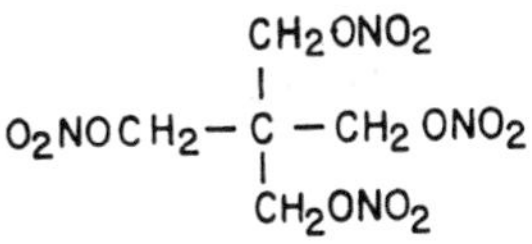

g. PETN

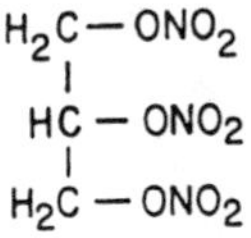

h. NG

Although some aromatic explosives possess strong absorption in
the UV region, this type of a detector provides little, if any,
selectivity or specificity for the compounds of interest. Thus one
must rely almost exclusively on chromatographic retention times,
capacity factors, coinjections with standards, and related, often
shaky approaches for confirmation of a compound in a complex matrix.
There have been very few, if any, reports wherein chemical derivatization has been utilized, together with strongly UV-absorbing deri-
vatizing reagents, for the analysis of explosive materials [38, 39].
This may be due to the fact that most explosives do not lend themselves readily to simple derivatization reactions, except perhaps
chemical/electrochemical reduction of the nitro group. That is,
there are no readily available handles for derivatization in most
explosives.

In discussing the use of HPLC for trace explosive analysis, it
must be remembered that modern HPLC, using microparticles as the
solid support, has only really been exploited for about a decade
(1970-1980). During that time frame, particle sizes have gone from
a range of 30-44 μm to 3-5 μm, while HPLC pumps have gone from an
operating range of 0-500 psi to ranges of 0-10,000 psi and above.
Injection has gone from stopped-flow, septum injections to continuous-
flow, valve-type injections, often automatically operated. Other
parts of the overall HPLC instrumentation have also changed drastically. Thus, to compare results obtained in the early 1970s with
those obtained more recently is a bit unfair to early researchers.
Limits of detection have improved, resolutions have been perfected
and reproducibility of analyses have gotten much better -- in large
part because of the advances made in the technology and instrumentation of HPLC.

Some of the earliest work in the HPLC-UV analysis of explosives and related materials was due to Chandler and co-workers at
Hercules Corp. [9, 40, 41]. In the earlier reports, Chandler et al.
[9] analyzed hexanitrobibenzyl (HNBB) and 3-methyl-2',4,4',6,6'-
pentanitrodiphenylmethane (MPDM) present in TNT samples. This

separation was performed on a column packed with Porasil A (37-75 µm), using methylene chloride (MC) as the mobile phase and UV detection at 254 nm. The concentrations of the impurities found in samples of commercially produced TNT ranged from 0.1 to 0.5% for HNBB and from 0.1 to 0.3% for MDPM. In a later publication, Dalton et al. [41] described a method for the HPLC determination of NG, diethyl phthalate (plasticizer), and ethyl centralite (EC) in various propellants. This work utilized a fixed-wavelength (254-nm) UV detector, together with a column packed with 30- to 44-µm Vydac adsorbent, and a mobile phase at 0.8 ml/min of 1,1-dichloroethane (DCE). The propellant samples were prepared by extracting these with technical-grade methylene chloride overnight in a Soxhlet apparatus, and then concentrating the final extracts on a steam bath to near dryness. The residue was then taken up in DCE containing the internal standard, and diluted to final volume for HPLC analysis. Figure 1 indicates one of the earliest HPLC-UV chromatograms for explosive and related material, using the conditions indicated above [41]. It should be compared with some of the later HPLC-detector chromatograms indicated later in this chapter. What is of interest in Fig. 1 relates to the total analysis time (about 23 min), the extreme broadness of each resolved peak, and the total time of elution for the longer-retained materials (2-4 min).

Walsh et al. have reported on the analysis for nitrotoluene and TNT in TNT wastewaters from various explosives-manufacturing sites [10]. The range of TNT measurements was 1-100 ppm in the original water samples, and quantities of less than 1 ppm could be readily determined using their methods. HPLC here involved the use of reversed-phase conditions with a C_{18} packing material, together with a mobile phase of 9:1 water:acetonitrile and dual UV-RI detection. It was also demonstrated here that certain polymeric resins, such as XAD-2, could be utilized to cleanup contaminated wastewaters, resulting in the removal of TNT and related materials.

Toward the mid-1970s, Doali and Juhasz at the Aberdeen Proving Grounds described the use of HPLC-UV-RI methods for the trace

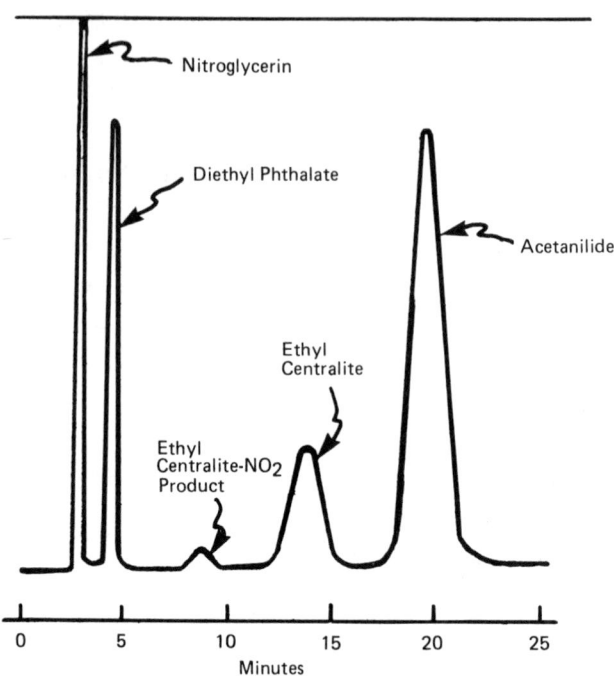

FIG. 1. Double-base propellant analysis by HPLC on a column of
Vydac adsorbent (30-44 μm) with mobile phase of 1,1-dichloroethane
at 0.8 ml/min, UV detection at 254 nm. (From Ref. 41. Reprinted
with permission of the *Journal of Chromatographic Science* and
Preston Publications, Inc.)

analysis of various explosives and propellants [42, 43]. In the

first report [42], they reported the HPLC-UV separation-detection

of a wide variety of nitrate esters, nitroaromatics, and nitramines,

such as p-nitrotoluene, 2,4-dinitrotoluene (DNT), 2,4,6-trinitro-

toluene (TNT), tetryl, RDX, HMX, NG, EC, diphenylamine (DPA),

dibutylphthalate, 2-nitrodiphenylamine, 2,6-dinitrotoluene, and N-

nitrosodiphenylamine [42]. Throughout this first systematic study

of explosives by HPLC, only one adsorbent was employed, Corasil II,

37-50 μm, but with a variety of mobile phases. Some of the mobile

phases and conditions are described in the Appendix. A typical

separation of TNT, tetryl, and RDX is indicated in Fig. 2, which

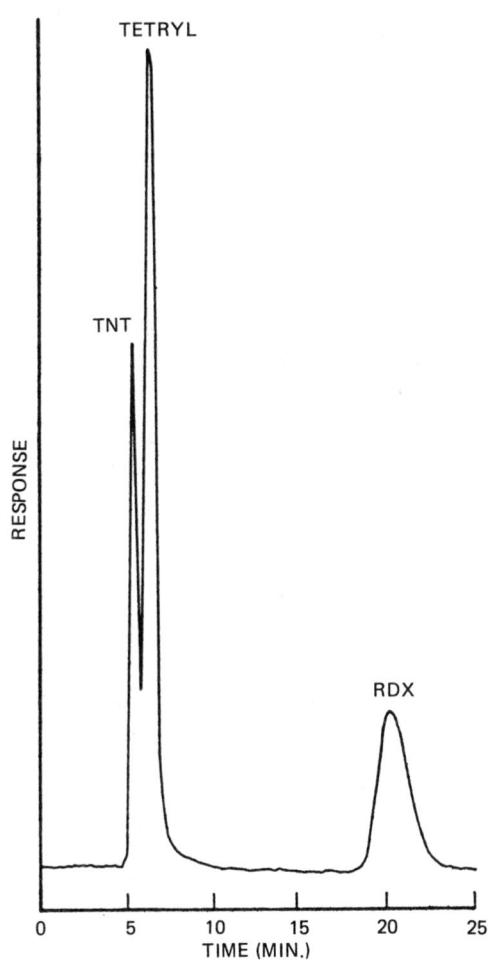

FIG. 2. HPLC separation of TNT, tetryl, and RDX on a Corasil II
(37-50 μm) column using a mobile phase of dioxane:cyclohexane (30:70)
at a flow rate of 0.6 ml/min with UV detection at 254 nm. (From
Ref. 42. Reprinted with permission of the *Journal of Chromatographic
Science* and Preston Publications, Inc.)

utilized a mobile phase of dioxane:cyclohexane (30:70) at a flow
rate of 0.6 ml/min with UV detection at 254 nm. In a subsequent
publication, Doali and Juhasz determined a common propellant component, 2-nitrodiphenylamine, in a composite modified double-base
propellant by HPLC-UV methods [43]. The analysis was performed
using a mobile phase of methylene chloride:cyclohexane (20:80) at a
flow rate of 0.5 ml/min, with UV detection at 254 nm, and a 100 cm ×
2.1 mm ID column packed with Corasil II (37-50 µm). The samples
were prepared by extracting them overnight with methylene chloride
in a Soxhlet apparatus, evaporating the MC to a few milliliters,
adding the internal standard, and then diluting in a volumetric
flask with 1,2-dichloroethane. The range of 2-nitrodiphenylamine
determined in the commercial propellant samples was 0.9-1.0% by
weight, with a high degree of precision evidenced by the final results [43]. Conditions for the HPLC separation of 2-nitrodipheny-
lamine, 2,4-dinitrotoluene, 2,2'-dinitrodiphenylamine, 2,4-dinitro-
diphenylamine, and 2,4'-dinitrodiphenylamine are indicated in the
appendix [43].

Farey and Wilson in 1975 studied the thermal stability of
tetryl, and analyzed the parent compound and its degradation products by HPLC-UV [44]. Their method utilized a µPorasil column
with a mobile phase of cyclohexane:tetrahydrofuran (97:3) at a flow
rate of 2.0 ml/min, as well as a µBondapak CN column. The latter
used a mobile phase of cyclohexane:chloroform:tetrahydrofuran
(77:20:3) at a flow rate of 2.0 ml/min, with UV detection at 254
nm. This system proved ideal for tetryl and its numerous degradation products, but trace analyses were not necessary here, as microgram amounts of materials were being injected. Picric acid was
determined using a phenyl-Corasil reversed-phase column and a mobile
phase of water:methanol (75:25) at a flow rate of 2.0 ml/min, again
with UV detection at 254 nm. Some of the tetryl degradation products
studied here were 2-nitroaniline, 2,4-dinitroanisole, 2,4,6-
trinitrobenzene, methylpicramide, picramide, 4-nitroaniline, 2,4-
dinitroaniline, and picric acid (2,4,6-trinitrophenol) [44] (appendix).

Poyet et al. have described the use of HPLC-UV for the analysis of 2-nitrodiphenylamine, a widely used stabilizer for various nitrate esters, together with a number of other aromatic nitro derivatives [45]. Some of these derivatives found often together with the parent compound were: 2-nitro-N-nitrosodiphenylamine, 2,2'-dinitrodiphenylamine, 2,4'-dinitrodiphenylamine, 2,4,2'-trinitrodiphenylamine, 2,4,4'-trinitrodiphenylamine, and 2,4,2',4'-tetranitrodiphenylamine. For the analysis of these compounds and 2-nitrodiphenylamine, the authors used a reversed-phase packing of µBondapak C_{18} with an eluent of methanol:water (67.5:32.5%) at a flow rate of 1.0 ml/min, with UV detection at 254 nm. For the analysis of RDX and HMX, the conventional liquid-solid approach was employed, with a column of LiChrosorb Si 60 and an eluent of cyclohexane:dioxan (55:45) at a flow rate of 1.5 ml/min and UV detection again at 254 nm [45]. Trace detection of RDX and HMX at levels of 0.01-0.02% in solution were observed and reported, with an overall HPLC analysis time of less than 20 min. Standards as well as actual explosive extracts were analyzed using both of these methods. Figure 3 illustrates the separation of RDX and HMX obtained under the conditions indicated.

Although not to be considered an HPLC approach, Freeman et al. have described the use of various adsorptive polymers for the separation of RDX and HMX explosives [46]. The mobile phase in this work was acetone or acetone:methanol (3:1), with polymers containing functional groups such as o-methyl pyridinyl, alkyl phenyl, carboxylic acid, phenyl sulfonamides, and N-vinylpyrrolidone. The separations and retentions observed with some of these polymers for both RDX and HMX was attributed to the formation of insoluble complexes.

Schaffer and Teter have recently described an HPLC-based assay for 1,3,5-triamino-2,4,6-trinitrobenzene (TATB) using a column of µBondapak NH_2 with a mobile phase of DMF:toluene:heptane (1.5:43:55.5) [47, 48]. TATB, as is the case for many other explosives, is a highly insoluble material, and injection solutions were prepared here in N,N-dimethylformamide (DMF), which must then be miscible and

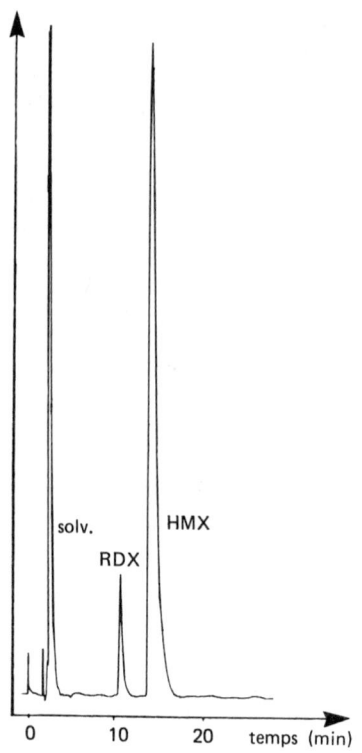

FIG. 3. Separation of RDX and HMX on a LiChrosorb Si 60 column
with a mobile phase of cyclohexane:dioxan (55:45) at a flow rate of
1.5 ml/min and UV detection at 254 nm. (From Poyet and Michaud,
*Analysis 4:*55.)

compatible with the mobile phase chosen. Also, it must be the case
that when injections are made, the explosives do not precipitate
out of the mobile phase before or during passage through the column-
detector. Some of the impurities found in TATB and fully resolved
from the parent compound using the foregoing conditions were:
1,2,3,5-tetrachloro-4,6-dinitrobenzene, 1,3,5-trichloro-2,4,6-
trinitrobenzene, 1,3,5-trichloro-2,4-dinitrobenzene, 1-chloro-3,5-
diamino-2,6-dinitrobenzene, 1-chloro-3,5-dinitro-2,4,6-triamino-
benzene, and related compounds. Figure 4 indicates a typical separation of TATB from its impurities, using this column and mobile phase

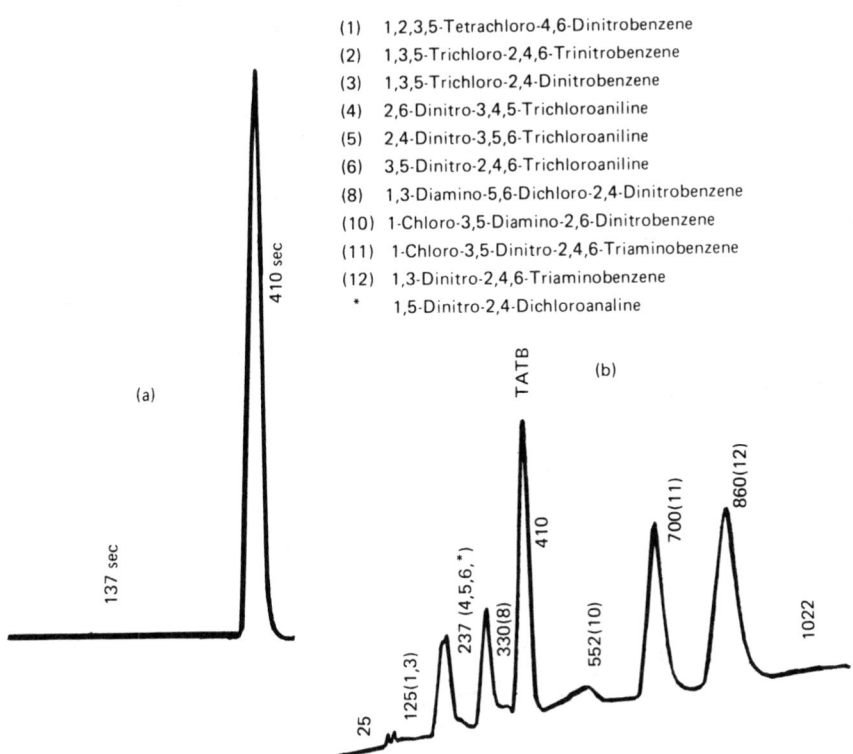

Impurities Found in TATB

(1) 1,2,3,5-Tetrachloro-4,6-Dinitrobenzene
(2) 1,3,5-Trichloro-2,4,6-Trinitrobenzene
(3) 1,3,5-Trichloro-2,4-Dinitrobenzene
(4) 2,6-Dinitro-3,4,5-Trichloroaniline
(5) 2,4-Dinitro-3,5,6-Trichloroaniline
(6) 3,5-Dinitro-2,4,6-Trichloroaniline
(8) 1,3-Diamino-5,6-Dichloro-2,4-Dinitrobenzene
(10) 1-Chloro-3,5-Diamino-2,6-Dinitrobenzene
(11) 1-Chloro-3,5-Dinitro-2,4,6-Triaminobenzene
(12) 1,3-Dinitro-2,4,6-Triaminobenzene
 *   1,5-Dinitro-2,4-Dichloroanaline

FIG. 4. Separation of TATB from various impurities found in commercial TATB samples. (a) Standard sample of TATB on a μBondapak NH$_2$
column with a mobile phase of DMF:toluene:heptane (1.5:43.5:55.5)
at a flow rate of 2.0 ml/min with UV detection at 355 nm. (b) Separation of commercial TATB from its commonly found impurities; conditions as in (a). (From Ref. 48.)

at 2.0 ml/min with UV detection at 355 nm. This is one of the few
times where UV methods were used at a wavelength other than 254 nm
[47, 48]. The figure indicates a typical chromatogram for a standard of TATB (a), as well as a chromatogram for TATB and most of the
impurities present in commercial samples (b).

Meier et al. have recently described the successful application
of HPLC-UV techniques to the trace (less than 100 ppb) analysis of

TNT, tetryl, RDX, and HMX [49]. The HPLC conditions employed a Zorbax
C_{18} column (25 cm × 4.6 mm ID) with two different sets of conditions,
depending on the compounds being analyzed. For the elution of TNT
and its degradation products, a mobile phase of acetonitrile:water
(40:60) was used at 1.0 ml/min. These same conditions could be used
for tetryl. For RDX and HMX, it was shown advantageous to use a
mobile phase of methanol:water (30:70) at 1.0-ml/min flow rate.
Using these conditions, each munition was eluted with a sufficient
retention time to permit its separation from both more and less
polar degradation products. The limits of detection were about
50 ng for each munition of interest, with UV detection set at 230
nm, where RDX, TNT, and tetryl have larger ε values than at 254 nm.
Some of the degradation products separated were 1,4-dinitrobenzene,
1,3,5-trinitrobenzene, and various dinitrotoluenes, in combination
with TNT and tetryl. All of these were readily resolved using the
acetonitrile:water mobile phase and conditions as above [49]. These
workers were also able to analyze trace amounts (50 ppb) of these
explosives in wastewater, by suitable concentration techniques,
prior to the HPLC-UV analysis.

Cattran et al. have described, in an unpublished report, a
method for the quantification of HMX, RDX, and TNT in wastewaters by
HPLC-UV methods [50]. This method is capable of detecting and quantitating munitions down to 10 ng, using UV detection with direct injection of water samples. The system described employs a reversed-
phase, μBondapak C_{18} column with a mobile phase of methanol:water
(40:60), together with UV detection at 230 nm, and a flow rate of
100 ml/h. Numerous wastewater samples were taken from munitions
production sites for determination of materials present using this
HPLC-UV method. Samples were transferred directly to the HPLC
column, by injecting large volumes, without the need for initial
sample concentration prior to HPLC. In addition to the detection
of the munitions of interest, a number of their possible degradation
products were also analyzed using this approach. Some of these
derivatives were: dinitrotoluene isomers, dinitrobenzene isomers,
p-aminodinitrotoluene isomers, and trinitrobenzene isomers.

Haag recently described the use of reversed-phase HPLC with
variable-wavelength UV detection for the analysis and comparison
of smokeless propellants [51]. Individual lots of the same powder,
aged samples of a single lot of propellant, and samples that could
not be distinguished on the basis of physical properties were all
capable of being discriminated using this approach. Specific details
of the analytical methods and conditions are not yet available.

Prime and Krebs have recently described a highly sensitive and
simple-to-apply recovery-HPLC analysis scheme for the identification
of EGDN and NG in explosion debris [52]. The method of volatile
sample collection involved heating the sample debris at carefully
controlled temperatures, under a stream of inert gas used to transfer the volatile explosive residues onto a collection tube packed
with a solid adsorbent. The method is quite similar to that of head-
space sample analysis, used for many years to trap volatile materials above a liquid or solid sample [53-56]. After transfer of
the volatile explosive compounds was complete, the adsorbent was
extracted with a small volume of 2-propanol. This was then injected, following filtration, onto the HPLC-UV arrangement directly,
or first concentrated under nitrogen to a small volume for reinjec-
tion. This was one of the few studies available to compare directly
the use of various wavelengths for the detection of trace amounts
of aliphatic organic nitrates, such as NG and EGDN. Although these
materials can be detected at high levels at 254 nm, this is less
useful for trace amounts of materials recovered from test blasts
[52]. These authors also recovered traces of volatile explosives
by air sampling at the site of a known explosion, again using solid
adsorbent trapping methods. For both EGDN and NG, a considerable
improvement in detection limits was observed by going from 254 nm
to 190 nm, as indicated in Fig. 5. However, even at 210 nm, there
is a significant improvement in the detection limit and sensitivity
of these compounds, thus allowing for a wider choice of HPLC-UV-
compatible mobile phases. The minimum detectable amounts of EGDN
and NG using these methods was found to be about 30 ng per injection, using a UV wavelength of 200 nm. Figure 6 is a gradient

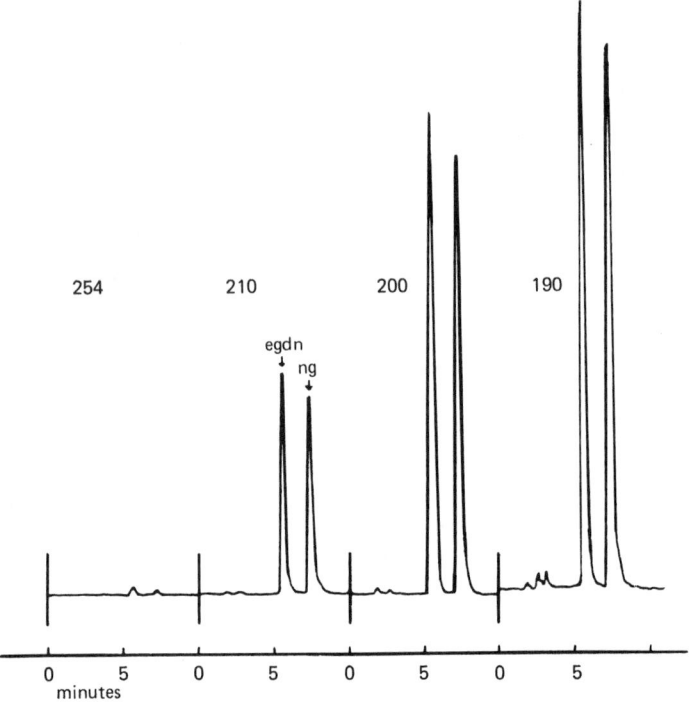

FIG. 5. Variation of response with various UV wavelengths for EGDN
and NG. HPLC conditions used isocratic elution with acetonitrile:
water (50:50) on a µBondapak C_{18} reversed-phase column at 1.0
ml/min. (From Ref. 52.)

elution HPLC-UV assay for EGDN and NG from an extract of dynamite,
using an initial mobile phase of methanol:water and a final eluant
of methanol alone. The particular gradient profile used here is
also indicated in Fig. 6. The method used a µBondapak C_{18} reversed-
phase column, a mobile-phase flow rate of 1.0 ml/min, and UV detection at 254 nm [52].

     Before concluding this section with a discussion of the work
of Alm et al., mention must be made of various references related
to the HPLC-UV determination of explosives, but found in unusual
literature sources or most readily from Chemical Abstracts or the
NTIS (National Technical Information Service). Thus Bissett and

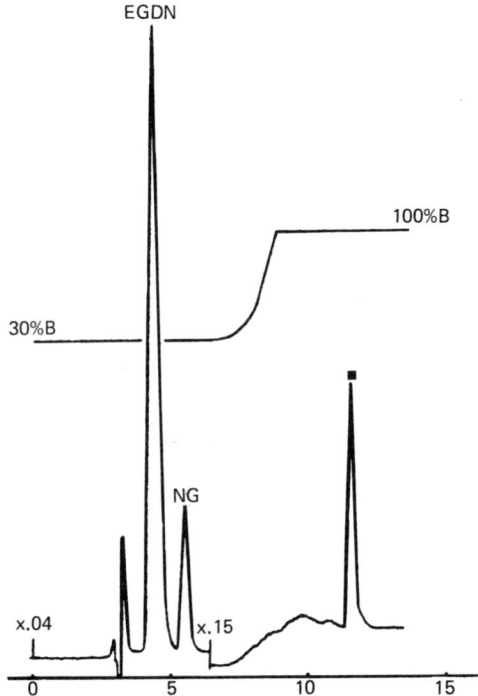

FIG. 6. Separation of EGDN and NG from an extract of dynamite using gradient elution HPLC with UV detection at 254 nm. HPLC conditions used a µBondapak C_{18} column with an initial mobile phase of 30% B and a final eluant of 100% B, all at 1.0 ml/min, solvent A = methanol:water (50:50), solvent B = methanol. (From Ref. 52.)

Levasseur have described an HPLC-UV assay for nitroguanidine and some of its degradation products [57]. Gorin et al. have discussed the HPLC-UV assay for diphenylamine, centralite, and a variety of derivatives in powders and propellants [58]. These methods utilized HPLC separations on 5-µm silica gel with a mobile phase of dioxane:hexane (20:80). Fariwar-Mosheni et al. have described the HPLC-UV analysis of NG, 2-nitrodiphenylamine, and diethyl phthalate via HPLC-UV methods [59], as well as the analysis of other common explosives and propellants [60]. An improved assay for HNS by HPLC was reported by Schaffer, this using a µBondapak NH_2 column [61].

An excellent summary of generally unpublished HPLC-UV analyses
for a wide variety of standard explosives and related materials has
been put together by Alm et al. in the National Defense Research
Institute of Sweden [5]. This report is concerned mainly with a
variety of aromatic nitro derivatives, and provides analytical data
such as MS, IR, NMR, GC, and of course HPLC. At least eight different sets of conditions are provided, all on μPorasil columns, with
a variety of mobile phases (see the appendix) [5]. The compounds
studied included TNT, EGDN, NG, TNB, PETN, TETR, HNS, RDX, and HMX.
All analyses were done at 254 nm with a fixed-wavelength type of
detector. Figure 7 illustrates a typical HPLC-UV chromatogram from
the Alm et al. report, here analyzing for various aromatic nitro
derivatives. The separation involved use of a μPorasil column with
a mobile phase of n-hexane:methylene chloride (87.5:12.5) at a flow
rate of 2.0 ml/min.

III. HPLC-MASS SPECTROMETRY ANALYSIS AND APPLICATIONS

Mass spectrometry (MS) has been a standard method of analysis for
volatile or semivolatile and thermally stable explosive materials
[1, 2, 5, 6]. MS has also been utilized very often for the characterization of various nitro compounds, other than explosives, but
the methods and techniques perfected are often immediately applicable to explosives analysis. However, very little has been described in the literature wherein MS has been combined on-line or
even off-line with HPLC separations for the trace analysis of explosives present in real-world samples. Only recently, Vouros et
al. [62] and Camp [63] have described an off-line combination of
HPLC with MS for the determination of various standard explosives.
This approach utilized an initial off-line HPLC separation and collection of the explosives of interest, via UV detection, followed
by individual sample concentration and chemical ionization mass
spectrometry (CIMS) using ammonia as the soft reagent gas. The
method was also shown to be effective when applied to the analysis
of simple residues from test explosions designed to simulate actual

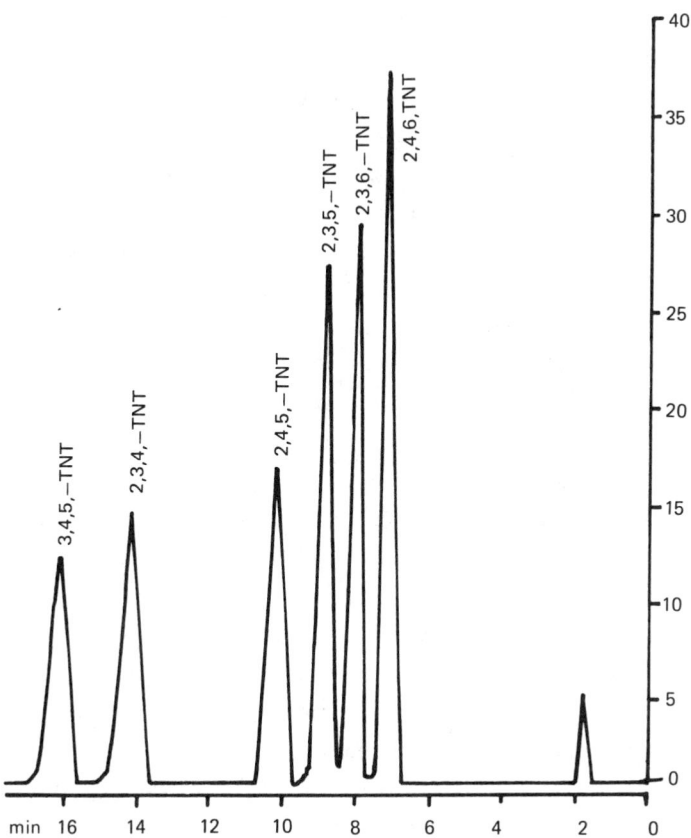

FIG. 7. Separation by HPLC-UV of various trinitrotoluene derivatives, using a μPorasil column with a mobile phase of n-hexane:
methylene chloride (87.5:12.5) at a 2.0-ml/min flow rate. (From
Ref. 5.)

bombings. Previous work using MS for explosive detection had employed primarily electron impact ionizatqon, but this approach resulted in extensive fragmentation of the parent compounds and an
undetectable molecular ion. Similarly, negative-ion MS was also of
limited success, because of the relatively weak molecular ion peaks
observed in these earlier studies.

    The advantages of this approach were many, especially in the
repeated use of ambient temperatures for both HPLC and CIMS. The

method provided virtually unambiguous identification in two ways,
on the basis of the HPLC retention times and capacity factors as
compared with standard explosives, and on the basis of the very
specific chemical ionization mass spectra obtained. In every case
studied, the CIMS spectra provided $(M + H)^+$ ions and fragmentation
patterns unique for the individual compounds analyzed. The limits
of detection for CIMS are routinely in nanogram or subnanogram
range, and thus the off-line approach described here is limited only
by the amount of sample that can be collected from the HPLC and/or
that originally available. The HPLC-UV methods for the initial collection of the explosives of interest involved a 10-µm Partisil
silica-type column with a mobile phase of 1,2-dichloroethane at 0.5
ml/min with UV detection at 254 nm.

It should be emphasized here that combined HPLC-CIMS methods
were successful in identifying materials such as EGDN, NG, and
PETN, all of which cannot be differentiated by MS methods alone.
Thus this method offers significant advantages over all other approaches for explosives analysis, in that it provides for the accurate and reliable qualitative and quantitative identification of
explosive materials present in very complex matrices at the trace
level. Using total ion monitoring, selected ion monitoring, single-
ion monitoring, and ancillary highly sensitive and specific techniques, together with on-line HPLC, it should be possible to provide
for the most sensitive and specific methods of trace analysis for
these materials [64-68]. The major technical difficulty in applying on-line HPLC-MS techniques in trace organic analysis has always been the design and routine operation of the interface itself.
However, recent instrumentation advances, notably by Finnigan Corp.,
have significantly resolved much of this problem [69-74]. A recent
Finnigan Corp. Application Report [74] indicates the application of
the latest HPLC-MS techniques for the on-line trace analysis of the
herbicide Oryzalin and its various photolysis products from environmental samples. Oryzalin is 3,5-dinitro-N^4,N^4-dipropylsulfanila-
mide, an aromatic dinitro compound very similar in structure to many

of the commonly encountered explosives (TNTs, DNTs, MNTs, Tetryl, etc.). Electron impact (EI) and negative- and positive-ion chemical ionization (NICI, PICI) approaches are described for Oryzalin and its photolysis products [74]. In view of these recent advances and their application to at least one aromatic dinitro compound analogous to the explosives of interest here, there is every reason to expect that on-line HPLC-NICI, HPLC-PICI, and of course HPLC-EIMS approaches will shortly be applied to a variety of explosives analyses.

One major problem in the direct interfacing of reversed-phase HPLC with any mode of MS operation has been the successful, rapid removal of the mobile phases containing varying amounts of water or alcohols. This problem has recently been resolved by the work of Kirby [75] and Karger et al. [76]. This approach utilizes a continuous, on-line, solvent-solvent extractor operating between the end of the HPLC column and the entrance to the LC-MS interface. In this manner, a large number of polar and nonpolar organic compounds can be quickly and efficiently recovered from the reversed-phase eluant, transferred to an organic, highly volatile solvent such as methylene chloride, and this final eluant is then introduced into the LC-MS interface system. Hence there is every reason to believe that by using this type of an extractor, all forms of reversed-phase and normal-phase HPLC, as well as ion-exchange and paired-ion HPLC, can now be interfaced with a commercially available LC-MS system.

IV. HPLC-THERMAL ENERGY ANALYSIS AND APPLICATIONS

The thermal energy analyzer (TEA) is a detector developed in the early 1970s at Thermo Electron Corporation, and thereafter extensively employed for the trace analysis of various N-nitroso derivatives [77-79]. During the late 1970s, various investigators demonstrated that with slight modifications to the original TEA, an instrument could be had which was then useful in the trace analysis of various aliphatic and aromatic nitro derivatives [80-83]. HPLC-TEA methods have now been applied in the analysis of various cardiac

vasodilators, compounds such as NG, isosorbide dinitrate, and isomeric isosorbide mononitrates, from a variety of biological matrices
[80]. In addition, these same approaches have been extensively
developed and applied for the trace analysis of a large number of
explosive materials in often complex sample matrices [81-83].
Whereas the UV detector depends on some degree of absorption of UV
energy by the compounds of interest, the TEA relies on the thermal
release of NO_2 from an explosive at elevated temperatures (550-650°C).
The initially formed NO_2 is converted within the detector to NO,
which then combines with ozone in a reaction chamber to generate
excited-state NO_2. This finally degrades back to its ground state
with the release of energy of a specific wavelength, and it is this
energy released in the form of IR radiation which is then measured
within the TEA. Combined HPLC-TEA methods are capable of distin-
guishing between structurally similar explosives, such as the TNTs,
DNTs, or MNTs, and those compounds containing structurally different
nitro groups, such as in PETN, RDX, and NG. Such differentiation,
however, relies completely on the HPLC separations obtained under
specific HPLC conditions. HPLC-MS methods identify individual explosives using retention times, as well as extensive structural information provided by the fragmentation patterns and ionization
methods employed (see above). Some typical results of HPLC-TEA
analyses are indicated in Fig. 8. This represents the HPLC separation of PETN, RDX, NG, and HMX using two different sets of conditions. In Fig. 8a the compounds are separated on a LiChrosorb Si
60 column using a mobile phase of ethanol:isooctane with gradient
elution, all at a flow rate of 1.5 ml/min. In Fig. 8b, the same
compounds have different elution parameters on a LiChrosorb μNH_2
column using a mobile phase of ethanol:isooctane with different
solvent programming conditions at 1.5-ml/min flow rates. Of interest here are the different retention times realized for these materials, depending on the particular column and solvent programming
conditions utilized. Such readily obtained chromatographic differences could serve as confirmatory evidence for the presence, for
example, of PETN in complex mixtures of explosives.

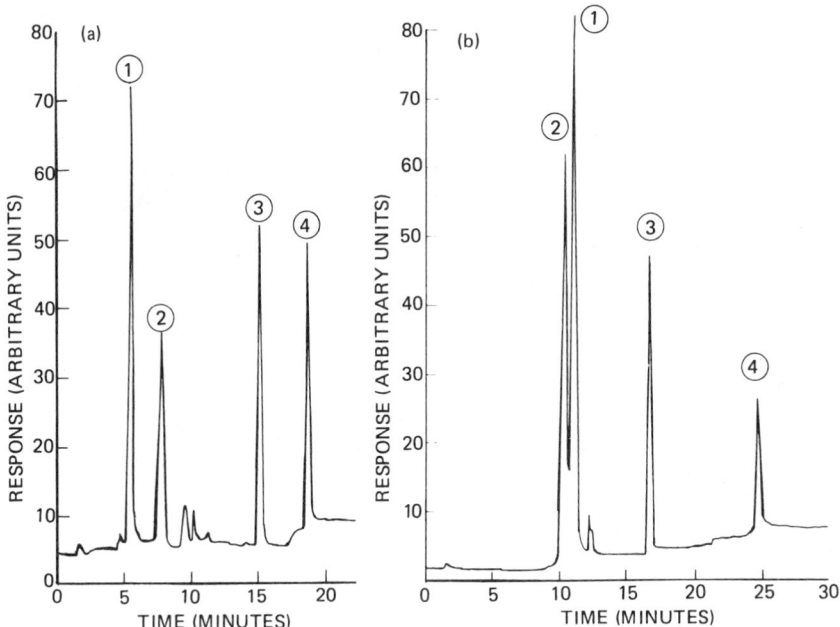

FIG. 8. The HPLC-TEA analysis of four common explosives: (1) NG, (2) PETN, (3) RDX, and (4) HMX. HPLC conditions: (a) LiChrosorb Si 60 column with ethanol:isooctane (1.5-90%) with solvent programming and 1.5-ml/min flow rates; and (b) LiChrosorb μNH_2 column using ethanol:isooctane (1.5-80%) with different solvent programming conditions and 1.5-ml/min flow rates. (Reprinted with permission from Lafleur and Morriseau, *Anal. Chem.* 52:1313; copyright 1980, American Chemical Society.)

HPLC-TEA methods are sensitive to nanogram amounts of materials, even in the presence of extraneous materials, such as plasticizers and stabilizers, often present in various propellant and explosive mixtures. All explosives containing the NO_2 and/or NO moiety are detectable using HPLC-TEA, with a high degree of reproducibility, accuracy, and precision already demonstrated by Lafleur and Morriseau [83]. These authors have compared retention data for eight common explosives -- EGDN, PETN, NG, Petrin, PEDN, RDX, NGU, and HMX -- all on both silica and μNH_2 columns.

However, as discussed previously, the TEA is not a completely specific detector, even when combined with HPLC, since it is known

that other types of organic compounds provide significant responses
[84, 85]. HPLC-TEA suffers from other serious drawbacks, including
the fact that water-based mobile phases are not routinely compatible
with the commercially available detector. This essentially excludes
from routine use any form of HPLC that would require water or other
solvents which have a freezing point near 0°C. Hence it is not possible with the TEA to utilize reversed-phase HPLC, ion-exchange, or
paired-ion methods. Unfortunately, these methods now comprise about
75% of all reported HPLC applications. However, as Lafleur and
Morriseau have demonstrated so well, satisfactory HPLC-TEA conditions
can be developed for a very large number of explosives, all of which
avoid the use of water-based mobile phases. Another serious drawback to be considered is the current selling price of HPLC-TEA instrumentation, now exceeding $35,000 for the detector alone. Fin-
ally, because the TEA is a relatively selective detector, it must
be dedicated to nitro and/or nitroso analyses; it cannot be turned
into a general HPLC detector, as is the case with UV, RI, and/or
MS detectors.

    The methods of analysis developed and described for HPLC-TEA
do provide for a minimum amount of sample handling and preparation
steps, compatible with the HPLC separation conditions. They also
provide a high degree of selectivity and sensitivity compared with
UV or EC methods, and an overall ease of analysis and data acquisition. There is no question but that this approach to explosives
analysis is of extreme interest, despite the fact that only a single
laboratory has thus far described their results in the literature
using the methods. Whether or not forensic-oriented laboratories
will be able to make use of the TEA for routine explosives determinations remains questionable, for some of the reasons indicated
above.

V. HPLC-ELECTROCHEMICAL ANALYSIS AND APPLICATIONS

HPLC has now been used quite extensively in combination with electrochemical detection (EC), and is especially suitable for the trace
analysis of nitro aromatics [86-90]. Aromatic C-nitro compounds

are generally suitable for HPLC-EC analysis, and many such derivatives have already been utilized in the analysis of other classes
of organic compounds, these having a wide variety of functional
groups [88]. The nitrophenyl or dinitrophenyl group has a unique
suitability for electrochemical detection in the reduction mode,
with a concomitant high sensitivity and low detection limits. Bioanalytical Systems, Inc., the firm that offers one of the more popular HPLC-EC systems and accessories, has already studied the reduc-
tive LCEC analysis of compounds such as 4-nitrobiphenyl, 2-nitro-
naphthalene, 2-nitrofluorene, 2,4,7-trinitro-9-fluorenone, 2,4-
dinitrofluorene, 2-nitro-p-phenylenediamine, and related materials
[90]. However, to date this method has not been reported for the
trace analysis of explosives present in environmental, real-world
samples. It is our understanding that these studies are currently
in progress, the results are most promising, and that a number of
publications should appear in this area during the coming year or
two [91]. It is expected that HPLC-EC will continue to play an important and ever-increasing role in trace organic analysis, and
this will shortly include explosives as well.

Graffeo and Riggin have also described the use of HPLC-EC for
the trace analysis of nonvolatile pollutants, including certain
nitro aromatics. However, specific details and results of these
methods are not yet generally available [89]. In most instances,
the limits of detection for nitro aromatics by HPLC-EC are in the
low-nanogram or subnanogram ranges, and specificity is considerably
better than via UV detection. Apparently, this approach is most
valuable for aromatic derivatives, and may not be employed as easily
with other types of nitro-containing compounds.

VI. HPLC-ELECTRON-CAPTURE DETECTION ANALYSIS AND APPLICATIONS

The electron-capture detector (ECD) is uniquely suited for the trace
analysis of organic nitro compounds, be they aliphatic or aromatic,
often at the picogram levels, and much of the initial work with explosives detection involved GC-ECD methods [1, 2, 4-6]. However,
the major restriction in the application of GC-ECD has always been

the thermal instability and volatility problems with many explosives
of interest. In recent years, during the 1970s, it became apparent
that the ECD can often be successfully interfaced with various forms
of HPLC [92-96]. Until quite recently, a commercial LC-ECD unit
was available in Europe from Pye Unicam Ltd., but this has now been
withdrawn from the market. Most of the published work dealing with
HPLC-ECD has involved the trace analysis of halogen-containing compounds, but there is at least one reference that describes the
successful analysis of certain standard nitro aromatics [96, 97].
Some minimum detection limits for various nitro compounds in HPLC-
ECD are presented in Table 2, all of these being aromatic nitro derivatives; none are explosives. Other than our own work described
below, this appears to be the only literature reference wherein
HPLC-ECD approaches have been described for explosive materials.

In general, there is often a difference in sensitivity of
several orders of magnitude between GC-ECD and HPLC-ECD for identical compounds. However, even for the HPLC conditions employed in
the original studies of Willmott and Dolphin [94], which were apparently not optimized, the limits of detection were in the low-ppb
(billion) or high-ppt (trillion) ranges. There is no question
that ECD is much more difficult to operate in the HPLC mode than it
is in GC, and various reports attest to this [92-97]. Both linearity and sensitivity are very dependent on the electron-capturing
properties of the solvent and any impurities that may be present.

TABLE 2. Minimum Detection Limits (MDL) for Certain Aromatic Nitro
Compounds by HPLC-ECD

Compound	MDL (g/ml)
Nitrobenzene	8×10^{-10}
2-Nitrotoluene	3×10^{-9}
3-Nitrotoluene	3×10^{-9}
4-Nitrotoluene	5×10^{-9}
2,4-Dinitrotoluene	9×10^{-10}
2,6-Dinitrotoluene	8×10^{-10}

Source: A. T. Chamberlain and J. S. Marlow, J. Chromatogr. Sci.
15:29 (1977). Reprinted with permission of the Journal of Chromatographic Sciences and Preston Technical Abstracts Company.

Extensive solvent purification and storage procedures are often necessary [92, 95]. Also, solvent flow rates and total volume affect
the linearity and sensitivity, as well as minimum limits of detection. An increased flow rate decreases signal height due to dilution effects and also to a decreased vaporization of the solute
present as the total eluent enters the ECD unit [92, 98, 99]. The
ECD also has a high degree of variability in its reponse to various
members of the same class of compounds, and can respond to a large
number of compounds other than just nitro derivatives.

Recently, Krull and Bushee have described the direct interfacing of HPLC with GC-ECD, using the GC-ECD only as a vaporization
oven, heated transfer interface, and detector-electronics arrangement. There is no GC column present within the interface, for the
HPLC eluent is passed directly into the GC oven and vaporized, then
split with a variable ratio splitter valve. A fraction of that
vaporized eluant is then passed directly into the ECD. Limits of
detection with this arrangement appear to be in the low-nanogram
range, at least for simple compounds such as nitrobenzene and dinitrotoluenes [98]. A variety of splitting ratios were analyzed for
each standard compound, and flow rates of about 10-80 μl/min into
the GC-ECD unit were found to be optimal for maximizing detection
limit capabilities (MDLs). In view of the fact that there is no
longer any commercial LC-ECD unit on the market, the ability to
interface HPLC inexpensively and rapidly with any commercial GC-ECD
unit is hoped to become a commonly employed research and/or development tool. It should be apparent that this approach will lend itself readily to trace explosives analysis. We have already emphasized that ECD is uniquely suited for the trace analysis of organic
nitro compounds, and thus in the detection of many commonly used
explosives [3].

In fact, we have recently undertaken certain preliminary
studies with a limited number of explosive standards, utilizing
here a combination of split eluant HPLC with the Pye Unicam LC-ECD
unit once commercially available [99]. Using a LiChrosorb Si 60

column with a mobile phase of 0.5-6.0% isopropanol:n-hexane at an
initial flow rate of 1.5 ml/min, with a split ratio of 0.3 ml/min
to the ECD unit, it has now been possible to detect 2,6-dinitrotolu-
ene, 2,4,6-trinitrotoluene (TNT), and tetryl in the range 1-6 ng per
injection. These analyses have been performed isocratically thus
far, changing the concentration of IPA-hexane as one progresses from
DNT to TNT to tetryl. Linearity plots for these compounds have been
prepared, with all exhibiting at least one-order-of-magnitude linearity (DNT = 10-100 ng, TNT = 10-60 ng, tetryl = 10-200 ng). It
is generally not possible to use reversed-phase HPLC conditions
with the HPLC-ECD arrangement described above, nor is it possible
to utilize IPA above about the 10%/hexane level. Attempts to use
absolute ethanol hexane with a LiChrosorb Si 60 or a μNH_2 column
were not successful; that is, the ECD loses all response to injected
nitro compounds when this mobile phase is present. Methanol, acetonitrile, water, acetone, ethyl acetate, ethanol, and related polar
solvents are simply not viable to utilize with any direct HPLC-ECD
arrangement, as previous workers have also reported. Hence one is
severely limited in the range of solvents that are compatible with
the ECD, especially not being able to utilize halogen-containing
materials. It may be possible to use ether-type solvents, such as
dioxane, THF, and diethyl ether, but this remains to be demonstrated.
The need to use HPLC on quite polar compounds, such as PETN, NG,
EGDN, RDX, and HMX, coupled with an inability to utilize reversed-
phase conditions or highly polar solvents with bonded-phase packings or silica gel, makes it extremely difficult to obtain suitable
HPLC-ECD conditions for the majority of explosive compounds commonly
encountered [99]. This apparent inability to employ some of the
more important types of HPLC separations is an inherent, significant
disadvantage and drawback in all HPLC-ECD work. Until this situation is fully rectified, the real potentials of this interfaced
system will never be realized or recognized. It is also true that
the overall performance of HPLC-ECD is highly dependent on the
quality of the solvents used as the mobile phase [92, 99].

The current situation with regard to HPLC-ECD interfacing is
not that different from what existed several years ago with HPLC-MS
systems. Thus the methods for overcoming removal of highly polar,
relatively nonvolatile mobile-phase solvents can often make use of
the continuous solvent-solvent extractor described by Kirby [75]
and Karger et al. [76]. This approach has thus far been described
for HPLC-MS, but there is every reason to expect it to be just as
practical and applicable to HPLC-ECD systems. This would then make
it possible to utilize virtually all modes of HPLC, with any type
of mobile-phase composition initially, as long as there is efficient extraction of the explosive materials of interest from the
initial HPLC eluant into the ECD-compatible organic solvent used
in the continuous extractor operating in an on-line mode. Brinkman
has also suggested the use of an extractor reactor that can be
interfaced between the HPLC and the ECD, here making use of hexane
instead of a heavier-than-water solvent in the phase-separator device [100]. This apparently does not cause any problems in operating the separation device in an on-line mode. However, extraction
efficiencies and the carryover of polar solvents in the hexane may
remain problems yet to be resolved. It is obvious that the whole
area of HPLC-ECD for explosives analysis is one requiring a significant amount of research and development in years to come. Nevertheless, it seems to offer a simple, inexpensive, and potentially
reliable method for the trace analysis of these particular compounds.

VII. CONCLUSIONS

Explosives are one of the more difficult classes of organic compounds to analyze for at the trace level, with a high degree of
specificity and sensitivity. Many of them are just not suitable
for GC separations, and TLC has long suffered from poor resolutions
and inaccuracy in the quantitative aspects of trace analysis. Thus
it seemed in the early 1970s that HPLC would offer a sure solution
to performing trace explosive analyses. To some extent this promise has now been realized, but there is much yet to be desired in

the existing instrumentation for performing HPLC-detector analyses.
It is obvious when one considers Figs. 1-8 that we have come a very
long way in improving the separations and reproducibility now possible with the latest advances in HPLC alone. The problem remains
in providing the average forensic laboratory with a complete HPLC-
detector system which is inexpensive to purchase and operate, simple
to maintain and repair, easy to automate, and provides a high degree
of accuracy, precision, and reproducibility in its day-to-day operations. HPLC-RI could be used for explosives analysis, but it will
just never make the degree of specificity and sensitivity required
in explosives analysis. HPLC-UV offers excellent sensitivity for
most explosives, especially at the now more widely usable wavelengths of 190-210 nm, but it suffers from a complete lack of specificity for the particular compounds of interest. Thus, to make use
of HPLC-UV, one really requires the use of certain confirmatory
techniques, which have not yet been adequately described for most
explosives. HPLC-MS provides both the specificity and sensitivity
required for trace explosives analysis, but it requires a high initial investment of money, plus highly trained and skilled operators
for routine, practical operations. It is clear that the average
forensic laboratory will not soon be able to utilize HPLC-MS approaches for its average explosives problem. HPLC-TEA suffers from
a high initial investment, commitment of a complete HPLC system for
the TEA analysis of explosives, an inability to be used for the
analysis of other classes of compounds, and an inability to function
with aqueous-based mobile phases. HPLC-EC may be the panacea for
explosive analysis, but there is too little published information
available to evaluate this system fully today. HPLC-ECD seems to
offer significant advantages, in that it can be constructed for
very little initial investment from existing HPLC and GC-ECD instrumentation. It is also easy to operate, and can be used for the
analysis of other classes of organic/organometallic compounds of
interest. Its problems are many, especially an inability, as with
TEA, to utilize fully and easily aqueous-based mobile phases in

reversed-phase, ion-exchange, or paired-ion chromatography modes. Nevertheless, it seems apparent that the method does lend itself to explosives analysis, but more involved research and development work is still needed here.

There is no ideal HPLC-detector arrangement for all explosive studies; some arrangements work very well with aromatic explosives and not as well with aliphatic derivatives. Significant advances have already been made, but many other advances remain to be described. There are still numerous HPLC detectors which have never been evaluated in explosives analysis [101]. If one had to choose the ideal HPLC-detector arrangement, it would have to be HPLC-MS, especially where this is operative in the positive- or negative-ion chemical ionization modes. However, this is an ideal situation, and not a very practical or affordable one for most forensic laboratories today. Thus, for the chemist involved in the development of trace analytical methods, this field remains one that calls for continued research and improvements.

ACKNOWLEDGMENTS

The author wishes to acknowledge the assistance and encouragement of Mike Camp, Diane Bushee, and David Young in various literature surveys and searches. He also acknowledges receipt of the valuable preprint of a chapter from the book *Analysis of Explosives* by J. Yinon and S. Zitrin. This book is due to appear in early 1981, published by Marcel Dekker. Arthur Lafleur devoted considerable time and effort in various discussions on the applications of the thermal energy analyzer to explosives analysis.

We indicate our appreciation to Ira Lurie for the invitation to contribute this chapter.

This is Publication 84 from the Institute of Chemical Analysis, Applications, and Forensic Science at Northeastern University.

APPENDIX: HPLC CONDITIONS FOR THE ANALYSIS OF VARIOUS EXPLOSIVES

1. *Nitroglycerin, diethyl phthalate, ethyl centralite, and ace-
   tanilide:* A stainless steel column packed with 30-44 μm Vydac
   adsorbent, mobile phase of 1,1-dichloroethane at 0.8 ml/min
   [41]. UV detection at 254 nm.

2. *Nitrotoluene and 2,4,6-trinitrotoluene:* A column of Bondapak
   C_{18} together with a mobile phase of 9:1 water:acetonitrile with
   UV detection at 254 nm [10].

3. *p-Nitrotoluene, 2,4-dinitrotoluene, and 2,4,6-trinitrotoluene:*
   A column of Corasil II (37-50 μm), with a mobile phase of
   hexane:methylene chloride (60:40) at a flow rate of 0.5 ml/min,
   RI detection [42].

4. *2,4,6-Trinitrotoluene and tetryl:* A Corasil II (37-50 μm)
   column with a mobile phase of dioxane:cyclohexane (10:90) at
   flow rate of 0.9 ml/min with UV detection at 254 nm [42].

5. *RDX and HMX:* A Corasil II (37-50 μm) column with a mobile
   phase of dioxane:cyclohexane (35:65) at flow rate of 1.2 ml/min
   with UV at 254 nm [42].

6. *TNT, tetryl, and RDX:* A Corasil II (37-50 μm) column with a
   mobile phase of dioxane:cyclohexane (30:70) at flow rate of
   0.6 ml/min with 254-nm detection [42].

7. *NG and ethyl centralite (EC):* A Corasil II (37-50 μm) column
   with chloroform:cyclohexane (16:84) at 1.0 ml/min with 254
   nm UV detection [42].

8. *Diphenylamine, dibutylphthalate, and NG:* A Corasil II column
   (37-50 μm) with a mobile phase of chloroform:cyclohexane (10:90)
   at a flow rate of 0.b ml/min with 254-nm UV detection [42].

9. *2-Nitrodiphenylamine and NG:* A Corasil II column (37-50 μm)
   with a mobile phase of methylene chloride:cyclohexane (10:90)
   at a flow rate of 1.8 ml/min and UV detection at 254 nm [42].

10. *Diphenylamine, 2-nitrodiphenylamine, 2,6-dinitrotoluene, 2,4-
    dinitrotoluene, and N-nitrosodiphenylamine:* A Corasil II
    (37-50 μm) column with a mobile phase of methylene chloride:
    hexane (10:90) at a flow rate of 0.9 ml/min and UV detection
    at 254 nm [42].

11. *2-Nitrodiphenylamine, 2,4-dinitrotoluene, 2,2'-dinitrodipheny-
    lamine, 2,4-dinitrodiphenylamine, and 2,4'-dinitrodiphenylamine:*
    A Corasil II (37-50 μm) column (100 cm × 2.1 mm ID) with a
    mobile phase of methylene chloride:cyclohexane (20:80) at a
    flow rate of 0.5 ml/min with UV detection at 254 nm [43].

12. *Tetryl, 2-nitroaniline, 2,4-dinitroanisole, 2,4,6-trinitroani-
    sole, methylpicramide, picramide, 4-nitroaniline, 2,4-dinitro-
    aniline, and picric acid:* (a) A 30 cm × 4.4 mm ID stainless
    steel column of μBondapak CN with a mobile phase of cyclohexane:
    chloroform:tetrahydrofuran (77:20:3) at 2.0 ml/min and UV

detection at 254 nm [44]. (b) A 60 cm × 2.2 mm ID column of phenyl Corasil with a mobile phase of water:methanol (75:25) at 2.0 ml/min and UV detection at 254 nm [44].

13. *RDX and HMX:* A 20-μm LiChrosorb Si 60 column with a mobile phase of cyclohexane:dioxane (55:45) at 1.5 ml/min, with UV detection at 254 nm [45].

14. *2-Nitrodiphenylamine, 2-nitro-N-nitrosodiphenylamine, 2,4,4'-trinitrodiphenylamine, 2,4'-dinitrodiphenylamine, 2,4,2'-trinitrodiphenylamine, 2,2'-dinitrodiphenylamine, 2-nitrodi-phenylamine, triphenylphosphate, and dibutyl phthalate:* A μBondapak C_{18} column with an eluant of methanol:water (67.5: 32.5) at a flow rate of 1.0 ml/min with UV detection at 254 nm [45].

15. *RDX and HMX:* A glass column operated at atmospheric pressure containing polymeric adsorbents, with a mobile phase of acetone or acetone:methanol (3:1) [46].

16. *TATB and impurities found in TATB samples:* A μBondapak NH_2 column (25 cm × 2.6 mm) with a mobile phase of DMF:toluene: heptane (1.5:43:55.5), at a flow rate of 2.0 ml/min, a column temperature of 45°C, and UV detection at 355 nm [48].

17. *TNT and tetryl from degradation products:* A Zorbax C_{18} column with a mobile phase of acetonitrile:water (40:60) at a flow rate of 1.0 ml/min with UV detection at 230 nm [49].

18. *RDX and HMX:* A Zorbax C_{18} column with a mobile phase of methanol:water (30:70) at a flow rate of 1.0 ml/min, with UV detection at 230 nm [49].

19. *HMX, RDX, and TNT, as well as various degradation products of these:* A μBondapak C_{18} column with a mobile phase of methanol: water (40:60) at a flow rate of 100 ml/hr, with UV detection at 230 nm [50].

20. *EGDN and NG:* (a) A μBondapak C_{18} column with gradient elution using 30-100% solvent B, the remainder solvent A, as below, at 1.0 ml/min, with UV detection at 254 nm. Solvent A = methanol:water (50:50); solvent B = methanol alone, [52]. (b) A μBondapak CN column with a mobile phase of hexane:2-propanol (90:10) at 1.0 ml/min with UV detection at 214 nm [52].

21. *Diphenylamine, centralite, and their derivatives:* A 5-μm silica gel column with mobile phase of dioxane:hexane (20:80), with UV detection [58].

22. *NG and EGDN:* A μPorasil column with a mobile phase of hexane: methylene chloride (70:30) at 2.0 ml/min with UV detection at 254 nm [5].

23. *HMX, RDX, and TNT:* A μPorasil column with a mobile phase of chloroform:acetonitrile (90:10) at a flow rate of 1.0 ml/min, with UV detection at 254 nm [5].

24. *TNT and tetryl:* A µPorasil column with a mobile phase of chloroform alone at 1.0 ml/min with UV detection at 254 nm [5].

25. *TNT and PETN:* A µPorasil column with a mobile phase of n-hexane:chloroform (80:20) at a flow rate of 2.0 ml/min with UV detection at 254 nm [5].

26. *TNT and TNB:* A µPorasil column with a mobile phase of n-hexane: chloroform (80:20) at a flow rate of 1.0 ml/min with UV detection at 254 nm [5].

27. *2,4,6-TNT, 2,3,6-TNT, 2,3,5-TNT, 2,4,5-TNT, 2,3,4-TNT, and 3,4,5-TNT:* A µPorasil column with a mobile phase of n-hexane: methylene chloride (87.5:12.5) at a flow rate of 2.0 ml/min with UV detection at 254 nm [5].

28. *2,5-DNT, 2,6-DNT, 3,5-DNT, 2,4-DNT, 2,3-DNT, and 3,4-DNT:* A µPorasil column with a mobile phase of n-hexane:chloroform (95:5) at a flow rate of 2.5 ml/min with UV detection at 254 nm [5].

29. *2-MNT, 3-MNT, and 4-MNT:* A µPorasil column with a mobile phase of n-hexane at a flow rate of 1.0 ml/min with UV detection at 254 nm [5].

30. *TNT, RDX, and HMX:* A µPartisil column with a mobile phase of 1,2-dichloroethane at 0.5 ml/min with UV detection at 254 nm [62].

31. *RDX separated from explosion residue components:* A µPartisil column with a mobile phase of heptane:1,2-dichloroethane (40:60) at 0.5 ml/min with UV detection at 254 nm [62].

32. *NG, PETN, RDX, and HMX:* A LiChrosorb Si 60 column with gradient elution using ethanol:isooctane from 1.5 to 80% over 20 min, then at 80% for rest of analysis, all at a flow rate of 1.5 ml/min with TEA detection [83].

33. *NG, PETN, RDX, and HMX:* A µNH$_2$ column with a gradient elution using ethanol:isooctane from 1.5 to 90% over a period of 20 min and held at 90% for the rest of analysis, all at a flow rate of 1.5 ml/min with TEA detection [83].

34. *PETN, Petrin, and PEDN:* Either a LiChrosorb Si 60 or a µNH$_2$ column with gradient elution and a mobile phase of ethanol: isooctane (1.5 to 80 or 90%) at a flow rate of 1.5 ml/min with TEA detection [83].

35. *NG and EGDN:* Either a LiChrosorb Si 60 or a µNH$_2$ column with gradient elution and a mobile phase of ethanol:isooctane (1.5 to 80 or 90%) at a flow rate of 1.5 ml/min with TEA detection [83].

REFERENCES

1. J. Yinon and S. Zitrin, *The Analysis of Explosives,* Pergamon Press, Elmsford, N. Y., 1981 (in press).
2. J. Yinon, *Crit. Rev. Anal. Chem.,* pp. 1-35 (Dec. 1977).
3. I. S. Krull and M. J. Camp, *Am. Lab.,* p. 63 (May 1980).
4. R. B. Moier, New Concepts Symp. Workshop Detect. Identif. Explos., Reston, Va., Oct. 30-Nov. 1, 1978. U.S. Depts. of Treasury, Energy, Justice, and Transportation, *Proceedings.*
5. A. Alm, O. Dalman, I. Frolen-Lindgren, F. Hulten, T. Karlsson, and M. Kowalska, *FOA Report C 20267-Dl,* National Defence Research Institute, Dept. 2, Stockholm, Oct. 1978.
6. A. Alm, *FOA Report C 20133-Dl,* National Defence Research Institute, Dept. 2, Stockholm, Sweden, Aug. 1976.
7. M. J. Camp, New Concepts Symp. Workshop Detect. Identif. Explos., Reston, Va, Oct. 30-Nov. 1, 1978. U.S. Depts. of Treasury, Energy, Justice, and Transportation, *Proceedings.*
8. Waters Associates, Milford, Mass.: Bioanalytical Systems, Inc., West Lafayette, Ind.; and Thermo Electron Corporation, Waltham, Mass., personal communications.
9. C. D. Chandler, J. A. Kohlbeck, and W. T. Bolleter, *J. Chromatogr.* 67:255 (1972).
10. J. T. Walsh, R. C. Chalk, and C. Merritt, Jr., *Anal. Chem.* 45:1215 (1973).
11. J. O. Doali and A. A. Juhasz, *J. Chromatogr. Sci.* 12:51 (1974).
12. P. Vestergaard, A. Bachman, T. Piti, and M. Kohn, *J. Chromatogr.* 111:75 (1975).
13. E. L. Johnson, R. Gloor, and R. E. Majors, *J. Chromatogr.* 149:571 (1978).
14. H. Engelhardt, *High Performance Liquid Chromatography.* Springer-Verlag, New York, 1979.
15. L. R. Snyder and J. J. Kirkland, *Introduction to Modern Liquid Chromatography.* Wiley-Interscience, New York, 1974.
16. P. H. Dixon, C. H. Gray, C. K. Lim, and M. S. Stoll (Eds.), *High Pressure Liquid Chromatography in Clinical Chemistry.* Academic Press, New York, 1976.
17. P. M. Rajcsanyi and E. Rajcsanyi, *High Speed Liquid Chromatography.* Marcel Dekker, New York, 1975.
18. N. A. Parris, *Instrumental Liquid Chromatography,* J. Chromatogr. Libr., vol. 5. Elsevier, Amsterdam, 1976.
19. J. F. K. Huber, *Instrumentation for High-Performance Liquid Chromatography,* J. Chromatogr. Libr., vol. 13. Elsevier, Amsterdam, 1978.
20. R. E. Major, *Proc. Anal. Div., Chem. Soc.* (Lond.), p. 25 (Jan. 1975).
21. N. A. Parris, *Am. Lab.,* p. 124 (Oct. 1978).
22. L. S. Ettre and C. Horvath, *Anal. Chem.* 47:422A (1975).
23. C. Horvath and W. Melander, *Am. Lab.,* p. 17 (Oct. 1978).
24. H. M. McNair, *Am. Lab.,* p. 33 (May 1980).
25. J. H. Knox, *J. Chromatogr. Sci.* 15:352 (1977).

396 Krull

26. N. H. C. Cooke and K. Olsen, *Am. Lab.*, p. 45 (Aug. 1979).
27. H. Kern and K. Imhof, *Am. Lab.*, p. 131 (Feb. 1978).
28. R. P. W. Scott, *Liquid Chromatography Detectors*, J. Chromatogr. Libr., vol. 11. Elsevier, Amsterdam, 1977.
29. P. T. Kissinger, L. J. Felice, D. J. Miner, C. R. Preddy, and R. E. Shoup, in *Contemporary Topics in Analytical and Clinical Chemistry*, vol. 2, D. M. Hercules, G. M. Hieftje, L. R. Snyder, and M. A. Evenson (Eds.). Plenum, New York, 1978, Chap. 3.
30. L. A. Carreira, L. B. Rogers, L. P. Goss, G. W. Martin, R. M. Irwin, R. Von Wandruszka, and D. A. Berkowitz, *Chem. Biomed. Environ. Instrum.* 10:249 (1980).
31. L. N. Klatt, *J. Chromatogr. Sci.* 17:225 (1979).
32. L. S. Ettre, *J. Chromatogr. Sci.* 16:396 (1978).
33. S. A. Wise and W. E. May, *Res./Dev. Mag.*, p. 54 (Oct. 1977).
34. W. A. McKinley, D. J. Popovich, and T. Layne, *Am. Lab.*, p. 37 (Aug. 1980).
35. D. W. Janzen and D. J. Farley, *Am. Lab.*, p. 43 (Mar. 1976).
36. D. Janzen, M. Munk, B. Leaver, and P. DeLand, *Am. Lab.*, p. 67 (Jan. 1979).
37. D. H. Rodgers, *Am. Lab.*, p. 133 (Feb. 1977).
38. D. R. Knapp, *Handbook of Analytical Derivatization Reactions*. Wiley-Interscience, New York, 1979.
39. K. Blau and G. Kind (Eds.), *Handbook of Derivatives for Chromatography*. Heyden & Son, London, 1977.
40. C. D. Chandler, J. A. Kohlback, A. W. Tiedemann, and W. T. Bolleter, JANNAF Combined Propul. Meet., Las Vegas, 1971.
41. R. W. Dalton, C. D. Chandler, and W. T. Bolleter, *J. Chromatogr. Sci.* 13:40 (1975).
42. J. O. Doali and A. A. Juhasz, *J. Chromatogr. Sci.* 12:51 (1974).
43. J. O. Doali and A. A. Juhasz, *Anal. Chem.* 48:1859 (1976).
44. M. G. Farey and S. E. Wilson, *J. Chromatogr.* 114:261 (1975).
45. J. M. Poyet, H. Prigent, and M. Vignaud, *Analysis* 4:7 (1976).
46. D. H. Freeman, R. M. Angeles, and I. C. Poinescu, *J. Chromatogr.* 118:157 (1976).
47. C. L. Schaffer, *Assay of TATB by HPLC*, Mason & Hanger-Silas Mason Co., Inc., Dept. of Energy, U.S. Government Contract DE-AC04-76DP-00487, Dec. 1978. Available from NTIS.
48. C. L. Schaffer and A. C. Teter, *Assay of TATB by HPLC: Precision and Accuracy Study*, Mason & Hanger-Silas Mason Co., Inc., Dept. of Energy, U.S. Government Contract DE-AC04-76DP-00487, June 1979. Available from NTIS.
49. E. P. Meier, L. G. Taft, A. P. Graffeo, and T. B. Stanford, 4th Int. Conf. Sensing Environ. Pollut., New Orleans, Nov. 1977, p. 487, Paper 132.
50. D. Cattran, T. B. Stanford, and A. B. Fraffeo, Quantification of the Munitions HMX, RDX, and TNT in Wastewaters by Liquid Chromatography, Battelle Columbus Labs, unpublished report, private communication, 1979.
51. L. Haag, Calif. Assoc. Criminalists (CAC) Meet., Nov. 1980. *SAFS Newslett.* 8:19 (1980).

52. R. J. Prime and J. Krebs, *Can. Soc. Forensic Sci. J.* 13:27 (1980).
53. J. Drozd and J. Novak, *J. Chromatogr. Chromatogr. Rev.* 165:141 (1979).
54. B. Versino, M. deGroot, and F. Geiss, *Chromatographia* 7:302 (1974).
55. L. D. Butler and M. F. Burke, *J. Chromatogr. Sci.* 14:117 (1976).
56. R. H. Brown and C. J. Purnell, *J. Chromatogr.* 178:79 (1979).
57. F. H. Bissett and L. A. Levasseur, NTIS, Report AD-A034455; Available from NTIS. From *U.S. Gov. Rep. Announe. Index* 77(7):96 (1977).
58. P. Gorin, M. Lebert, M. Stephan, and B. Zeller, *Inf. Chim.* 158:209 (1976).
59. M. Fariwar-Mosheni, E. Ripper, and J. Zierath, *GIT Fachz. Lab.* 22:781 (1978).
60. M. Fariwar-Mosheni, E. Ripper, and K. H. Habermann, *Fresenius Z. Anal. Chem.* 296:152 (1979).
61. C. L. Schaffer, Report MHSMP-77-51, 1977. Available from NTIS. From *Energy Res. Abstr.* 3(8):Abstr. 19661 (1978).
62. P. Vouros, B. A. Petersen, L. Colwell, and B. L. Karger, *Anal. Chem.* 49:1039 (1977).
63. M. J. Camp, New Concepts Symp. Workshop Detect. Identif. Explos. Reston, Va., Oct. 30-Nov. 1, 1978. U.S. Depts. of Treasury, Energy, Justice, and Transportation, *Proceedings,* p. 579.
64. R. A. Hites and V. Lopez-Avila, *Anal. Chem.* 51:1452A (1979).
65. H. Brandenberger, in *Recent Developments in Mass Spectrometry in Biochemistry and Medicine,* vol. 2, A. Frigerio (Ed.). Plenum, New York, 1979, p. 227.
66. D. H. Russell, E. H. McBay, and T. R. Mueller, *Am. Lab.,* p. 50 (Mar. 1980).
67. C. Chang, *Am. Lab.,* p. 49 (Nov. 1980).
68. R. T. McIver, Jr., *Am. Lab.,* p. 18 (Nov. 1980).
69. J. D. Henion and G. A. Maylin, *Biomed. Mass Spectrom.* 7:115 (1980).
70. W. H. McFadden, D. C. Bradford, D. E. Games, and J. L. Gower, *Am. Lab.,* p. 55 (Oct. 1977).
71. P. J. Arpino and G. Guiochon, *Anal. Chem.* 51:682A (1979).
72. W. H. McFadden, *J. Chromatogr. Sci.* 18:97 (1980).
73. Y. Yoshida, H. Yoshida, S. Tsuge, and T. Takeuchi, *J. High Resolut. Chromatogr. Chromatogr. Commun.* 3:16 (1980), and references therein.
74. R. F. Skinner, Q. Thomas, J. Giles, and D. G. Crosby, *Application Report No. 8 LC/MS PPINICI.* The Finnigan Corporation, Sunnyvale, Calif., Aug. 1979.
75. D. Kirby, Ph.D. thesis, Northeastern University, Nov. 1980.
76. B. L. Karger, D. P. Kirby, P. Vouros, R. L. Foltz, and B. Hidy, *Anal. Chem.* 51:2324 (1979).
77. I. S. Krull and M. Wolf, *Am. Lab.,* p. 84 (May 1979).
78. I. S. Krull, K. Mills, G. Hoffman, and D. H. Fine, *J. Anal. Toxicol.* 4:260 (1980).

79. I. S. Krull and D. H. Fine, in *Handbook of Carcinogens and Other Hazardous Substances. Chemical Trace Analysis*, M. C. Bowman (Ed.). Marcel Dekker, New York, 1981, (in press), Chap. 6.

80. A. P. Silvergleid, W. Yu, and B. Morriseau, Pittsburgh Conf. Analy. Chem. Appl. Spectrosc., Atlantic City, N.J., Mar. 1980, Paper 170.

81. A. L. Lafleur, Pittsburgh Conf. Anal. Chem. Appl. Spectrosc., Atlantic City, N.J., Marh. 1980, Paper 285.

82. A. L. Lafleur, B. D. Morriseau, and D. H. Fine, New Concepts Symp. Workshop Detect. Identif. Explos., Reston, Va., Oct. 30-Nov. 1, 1978. U.S. Depts. of Treasury, Energy, Justice, and Transportation, *Proceedings,* p. 597.

83. A. L. Lafleur and B. D. Morriseau, *Anal. Chem.* 52:1313 (1980).

84. I. S. Krull, E. U. Goff, G. G. Hoffman, and D. H. Fine, *Anal. Chem.* 51:1706 (1979).

85. I. S. Krull, T. Y. Fan, and D. H. Fine, *Anal. Chem.* 50:698 (1978).

86. P. T. Kissinger, *Anal. Chem.* 49:447A (1977).

87. P. T. Kissinger, L. J. Felice, D. J. Miner, C. R. Preddy, and R. E. Shoup, in *Contemporary Topics in Analytical and Clinical Chemistry*, D. M. Hercules, G. M. Hieftje, L. R. Snyder, and M. A. Evenson (Eds.). Plenum, New York, 1978, Chap. 3.

88. P. T. Kissinger, K. Bratin, G. C. Davis, and L. A. Pachla, *J. Chromatogr. Sci.* 17:137 (1979).

89. A. P. Graffeo and R. M. Riggin, 4th Joint Conf. Sensing Environ. Pollut., New Orleans, Nov. 1977, *Proceedings,* p. 637.

90. Bioanalytical Systems, Inc. *LCEC Application Notes,* Nos. 11 and 30.

91. P. T. Kissinger and K. Bratin, personal communications, 1980.

92. U. A. Th. Brinkman, in *HPLC Detectors,* A. Zlatkis et al. (Eds.). Marcel Dekker, New York, 1981, (in press), Chap. 12.

93. F. J. Conrad and P. K. Peterson, New Concepts Symp. Workshop Detect. Identif. Explos., Reston, Va., Oct. 30-Nov. 1, 1978. U.S. Depts. of Treasury, Energy, Justice, and Transportation, *Proceedings.*

94. F. W. Willmott and R. J. Dolphin, *J. Chromatogr. Sci.* 12:695 (1974).

95. U. A. Th. Brinkman, P. M. Onel, and G. DeVries, *J. Chromatogr.* 171:424 (1979).

96. *Liquid Chromatography Applications,* No. 13, Pye Unicam, Ltd. (Philips Electronics Instruments), Cambridge, England.

97. A. T. Chamberlain and J. S. Marlow, *J. Chromatogr. Sci.* 15:29 (1977).

98. I. S. Krull and D. Bushee, *Anal. Lett.* 13(A14):1277 (1980).

99. I. S. Krull, E. A. Davis, C. Santasania, S. Kraus, A. Basch, and Y. Bamberger, *Anal. Lett.* 14(A16):1363 (1981).

100. U. A. Th. Brinkman, personal communication, 1980.

101. K. R. Hill, *J. Chromatogr. Sci.* 17:395 (1979).

8 Analysis of Writing Inks by High-Performance Liquid Chromatography

Albert H. Lyter, III*

Bureau of Alcohol, Tobacco, and Firearms
Rockville, Maryland

I. INTRODUCTION

Although not widely practiced, the analysis of writing inks can play
an important part in the successful investigation and adjudication
of both criminal and civil cases. Many times an entire case will
revolve around a single document or bit of writing. One such case
was that of the so-called "Mormon will" of the late Howard R. Hughes.
In this instance the whole question before the court was: Is this
authentic? The unaware forensic scientist would have merely thrown
up his or her hands and asked; What next?

Ink, like any mixture, can be separated and therefore characterized and this separation and characterization is in essence what
every forensic scientist performs. Document examiners have long
been the keepers of document evidence and have tried over the years
with some success to analyze inks on paper. Their attempts usually
involved optical and photographic methods of differentiation, of
which many such techniques are the first step in any analysis scheme
[1-6]. Techniques involving the actual chemical analysis of the

---

*
Present affiliation: Federal Forensic Associates, Los Angeles,
California.

ink, however, provided some difficulty until forensic chemists got
involved [7-18]. These chemists evolved the analysis of writing ink
to the point at which it is today. This is, the use of state-of-
the-art scientific instrumentation such as high-performance liquid
chromatography (HPLC). Although the analysis of writing ink has
evolved through the years, and has come under criticism [19, 20],
the forensic community must continue to develop this sometimes neglected and often important specialty.

II. BACKGROUND

To appreciate fully the problems confronting the forensic scientist
in his or her attempt to analyze writing ink it is necessary to have
some understanding of the product. Writing ink has progressed
through the years from naturally occurring colored liquids to metal
compounds to the present day, when two basic categories of ink
exist: ball pen and non-ball pen [21-24].

   Although ball pen inks are a rather recent development, circa
1945, they too have undergone change. Starting as a mixture of
dyes and pigments in an oil-based vehicle, they have progressed to
glycol-based systems that are extremely stable and easy to dispense.
The first ball pends were said to produce six carbons and no ori-
ginals eluding to the problem of bleeding, which easily distinguishes
this early product from its offspring. The present-day ball pen is
a mechanical marvel in that it is sometimes a closed system using
pressure to dispense the viscose ink through the ball mechanism.
These newer ball pen ink formulations contain acidic and basic dyes
as well as premetalized dyes and many other minor components, such
as surfactants, viscosity adjusters, corrosion control additives,
and many varied resinous materials [25]. No discussion of solvent
components will be undertaken since the written line is most often
depleted of these constituents. As one can see, any attempt to
characterize a product as varied and complex as ball pen ink is
extremely difficult. Although the complexity of ball pen ink is
the principal problem in characterization, other problems do exist
and I will elucidate these as they present themselves.

Non-ball pen inks are primarly water based and are found in
many different types and designs of writing instruments. As with
ball pen ink, non-ball pen ink is a mixture of colored and noncolored components. Because the majority of non-ball pen inks use
water as a base, the colored components used are the water-soluble
salts of acidic and basic dyes. Most often, organic solvents are
also present, to minimize drying of the ink at the pen tip, as well
as other ingredients, such as resins and surfactants, added to impart desired characteristics to the ink [25]. The same complexity
that plagues the examiner of ball pen ink is again present when one
confronts non-ball pen ink. This is, however, a surmountable task
faced by many forensic scientists.

In analyzing writing ink of any kind, the ultimate goal is
usually: Did this pen prepare that writing? In addition, it is
asked: When was it written? These two questions serve to provide
the challenge and therefore the goal of the forensic scientist:
Characterize this substance to the point of individualization.
Needless to say, this is a very difficult task, but many advances
have been made and more will be made through the aid of new techniques and advanced instrumentation.

Existing analysis techniques fall into three basic categories:
optical (including photographic), spectroscopic, and chromatographic.
Although the results of optical analyses are often attributable to a
single component of the ink, these techniques yield results indica-
tive of the ink as an entity. Optical methods of differentiation,
the results of which are usually displayed through photographic
development, were the first to be developed and have continued to
be extremely useful. Among these techniques are infrared (IR) absorbance, IR luminescence, ultraviolet (UV) fluorescence, and the
use of dichroic filters [1-6]. These techniques, especially IR
absorbance and IR luninescence, have become widely used in document
laboratories as an efficient method for detecting alterations and
obliterations, and as the first-line screening for ink comparisons.
There are certain pitfalls, however, which should be known before
routine use of these techniques [26].

As with the optical methods, the sepectroscopic techniques involve analysis of the ink in its entirety, even though often, the
ink is dissolved in a solvent. This being the case, any results
must be related to a summation of the ink's components -- and therefore characteristic, but not individualistic. Infrared (IR) spectroscopy has been used sparingly in ink analysis by the author, but
is of very little value due to the complexity of the spectra. IR,
however, may be applicable as a mode of detection when coupled with
chromatographic separation. Ultraviolet (UV) spectroscopy has also
been used, but because of its treatment of ink as a single component rather than a complex mixture, it can only characterize. One
spectroscopic technique that has been found to be useful on limited
occasion is spectrophotofluorometry (SPF) [27-29]. In cases where
a simple one- or two-dye system is encountered, it is possible to
quantitate relative amounts of the components with the SPF and therefore differentiate many inks.

The most widely used techniques in ink analysis are chromatographic in nature. The first attempts at chromatographic analysis
of the written line were actually no more than spotting a solvent
on the ink line and watching the ink bleed or separate into its components. This was in effect a form of paper chromatography, which
became the next step in the sophistication of ink analysis [7, 12,
15]. These early attempts at chromatography were hindered by the
attitude "don't destroy the evidence," which pervaded the forensic
community. Advances in ink analysis were made despite this attitude by first doing and then discussing the benefits versus the
destruction of the document. Allowing for some destruction of the
document, other chromatographic techniques such as thin-layer
chromatography (TLC) [10, 11, 13, 16, 17] and high-performance
liquid chromatography (HPLC) [30-32] were developed and are used
today as a part of an analysis scheme for writing inks.

III. BALL PEN INK

At least 80% of all evidence requiring ink analysis contains ball
pen ink. It is therefore not surprising that the majority of work

done in the area of HPLC analysis of ink has been limited to ball pen ink. As has been mentioned previously, this undertaking is not without its problems, the first of which is sample complexity. Ball pen ink contains dyes of several classes, including premetalized dyes and basic dyes. As seen in Fig. 1, these dye components could possibly be present in either ionic or neutral form, and in some instances contain both acidic and basic characteristics. Together with the dyes, which may constitute 40-50% of the ball pen ink, the other major constituent is the resinous fraction. Resins are present in ball pen ink to impart higher viscosity and therefore vary in concentration from about 5 to 40%. As you might well have guessed all types of resins are used in ball pen ink, ranging from natural to modified natural to synthetic polymers. The presence of resins in ball pen ink further complicates any chromatographic separation because of their molecular size and solubility characteristics. Other minor components do exist in most ball pen ink formulations, but they are of little consequence in chromatographic separations due to their concentrations (< 1%), and the ability to extract them from paper. This brings up the problem of the extracting solvent. It is evident that this problem stems from the mixture of components that make up ball pen ink. The majority of the dye components are soluble in alcohols, glycols, and several other organic solvents, while the resins have limited solubility in alcohol and better solubility in the glycols. The resins, unlike most of the dyes, also have an affinity for the cellulose fibers of paper which makes their extraction extremely difficult. Literature describing thin-layer chromatography (TLC) procedures have used pyridine with success as an extracting solvent, but these procedures have usually dealt only with the colored fraction of ball pen ink [10, 13]. It is evident that considering only these two problems -- sample complexity and sample extractability -- one would have an extremely difficult time in characterizing ball pen ink with a single chromatographic system.

Colwell and Karger [30] attempted to characterize ball pen ink

74350   C.I. Solvent Blue 25 (*Greenish blue*)

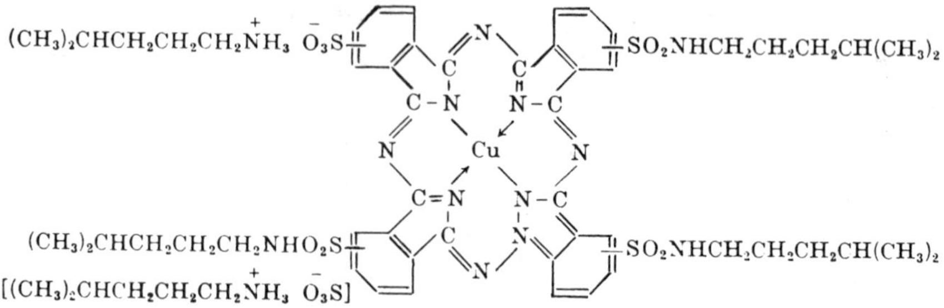

$(CH_3)_2CHCH_2CH_2CH_2\overset{+}{N}H_3 \ \overset{-}{O_3}S$

$SO_2NHCH_2CH_2CH_2CH(CH_3)_2$

$(CH_3)_2CHCH_2CH_2CH_2NHO_2S$

$[(CH_3)_2CHCH_2CH_2CH_2\overset{+}{N}H_3 \ \overset{-}{O_3}S]$

$SO_2NHCH_2CH_2CH_2CH(CH_3)_2$

(Substituents in the 3-positions)

44045   C.I. Basic Blue 26 (*Bright blue*)
             C.I. Pigment Blue 2 (*Bright blue*)*
44045B   C.I. Solvent Blue 4 (*Bright blue*)

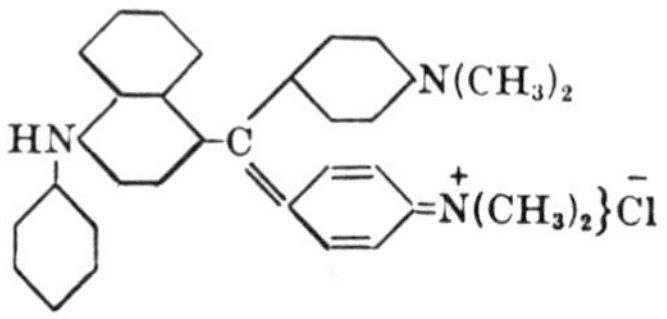

42500   C.I. Basic Red 9 (*Bright bluish red*)
Classical names **Para Magenta, Para Rosaniline**

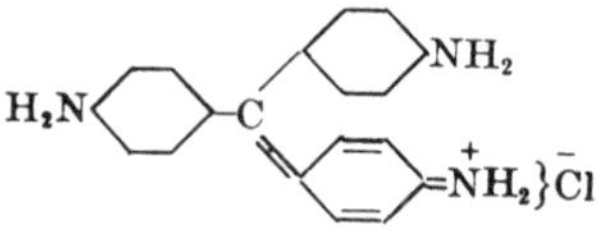

FIG. 1.   Chemical structures of three dyes used in making writing inks.   (From Ref. 32.)

with normal-phase HPLC conditions. Using a 25 cm × 3 mm ID column
packed with 10-μm Partisil silica and a methanol:formamide (98:2)
mobile phase, they were able to distinguish 25 different inks by
noting the presence, absence, and relative peak heights of components detected at 580 nm. These separations were carried out in
about 30 min using 10-μL injections of inks extracted with the mobile
phase and rather rigorous column conditions (silica gel and methanol:
formamide). This author questions the use of such rigorous conditions from an economic standpoint. It must be illustrated, however,
that the chromatographic separations obtained are quite adequate
and, given no alternative, would be acceptable. In addition to observation at 580 nm with the conditions above, Colwell and Karger
also used detection at 254 nm so as to include the "vehicle" components of the ink. Under these chromatographic conditions the
"vehicle" components (noncolored fraction) eluted as unretained
peaks and their peak heights were ratioed with peak heights from
the chromatogram of detection in the visible region. Through this
computation they differentiated inks that were similar in dye content and different in vehicle content. In the instance illustrated
by Figs. 2 and 3, there is a definite difference in appearance of the
chromatograms at 254 and 580 nm. This is due to differences in resin
type and content between the two ink samples, however; this type of
computation and differentiation may lead to false results for several reasons. First, the peaks of interest are unretained and therefore not specific. Second, the dye components do have absorbance
at 254 nm, which might interfere with these computations. Third,
given the situation of ink entries on different paper types, the
problem of preferential extraction of either dyes or noncolored components may lead to a false result in comparison.

   Another normal-phase chromatographic system developed by Colwell
and Karger is a 25 cm × 3 mm ID column packed with μPartisil silica
particles with a mobile phase of heptane:isopropanol (98:2). This
system appears to differentiate the vehicle components in inks in

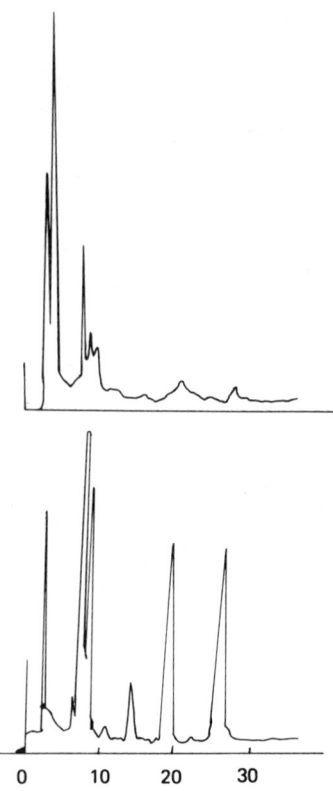

FIG. 2. Separation of components of an ink with a high-UV absorbing
vehicle content. Top: 254 nm, 0.02 absorbance for UV; bottom: 580
nm, 0.04 absorbance for visible. (From Ref. 30.)
Ref. 30.)

about 20 min at a flow of 0.5 ml/min using 10-µL injections of inks
extracted with the mobile phase. The detector wavelength in this
case was also at 254 nm and the chromatographic separations are quite
adequate for both qualitative and quantitative analyses. This variation in chromatographic conditions to allow for separation of the
vehicle components illustrates the difficulty in developing a single
chromatographic system for differentiation of ball pen inks.

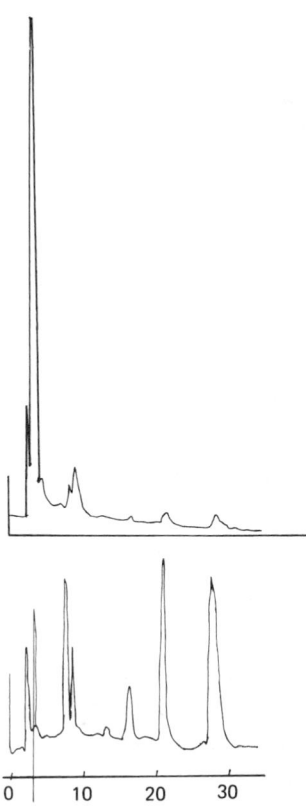

FIG. 3. Separation of components of an ink with a low-UV-absorbing vehicle content. Top: 254 nm, 0.02 absorbance for UV; bottom: 580 nm, 0.04 absorbance for visible. (From Ref. 30.) (From Ref. 30.)

The thin-layer chromatographic conditions reported by Brunelle and Pro [10] are also normal phase in their use of E. M. Merck silica gel plates, but contain some water in the mobile phase. These conditions adapted to HPLC have proven to be inadequate in separating the dye components of a selected group of ball pen inks when analyzed by the author using a Partisil column of 10-μm silica particles and a mobile phase of ethyl acetate:ethanol-water (14:7:6) [33]. Changing this system to exclude the water, and varying the amounts of ethanol and ethyl acetate, have resulted

in poor separation of the dye components at best, not to mention a
severe mobile phase, 90-100% ethanol. The separation of the vehicle
components was not possible under these conditions and any additional work with similar conditions was abandoned.

It is apparent that the use of normal-phase chromatographic
conditions can separate the various components of ball pen ink, but
with some problems. A single normal-phase system that will separate both colored and noncolored components does not exist at this
time. Because of these circumstances the evaluation of reversed-
phase HPLC was undertaken.

The author [32] and Kelley et al. [31] have used reversed-phase
HPLC in an attempt to characterize ball pen ink. Kelley et al.
used a reversed-phase system consisting of a μBondapak C_{18} column
of 10 μm particles and a mobile phase of 80% acetonitrile and 20% water
with a modifier of 0.005 M heptanesulfonic acid and 0.02% acetic
acid. Because of the ionic nature of the dye components in ball
pen ink the use of the modifier was considered necessary. This
modifier acts in two ways: by ionic suppression and by providing
a counterion to bind with the basic components. This is considered
ion-pairing chromatography. Kelly et al. report separation of the
dye components of several ball pen inks with this system in about
10 min, using 10-μL injections of ink extracted with the mobile
phase and a detection wavelength of 580 nm. Kelly et al. point out
that several of the inks studied could only be differentiated by
studying the color of the written line (1 cm) after extraction with
30-40 μL of the mobile phase. This point was also encountered by the
author in comparing the preferential extraction of the dye component in ink by several solvents. The author has found that several
dyes are extracted in varying amounts by methanol, ethanol, and
acetonitrile:water (80:20) modified with 0.005 M heptanesulfonic
acid and 0.02% acetic acid. Pyridine was therefore used as an extraction solvent and all inks were totally extracted. Kelly et al.
does address the question of the analysis of resinous or vehicle
components in the ink, but still reported being able to identify 10

of the 16 blue ball pen inks he examined. Kelly et al. also noted
a phenomenon that occurs related to the amount of time the ink has
been exposed to air. This phenomenon is manifested as either an
additional peak, usually a shoulder, or as a decrease in peak height
of one of the constituents. This phenomenon, as hypothesized by
Kelly, is due to a deterioration of one of the components by ex-
posure to the air. This may be true considering the nature of some
dyes but [34] this phenomenon did not occur in multiple analyses of
similar inks under similar chromatographic conditions. This author
did, however, use pyridine as an extraction solvent, which was found
to have very consistent and reproducible results. Using the chromatographic conditions of 80:20 acetonitrile:water with 0.005 M heptane
sulfonic and 0.02% acetic acid, a flow rate of 2 ml/min, detection at
546 nm, and a µBondapak C_{18} 10-µm particle column this author was able
to differentiate 10 blue ball pen ink formulations with a sample
size as small as 0.5 µg. This differentiation, like earlier TLC
procedures and Kelly's HPLC work, relies only on the colored or dye
components. Figure 4 shows the type of separation to be expected
with these chromatographic conditions and Table 1 illustrates the
quantitative differences that allow for differentiation. The choice
to use the above-mentioned reversed-phase system [µBondapak C_{18}
column, acetonitrile:water (80:20), 0.005 M heptanesulfonic acid,
0.02% acetic acid mobile phase] and rely on differentiation by
colored components alone was based on several circumstances. First,
the normal-phase systems previously described, [10-µm Partisil
silica column, methanol:formamide (98:2) mobile phase] are rather
rigorous or not sufficiently descriminatory. Second, the reversed-
phase system using detection at 254 nm was extremely complex. Third,
the reproducible extraction of the resinous fraction was very diff-
difficult with all solvents available. It was pointed out by Colwell
and Karger that some inks differ only in their resinous components
or in the amount of resin in relation to the amount of dye. For
this reason they thought it necessary to analyze the resins so as
to differentiate these inks. Granted, it would be preferable to

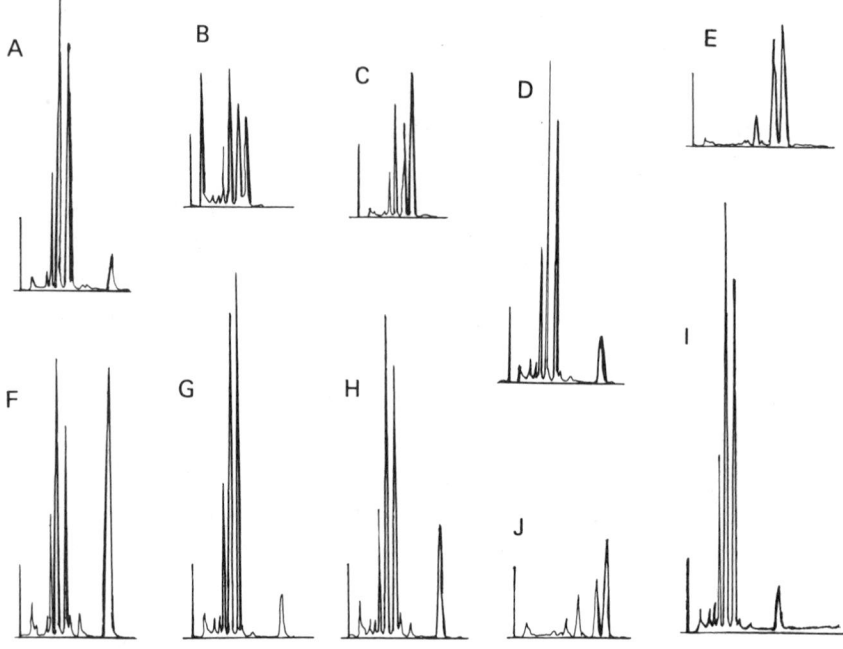

FIG. 4.   Chromatograms of 10 different ink formulations (A-J).
(From Ref. 32.)

analyze all fractions of ball pen ink, but this author was able,
using the previously mentioned chromatographic conditions [μBonda-
pak C_{18} column, acetonitrile:water (80:20), 0.005 M heptane sulfonic
acid, 0.02% acetic acid mobile phase] to differentiate between
different batches of the same ball pen ink. If this is the case
in all ball pen inks, the analysis of the resin fraction is unnec-
essary. Figure 5 shows the chromatographs of four batches of a
sinlge ball pen ink formulation and Table 2 illustrates the quantitative measurements allowing for the differentiation. It should,
however, be pointed out that these data are for a limited sample
population and should be expanded before they are used routinely.
Another parameter that was examined by this author was the effect
of different paper types on this type of analysis. By using three
types of paper -- bond, chromatographic, and tablet -- this author

TABLE 1. Normalized Average Peak Heights (%) ± Standard Deviation of Three 10-μl Injections of Each Ink (A–J) on Each of Three Papers (W, B, and T)

k	A	B	C	D	E	F	G	H	I	J
0.25	4.4 ± 2	99.7 ± 0.5	8 ± 0.5	5.6 ± 0.1	5.9 ± 3	13.1 ± 0.2	11.4 ± 1	10.4 ± 1	5.5 ± 0.1	6.7 ± 0.5
0.35	--	6.7 ± 0.5	--	6.9 ± 0.5	--	--	3.4 ± .2	5.4 ± 0.2	4.7 ± 0.2	--
0.55	--	7.2 ± 1	4.6 ± 0.1	6.5 ± 0.1	--	--	--	--	--	--
0.80	6.4 ± 1	--	--	--	--	7.7 ± 0.2	5.9 ± 1	5.8 ± 0.3	5.9 ± 0.3	--
1.00	40.3 ± 2	41 ± 2	30.4 ± 0.5	41.3 ± 0.1	--	44.4 ± 0.2	38.9 ± 1.5	38.9 ± 1	38.7 ± 2.5	--
1.20	100	95.9 ± 5	77.4 ± 0.5	100	--	99.5 ± 0.5	100	100	100	--
1.60	84.5 ± 2	71 ± 1.5	64.4 ± 0.3	2.3 ± 0.3	--	73.7 ± 2	79.4 ± 3	77.4 ± 3	84.5 ± 2	--
2.05	2.6 ± 0.5	65.2 ± 2	100	2.4 ± 0.2	3.6 ± 0.5	6.2 ± 0.3	2.5 ± 0.5	6.3 ± 0.1	1.7 ± 0.2	--
2.40	1.45 ± 0.5	--	--	2 ± 0.3	23.4 ± 3	8.9 ± 0.1	4.3 ± 0.5	4.9 ± 0.2	1.2 ± 0.2	16.7 ± 0.1
2.80	1.45 ± 0.5	--	--	--	2.7 ± 0.5	--	--	--	--	--
3.05	--	--	--	--	84.1 ± 5	. --	--	--	--	--
3.60	--	--	--	--	100	--	--	--	--	44.4 ± 0.3
4.15	13 ± 0.5	--	--	15.4 ± 0.2	--	99.3 ± 1	63.3 ± 1.5	32.8 ± 1	10.3 ± 0.2	--
4.35	--	--	--	--	--	--	--	--	--	50 ± 0.5
4.95	--	--	--	--	--	--	--	--	--	100

Source: Ref. 32.

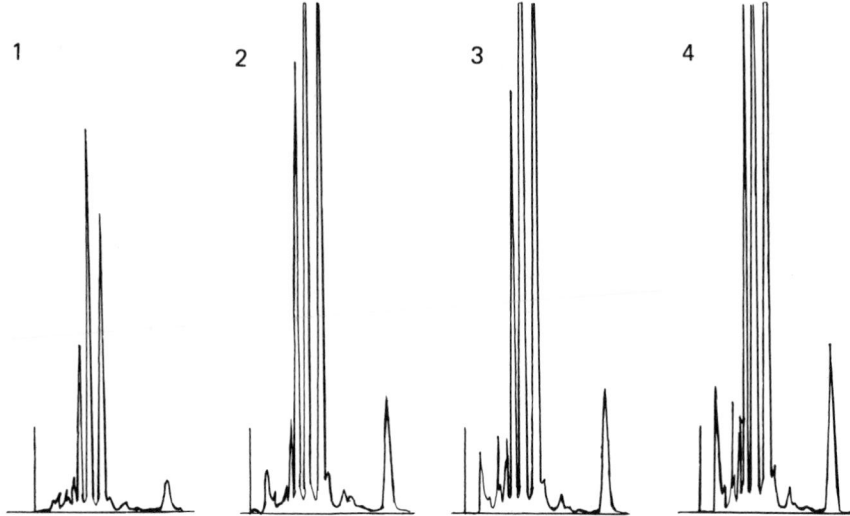

FIG. 5.   Chromatograms of four different batches of ink formulation
D.   (From Ref. 32.)

TABLE 2.   Normalized Average Peak Heights (%) of Three 10-μl Injections for Four Batches of Ink Formulation D

	Batch			
k'	1	2	3	4
-0.25	2.9	6.3	7.8	8.6
0.80	8.5	12.2	8.3	5.9
1.00	42.7	63.8	51.4	43.3
1.20	100	100	8.25	93.1
1.60	77.5	86.1	100	100
2.05	1.2	2.6	1.1	1.5
2.40	2.0	2.2	2.2	1.5
2.80	--	1.3	0.4	0.2
4.15	8.8	16.8	15.8	11.6

Source: Ref. 32.

found that the chromatographic separation was not effected, but that certain paper types did affect extractability of the ink. The paper types were also found not to impart any characteristics of their own to the chromatogram. Figure 6 illustrates the effect of paper on the amount of extracted ink by comparison of peak heights for ink extracted from 5 and 10 plugs of ink on two different paper types. The line graphs onto which these values are plotted were generated by using standard solutions of ink with known absorbance and plotting peak height versus absorbance for the resulting chromatograms (Fig. 7).

Additional reversed-phase work has been done by this author using a μBondapak CN column (10 μm), and a mobile phase of acetonitrile:water (80:20) with modification by 0.005 M heptanesulfonic acid and 0.02% acetic acid [35]. The results of this type of

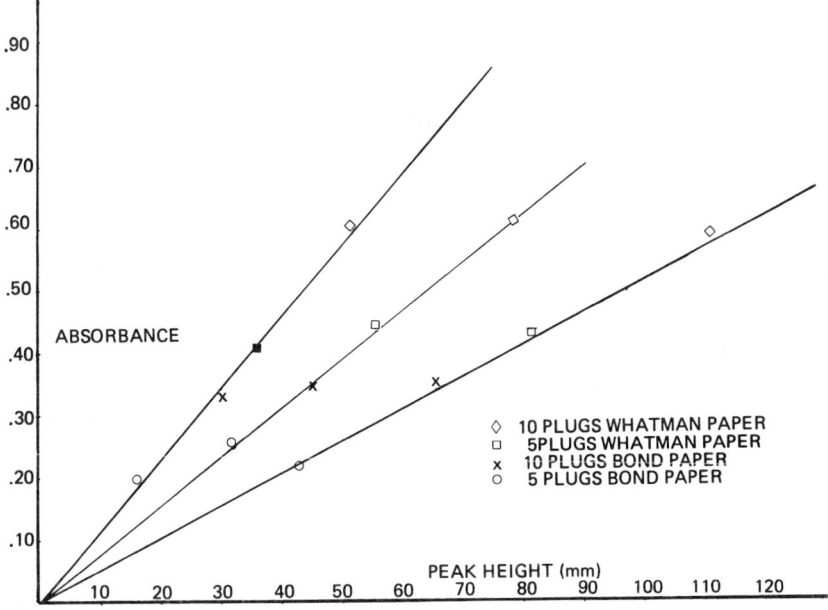

FIG. 6. Line graphs of Fig. 7 onto which peak heights from samples of 10 plugs B, 10 plugs W, 5 plugs B, and 5 plugs W have been plotted. (From Ref. 32.)

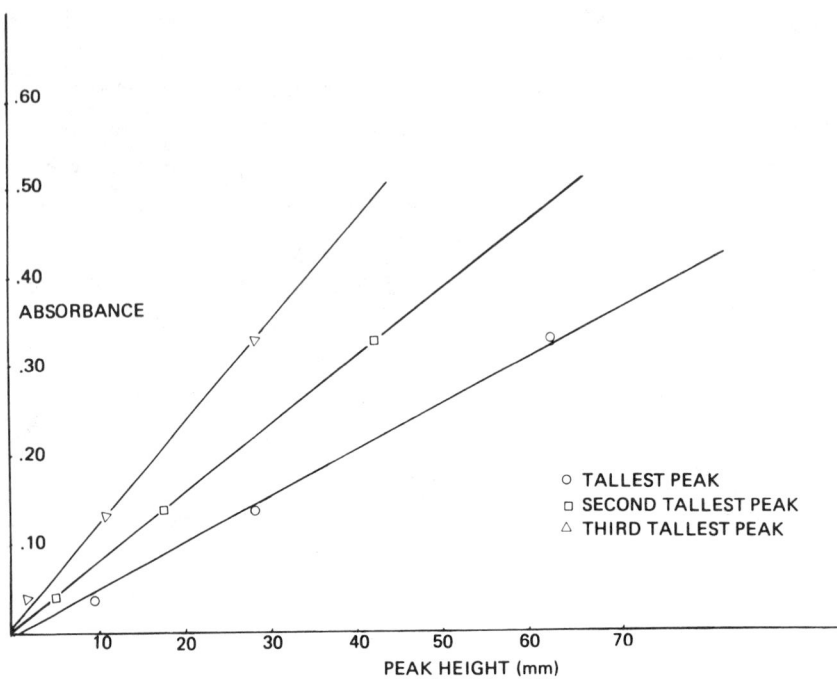

FIG. 7. Line graphs of average peak heights versus absorbance for
the three largest peaks of three 10-µl injections each, from three
solutions of ink formulation D. (From Ref. 32.)

analysis on a population of 36 blue ball pen inks shows good reproducibility but with sensitivity not as good as that of the previously mentioned reversed-phase system. Further work with this
µBondapak CN system might be useful as a means of identifying inks
through a computer search program.

The analysis of ball pen ink has progressed sufficiently from
thin-layer chromatography. Both normal- and reversed-phhase chromatographic conditions have been reported which were able to differentiate a limited number of ball pen ink formulations. The
normal-phase system of Colwell and Karger was able to differentiate
ball pen inks qualitatively by both colored and noncolored components, provided that the questions of preferential extraction, absorbance of the dyes at 254 nm, and identification of the unretained

peaks are answered. Semiquantitative differentiation was also reported by Colwell and Karger in cases where the inks were qualitatively similar. The same results were reported by this author using
a reversed-phase system [μBondapak C_{18} column, acetonitrile:water
(80:20), 0.005 M heptanesulfonic acid, 0.02% acetic acid mobile
phase] and analyzing only the colored components. In addition,
data indicate that quantitative analysis may be able to differentiate between batches of the same or similar ink formulations.

IV. NON-BALL PEN INK

Although non-ball pen inks make up a rather small percentage of the
ink entries examined, the need to expand technology beyond the
present TLC techniques exists. Non-ball pen inks have in the past
been used mainly in broad-tip markers not normally found on document evidence, but with the advent of new instruments such as the
rolling ball marker, their increased usage is inevitable. As with
ball pen ink, the non-ball pen ink contains two fractions: colored
and noncolored. The colored portion of non-ball pen ink is made
up of many different types of acidic and basic water-soluble dyes
that have been adapted from the textile industry [25]. Because of
this larger selection of dyestuffs, the characterization of non-
ball pen ink can be targeted at the colored portion of the ink.
Adding to the necessity to characterize by colored components is
the type of material present in the noncolored fraction of non-ball
pen ink. This noncolored fraction is made up mostly of water-based
solvents and several other organic solvents, with additional components such as surfactants and wetting agents usually present in
very small amounts. One can see that although non-ball pen ink is
a complex mixture, the analyst will more than likely be left with
just a mixture of dyes. For this reason any techniques that are
developed should be directed toward characterization of the dyestuffs.

The TLC procedures and conditions described earlier [E. M.
Merck silica gel plates, ethyl acetate:ethanol:water (14:7:6)] are

also applicable to the analysis of non-ball pen inks. As with ball
pen inks, however, when these conditions are adapted to HPLC, the
separations are not adequate. Normal-phase conditions for analysis
of non-ball pen ink have been suggested by Colwell and Karger --
Partisil 10-µm silica gel column, mobile phase of dichlorethane:
ethanol:formamide (89:10:1). Colwell and Karger did limited work
in this area and their results are encouraging and should be pursued. Very little other work has been done in this area, althrough
this author believes many of the conditions used for ball pen ink
can be applied to non-ball pen ink. Many of the same dyes or dye
bases are used in both ball pen and non-ball pen inks, and therefore similar chromatographic conditions should provide for similar
qualitative and quantitative results. As of this time, however, no
chromatographic conditions have been examined thoroughly so as to
permit routine analysis of non-ball pen inks by HPLC.

V. GEL PERMEATION CHROMATOGRAPHY

There were no references found to the use of gel permeation chromatography (GPC) in the analysis of writing ink. However, this author
feels that because of the complexity of writing ink, ball pen ink
in particular, and the nature of the constituents (dyes and resins
of rather large and diverse molecular weights), GPC could be an extremely useful technique in characterizing writing ink. The available GPC column technology, which allows for separation of components with as little as 30 molecular weight units of difference, is
ideally suited to analysis of writing ink. This technique could
also provide an excellent means by which to catelog standard ink
formulations for possible computer search programs.

VI. SUMMARY AND FUTURE TRENDS

It is evident that HPLC can be used to analyze writing ink from
forensic evidence. The conditions vary from normal phase to reverse
phase, with the results not quite answering the question "Did this
pen prepare that writing?" Although the goal has not been reached,

the technique of HPLC has moved us closer by allowing characterization of ball pen ink to the point of differentiating batches.

The future of ink analysis may not hold the answer to the question "Did this pen prepare that writing?" but HPLC will be in the future. This author believes the future of ink analysis to include the routine use of some of the techniques mentioned above in differentiating batches of ball pen ink, as well as the adaptation of these techniques to the analysis of non-ball pen ink. GPC or another HPLC mode will more than likely be adopted as a method for computer searching a standard ink collection and matching results to unknown inks. This will allow increased access to some of the larger standard ink collections, and therefore the increased usage of a valuable tool in forensic sciences.

ACKNOWLEDGMENTS

The author wishes to thank the U.S. Treasury's Bureau of Alcohol, Tobacco and Firearms for valuable support in research over the past several years and to Dr. A. A. Cantu for many valuable discussions. In addition to the above, a special thank you to all those persons within the ink manufacturing industry who donated both time and samples in the pursuit of my research.

REFERENCES

1. R. Chowdhry, S. K. Gupta, and H. L. Bami, *J. Forensic Sci.* 18:418-433 (1973).
2. R. M. Dick, *J. Forensic Sci.* 15:357-363 (1970).
3. L. Gowdown, *J. Crim Law Criminol. Police Sci.* 55:280-286 (1964).
4. R. M. Levern, *J. Forensic Sci.* 13:25-28 (1973).
5. G. B. Richards, *J. Forensic Sci.* 22(1):53-60 (1977).
6. U. Von Bremen, *J. Forensic Sci.* 10(3):368-375 (1965).
7. J. W. Brackett and L. W. Bradford, *J. Crim. Law Crimonol. Police Sci.* 43(4):530-539 (1952).
8. C. Brown and P. L. Kirk, *J. Crim. Law Crimonol. Police Sci.* 45(3):334-339 (1954).
9. C. Brown and P. L. Kirk, *J. Crim. Law Criminol. Police Sci.* 45(4):473-480 (1954).
10. R. L. Brunelle and M. J. Pro, *J. Assoc. Off. Anal. Chem.* 55:823-826 (1972).

11. D. A. Crown, *J. Crim. Law Crimonol. Police Sci.* 52:338-343 (1961).
12. L. Gowdown, Ann. Meet. Am. Soc. Questioned Doc. Exam., Rochester, N.Y., 1951.
13. J. D. Kelly and A. A. Cantu, *J. Assoc. Off. Anal. Chem.* 58(1): 122-125 (1975).
14. H. W. Moon, 31st Annu. Meet. Am. Acad. Forensic Sci., Atlanta, 1979.
15. A. W. Somerford and J. L. Souder, *J. Crim. Law Criminol. Police Sci.* 43(1):124-127 (1952).
16. J. Tholl, *Police* 2:55 (1966).
17. J. Tholl, *Police* 1:6-16 (1970).
18. A. H. Witte, *Methods of Forensic Science,* vol. 2. Wiley-Interscience, New York, 1963.
19. P. G. Tunstall, 30th Ann. Meet. Am. Acad. Forensic Sci., San Diego, 1978.
20. P. G. Tunstall, *Secur. Manage.,* pp. 69-72 (Oct. 1979).
21. D. Cary, *Can. Off. Prod. Stationery* 10(2):44-47 (1976).
22. C. H. Lindsly, R. B. Schmidt, and R. S. Casey, *Chem. Ind.,* pp. 50-51 (July 1948).
23. C. A. Mitchall, *Inks, Their Composition and Manufacture.* Lippincott, New York, 1937.
24. C. E. Waters, *Inks.* Nat. Bur. Stand. Circ. C426, Washington, D.C., 1940.
25. P. M. Daugherty, First Georgetown Univ. Conf. Surface Anal., Washington, D.C., 1969.
26. C. Sensi and A. A. Cantu, Veterans Administration and U.S. Treasury, Bureau of Alcohol, Tobacco and Firearms, to be published in *J. Forensic Sci.* 27(1):196-199 (1982).
27. B. Hanna, First Georgetown Univ. Conf. Surface Anal., Washington, D.C., 1969.
28. A. H. Lyter III, U.S. Treasury, Bureau of Alcohol, Tobacco, and Firearms, Rockville, Md., unpublished results, 1979.
29. J. H. Kelly, *J. Police Sci. Admin.* 1(2):175-177 (1973).
30. L. F. Colwell and B. L. Karger, *J. Assoc. Off. Anal. Chem.* 60(3):613-618 (1977).
31. J. Kelly, J. Reinstein, and K. Kempfert, Wisconsin State Crime Laboratory, Madison, Wis., unpublished results, 1978.
32. A. H. Lyter III, U.S. Treasury, Bureau of Alcohol, Tobacco, and Firearms, Rockville, Md., 1981; to be published in the *J. Forensic Sci.* 27(1):154-160 (1982).
33. A. H. Lyter III, U.S. Treasury, Brueau of Alcohol, Tobacco, and Firearms, Rockville, Md., unpublished results, 1981.
34. O. Menis, *Analysis of Ballpoint Ink.* Nat. Bur. Stand. Study, Aug. 1970.
35. A. H. Lyter, III, U.S. Treasury, Bureau of Alcohol, Tobacco, and Firearms, Rockville, Md., unpublished results, 1981.

Author Index

Beasley, T. H., 136(98), 138(98), 159, 205(60), 263
Bechell, H. C., 79(39), 158
Benoit, H., 283(24), 303
Berkersky, I., 320(20), 352
Berkowitz, D. A., 361(30), 363(30), 396
Bethke, H., 183(28), 262
Beyerman, H. C., 202(56), 262
Bidlingmeyer, B. A., 174(19), 261
Biggs, J. T., 331(60), 353
Bijster, P., 332(68), 336(68), 354
Billmeyer, F. W., 278(20, 21), 303
Bissett, F. H., 377(57), 397
Blank, C. L., 343(119), 355
Blasof, S., 199(51), 201(51), 202(51), 241(51), 242(51), 255(85), 262, 263
Blau, K., 366(39), 396
Blum, K., 331(51), 353
Bly, D. D., 267(4), 271(4), 273(4), 274(4), 277(4), 278(4), 279(4), 280(4), 284(26), 285(4, 26, 28), 286(4), 287(4), 291(4), 303
Bodnar, J. A., 200(53), 262
Boeme, W., 86(62), 158
Bogaert, M. G., 331(56), 353
Bollet, C., 343(120), 355
Bolleter, W. T., 359(9), 366(9, 40, 41), 367(41), 368(41), 392(41), 395, 396
Bommer, P., 320(21), 352
Boogt, W. H., 195(43), 262
Borga, O., 331(53), 353
Bos, P., 195(43), 262
Bounine, J. P., 69(22, 23), 98(23), 99(23), 100(23), 101(23), 102(23), 157
Bowers, G. N., 337(97), 338(97), 355
Bowman, L. M., Jr., 278(22), 303
Brackett, J. W., 400(7), 402(7), 417
Bradford, D. C., 380(70), 397
Bradford, L. W., 400(7), 402(7), 417
Braithwaite, R., 332(69), 336(69), 354

Braithwaite, R. A., 331(59), 353
Brandenberger, H., 380(65), 397
Bratin, K., 384(88), 386(88, 91), 398
Bredeweg, R. A., 116(88), 159
Bridges, R. R., 337(80), 354
Brinkman, U. A. Th., 386(92, 95), 387(92, 95), 388(92), 389(100), 398
Bristow, P. A., 103(76), 159
Brodie, R. R., 320(24), 324(24), 332(67), 352, 354
Brooks, M. A., 320(20), 352
Brown, C., 309(10), 310(10), 352, 400(8, 9), 417
Brown, R. H., 375(56), 397
Brugman, W. J. T., 68(20), 157
Brun, A., 69(22, 23), 98(23), 99(23), 100(23), 101(23), 102(23), 157
Brunelle, R. L., 400(10), 402(10), 403(10), 407(10), 417
Bruzzone, A. R., 278(18), 303
Bugge, A., 320(26), 352
Bunger, W. B., 90(69), 158
Burke, D., 196(44), 262
Burke, L., 337(83), 354
Burke, M. F., 375(55), 397
Bushee, D., 387(98), 398
Butler, L. D., 375(55), 397

Cahnmann, H. J., 80(57), 158
Camp, M. J., 357(3, 7), 358(3, 7), 378(3, 63), 387(3), 395, 397
Cantow, M. J. R., 271(9), 303
Cantu, A. A., 400(13), 401(26), 402(13), 403(13), 418
Carr, K., 337(87), 354
Carreira, L. A., 361(30), 363(30), 396
Carroll, I., 126(94), 128(94), 159
Cartoni, G. P., 125(93), 127(93), 159
Cary, D., 400(21), 418
Casey, R. S., 400(22), 418
Cashman, P. J., 123(91), 124(91), 129(95), 130(95), 150(91), 159
Cassiday, R. M., 79(43), 158
Cattran, D., 374(50), 393(50), 396

Subject Index

Refractive index (RI) detectors,
  38-40
Refractometers, 38-40
Retention indexes, 216-219
Retention times, 150-156
  tranquilizers, 191-192
  various drugs, 155-156
Retention volume data of basic
  drugs, 221-224
Reversed-phase chromatography,
  163-175
  relevant properties in, 163-
  167
Reversed-phase ion-pairing
  chromatography, 225-245

Sample retention, 84-85
Scott-Kucera model, 70, 71
Scott model, 61-68
Screening technique, 144-150
Secobarbital, 211, 228
Secondary solvent effects, 59-60
Sedative-hypnotics, 325-329
Septum injectors, 26-27
Short-term noise, 38
Silanols, 71-74
Silica
  elution order on, 77
  irregular, 81
  as most widely used adsorbent,
  53
  spherical, 82
  suppliers of, 83
  See also Adsorption chromato-
  graphy, use of silica
  in
Silica dissolution, 113-114
Silica packings, 78-80
  commerical availability of,
  80, 81, 82
  procedures of, 79
Size-exclusion chromatography, 5
Slatts model, 68
Snyder model, 54-55, 69, 70
Soczewinski model, 60-61, 69
Solute dispersion-zone spreading,
  274, 276
Solute retention, 57
Solvent delivery systems, 21-26
Solvent purity, 89-90

Solvents, properties, of, 93
Solvent strength, 57-59
Solvent viscosity, 92
Specific surface area, 78
Spherical silicas, 82
Step gradient system, 184
Stimulants, 123-129
Stop-flow injectors, 27
Strong polar solvents, 59
Sugars, 248-250

Tailing peak, 172-173
Ternary mobile phases, 98-102
Test mixtures, use of, 103-106
Tetracaine, 226
Tetrahydrofuran, 66, 161, 165
Thebaine, 136, 138, 139, 140,
  227
Thermal deactivation, 88-89
Thermal energy analyzer detectors,
  49
Thin-layer chromatography (TLC),
  91-92
Thin-layer chromatography scout-
  ing, 92-94
Thomas model, 69
Tranquilizers, 187-196
Triangulation methods, 50-51
Tricyclic antidepressants, 329-
  336
Trimethylsilyacetates, 179-180
Triton X-100, 18
Tryptamines, 180-181
Two-pump gradient systems, 184

Ultraviolet (UV) absorbance detectors, 40-44
Ultraviolet (UV) chromophore, 163

Valium, 118
Valve injectors, 27-30
  differences between, 28-29
Variable wavelength detectors,
  41, 43
Volatile nonpolar solvents, 48

Water